Extreme Weather Events and Human Health

Rais Akhtar
Editor

Extreme Weather Events and Human Health

International Case Studies

Springer

Editor
Rais Akhtar
International Institute of Health Management
Research (IIHMR) Delhi
New Delhi, India

ISBN 978-3-030-23772-1 ISBN 978-3-030-23773-8 (eBook)
https://doi.org/10.1007/978-3-030-23773-8

This Springer imprint is published by the registered company Springer Nature Switzerland AG
The registered company address is: Gewerbestrasse 11, 6330 Cham, Switzerland

Foreword

As I write this foreword, three events have occurred that have indicated the extreme importance of extreme weather events. The Intergovernmental Panel on Climate Change has just produced its latest analysis of the consequences of a 1.5 °C rise in temperature (http://report.ipcc.ch/sr15/pdf/sr15_spm_final.pdf) (accessed 12 November 2018). The UK Meteorological Office has just produced a detailed analysis of recent changes in the occurrence of extreme events in the UK (https://www.metoffice.gov.uk/binaries/content/assets/mohippo/pdf/uk-climate/state-of-the-uk-climate/soc_supplement-002.pdf) (accessed 12 November 2018). Thirdly, and, tragically, there has been huge loss of life in California as a result of severe fires associated with a long continued drought that may be a consequence of climate change.

Floods, dust and sand storms, thunderstorms, droughts, hurricanes, tornadoes, landslides and mudflows, extreme warm events, severe freezing, rapid ice and snow melting, rising storm surges and high tides and various other phenomena have severe impacts on human health. This is partly because of the adverse effects of global climate change brought about by human actions, partly because of the normal occurrence of extreme events, the expansion of urban environments (with heat island and other effects), burgeoning human populations living in potentially dangerous locations and land cover changes such as deforestation. The range of factors involved is well shown by the case of Brazil (chapter "Extreme Weather Events Under a Changing Climate: A Brief Context for Brazil and the Role of the Health Sector"). Among the factors that help to explain the increasing frequency of extreme events in that country, in addition to the threat of global environmental change itself, are increasing unplanned urbanization, environmental degradation, weak responsiveness of institutions and the population, the lack of infrastructure, and poverty. These conditions, combined with the occupation of locations with high exposure to natural hazards, have generated places with intense vulnerability and low response capacity.

This book takes a welcome international view and comprises case studies from 18 countries covering all continents: Taiwan, East Asia, Hong Kong, Fiji, Thailand, Indonesia, Australia, India, Malaysia, Horn of Africa, South Africa, Italy, France,

Canada, USA, Mexico, Brazil and Argentina. It also contains some more systematic chapters as, for example, on the health implications of dust storms in areas downwind from the world's great deserts. It also draws attention to especially vulnerable communities, including the elderly, the very young and those with pre-existing health conditions. As chapter "Climate Change, Wildfires, Heatwaves and Health Impacts in Australia" shows, in Australia, deaths attributable to heatwaves and fire smoke pollution are more commonly due to exacerbations of pre-existing health conditions, than to specific direct impacts such as heat stroke. Some groups, such as the elderly, infants and those with pre-existing conditions, tend to be more vulnerable to these impacts. In Thailand (chapter "Extreme Weather and Climate Events and Occupational Health in Thailand"), research revealed locations where that sunstroke, muscle cramps and/or heat exhaustion were possible with prolonged exposure. It also stressed that a heat health warning system is essential to reduce the negative impacts of such extreme weather. In neighbouring Malaysia (chapter "Climate-Related Disasters and Health Impact in Malaysia"), heat stress is exacerbated by the urban heat island effects, but also by fire hazes associated with deforestation. Fire effects are also severe in California (chapter "Wildland Fire, Extreme Weather and Society: Implications of a History of Fire Suppression in California, USA"), where there has been much debate about the role played by fire suppression and forest management policies. Thunderstorms have been implicated in asthma attacks, as shown in a study (chapter "Thunderstorms During Pollen Season as Risk Factors for Allergic Respiratory Diseases and Severe Asthma"). Increasing thunderstorm incidence may be a feature of urban growth. In northern Canada, Inuit communities and much of their infrastructure will be greatly impacted by changes in ice extent and thickness (chapter "The Impacts of Climate Change on Health and Development in Canadian Arctic and Sub-arctic Communities in the Twenty-First Century: A Systematic Review"). Desert dust (chapter "Dust Storms and Human Health") has been implicated in cardiovascular problems, asthma, allergy-related diseases, accidents and many other health effects.

Indeed, the nature and range of health effects as a result of extreme weather events are substantial. For example, if one takes floods, there are direct effects: drowning, injuries, health implications due to contact with cold and polluted water and cardiovascular incidents. There are also indirect effects: waterborne infections, vector-borne diseases (e.g. cholera), food shortages, health effects of chemical pollution, decrease of health care and emergency services, and psycho-social disturbances. During and after a flood, food, clean water and shelter can become scarce, which can increase the risk of starvation, malnutrition and dehydration.

One research finding of recent decades is that substantial changes in the frequency and intensity of extreme events can result from a relatively small shift in temperatures. As the Earth's climate has warmed, some types of extreme weather have become more frequent and severe, with increases in extreme heat, intense precipitation and drought. Heatwaves are longer and hotter. Heavy rains and flooding are more frequent. Drought, too, is more intense and more widespread. It is, therefore, imperative that the international community tries to reduce the growth in greenhouse gas loadings in the atmosphere and to keep the amount of global warming to less than 1.5 °C. It is also important that we monitor the changes in extreme events that are taking place so that we have firm evidence upon which to base mitigation and adaptation policy, including the appropriate design of engineering schemes. Equally, it is vital to develop forecasting systems so that vulnerable populations are given more accurate warnings of potential catastrophes. Similarly, planners need to make sure that fewer people live in vulnerable locations. For that we need to have a clear view of where such locations occur.

Andrew Goudie
University of Oxford
Oxford, UK

Preface

Since 1970, the earth's average temperature has risen by 0.8 °C. During this span, the rise in temperature each decade was greater than in the preceding one. The IPCC in the 2007 Fourth Assessment Report clearly showed that since 1880, both annual and five years' average temperatures continue to rise.

Globally, we are witnessing more hot days, heatwaves, increased frequency and intensity of hurricanes/typhoons and increased occurrence of forest fires. The past four years since 2015 have been the hottest on record, and we are seeing the effects. Rising sea level along the coast of most developed and developing countries are causing great concern for real estate values. As McMichael asserted, 'Extreme weather events are projected to increase further with the advance of human-driven climate change. Both recent and historical experiences indicate that infectious disease outbreaks very often follow extreme weather events, as microbes, vectors and reservoir animal hosts exploit the disrupted social and environmental conditions of extreme weather events' (McMichael 2015). It has been estimated that losses from anomalous weather events were more than three times higher in 2016 than in 2010, and as a proportion of GDP, much greater in poor than in rich nations.

From the Paris Climate Agreement to Katowice, scientists have issued warnings about human-induced climate change and in the Katowice conference, in December 2018, David Attenborough opined that the climate change may precipitate the collapse of our civilizations. Consequently, in accordance with the Paris Climate Agreement, developed and developing countries must fulfil the commitment under their own legally binding NDCs in order to keep global warming well below 2 °C above pre-industrial levels and if possible, below 1.5 degrees, encompassing rigorous adaptation and mitigation measures.

Thus, the climate change-induced extreme weather events are worsening the frequency, intensity and impacts of some types of extreme weather events. For example, sea level rise increases the impacts of coastal storms, and warming can place more stress on water supplies during droughts. The increased intensity of forest fires and droughts cause much distress to populations.

Forecasts, early warning systems and effective national weather services play an essential role in protecting local communities from weather and climate impacts such as flooding, hurricanes/typhoons and heatwaves.

Andrew Goudie in his Foreword has rightly stated, 'Equally, it is vital to develop forecasting systems so that vulnerable populations are given more accurate warnings of potential catastrophes. Similarly, planners need to make sure that fewer people live in vulnerable locations. For that we need to have a clear view of where such locations occur'.

At the same time, 'the pressures of population and economic growth need to be recognized and addressed. Housing and infrastructure constructed in danger zones such as flood plains or at the foot of unstable hillsides can put lives at risk. Alleviating poverty and strengthening public institutions is vital to reducing this risk' (https://www.c2es.org/content/extreme-weather-and-climate-change).

In order to achieve this, we need to focus not just on development in a country but also to lay emphasis on the spatial distribution of development in all regions in the country. This process may lead to reduced inequalities and may also help in counter-urbanization. The need of the hour is to develop holistic development policies by reducing vulnerability and strengthening the resilience of people as well as regions to the climate change-induced extreme weather events.

New Delhi/Aligarh, India Rais Akhtar

References

Center for climate and energy solutions, Arlington, USA. Extreme weather and climate change. (https://www.c2es.org/content/extreme-weather-and-climate-change).

McMichael, A. (2015). Extreme weather events and infectious disease outbreaks. *Virulence*, *6*(6): 543–7. doi: 10.4161/21505594.2014.975022.

Acknowledgements

In the process of writing, editing and preparing this book, there have been many people who have encouraged helped and supported us with their skill, thoughtful evaluation of chapters and constructive criticism.

First of all, we are indebted to all the contributors of chapters from both developed and developing countries for providing the scholarly and innovative scientific piece of research to make this book a reality. We are also thankful to the reviewers who carefully and timely reviewed the manuscripts.

We are also grateful to Prof. Andrew Goudie, University of Oxford for writing Foreword, which adds greatly to the book with his thoughtful insights.

I am thankful to my family—wife Dr. Nilofar Izhar, daughter Dr. Shirin Rais and son-in-law Dr. Wasim Ahmad—who encouraged and sustained me in developing the structure of the book and editing tasks, and I am deeply grateful for their support and indulgence.

Finally and most essentially, we are deeply obliged to the Springer and the entire publishing team, without whose patience, immense competence and support this book would not have come to fruition. We specially thank Dr. Robert K. Doe whose energizing leadership ensured that this book would indeed translate to reality. I am also grateful to Ms. Carmen Spelbos for her constant support.

We are also thankful to Mr. Sriram Srinivas and Mr. Muruga Prashanth Rajendran for constant cooperation.

New Delhi/Aligarh, India Rais Akhtar

Contents

About the Editor

Rais Akhtar is presently Adjunct Faculty, International Institute of Health Management and Research, New Delhi, formerly National Fellow and Emeritus Scientist (CSIR), CSRD, Jawaharlal Nehru University, New Delhi, and Visiting Professor Department of Geology, AMU, Aligarh. He has taught at the Jawaharlal Nehru University, New Delhi; University of Zambia, Lusaka; and The University of Kashmir, Srinagar. He is recipient of a number of international fellowships including Leverhulme Fellowship (University of Liverpool), Henry Chapman Fellowship (University of London), Visiting Fellowship (University of Sussex), Royal Society Fellowship (University of Oxford) and Visiting Professorship (University of Paris-10). He was Lead Author (1999–2007), on the Intergovernmental Panel On Climate Change (IPCC), which is the joint winner of Nobel Peace Prize for 2007. He is the recipient of Nobel Memento. He has to his credit 91 research papers and 19 books published from India, UK, USA, Germany and the Netherlands. His book entitled Environmental Degradation and Human Health was published in 2014 by Springer (with A. Malik and E. Grohmann). His book is entitled Climate Change and Human Health Scenario in South and Southeast Asia, published in 2016 by Springer, and his latest books include Climate Change and Air Pollution (with C. Palagiano) and Geographical Aspects of Health and Disease in India (with Andrew Learmonth) published in 2018 by Springer (the Netherlands) and Concept (New Delhi), respectively. He is Member of Expert Group on

Climate Change and Human Health of the Ministry of Health and Family Welfare, Government of India. He has been appointed as Corresponding Member of Italian Geographical Society in October 2018.

Introductory

Introduction: Extreme Weather Events and Human Health: A Global Perspective

Rais Akhtar

Abstract Flash floods, heatwave, coldwave, droughts, strong winds and storms, etc. are few examples of extreme weather conditions. 2003 heatwaves in Western and Central Europe resulted in the death of more than 70,000 people, and 2005 Katrina hurricane was considered the costliest disaster that devastated southern USA. It caused $160 billion worth of damages to the Gulf Coast. Hurricanes Harvey, Irma and Maria combined with devastating California wildfires and other natural catastrophes to make 2017 the most expensive year (about $306 billion), on record for disasters in the USA. 2018 brought Hurricane Michael, which struck the Florida coast in October, and the Camp Fire is California's deadliest and most destructive fire on record. The 1999 Orissa (now Odisha) cyclone in India was the strongest recorded tropical cyclone in the North Indian Ocean and among the most destructive in the region. Devastations as a result of unusual flooding in Uttarakhand in 2013, Kashmir in 2014, Kerala in 2018 and typhoon Mangkhut in Philippines in 2018 are examples of extreme weather events. Scientists assert that Europe's death toll from weather disasters could rise 50-fold by the end of twenty-first century, with extreme heat alone causing deaths of more than 1,50,000 a year if nothing is done to curb the effects of climate change. In July 2019, Europe has been confounded by another massive heat wave that could break records again.

Keywords Hippocrates · 2003 European heatwaves · Katrina hurricane · Hurricane Harvey · Typhoon Mangkhut · Forest fires · Kashmir · Kerala floods · Respiratory diseases · Donald Trump · Paris climate agreement · IPCC · Global Warming of 1.5 °C

1 Introduction

The understanding of impacts of extreme weather events on human health and well-being dates back to Hippocrates' fifth century treatise ON AIRS WATERS AND PLACES. The definition of extreme weather is when a weather event is significantly

R. Akhtar (✉)
International Institute of Health Management Research, New Delhi, India
e-mail: raisakhtar@gmail.com

R. Akhtar (ed.), *Extreme Weather Events and Human Health*,
https://doi.org/10.1007/978-3-030-23773-8_1

different from the average or usual weather pattern. Flash floods, heatwave, cold-waves, droughts, strong winds and storms etc. are a few examples of extreme weather conditions. The 2003 heatwaves in western and central Europe resulted in the death of more than 70,000 people, and the 2005 Katrina hurricane, was considered the costliest disaster that devastated the southern USA. It caused $160 billion worth of damages to the Gulf Coast. However, the disasters caused by hurricane Harvey in September 2017 proved to be even costlier with economic losses of about $190 billion. The 1999 Orissa (now Odisha) cyclone in India was the strongest recorded tropical cyclone in the North Indian Ocean and among the most destructive in the region. The surge combined with heavy rains to produce widespread flooding, damaging around 1.6 million homes. It has been estimated that 2017 natural disasters and climate change resulted in economic loss of $309 billion to the US. Scientists assert that Europe's death toll from weather disasters could rise 50-fold by the end of the twenty-first century, with extreme heat alone causing deaths of more than 152,000 a year if nothing is done to curb the effects of climate change (Davis 2017).

2 IPCC Report 2014

The views on the IPCC assessment report 2014 by some experts provide another opportunity for industrialized nations to create a fear psychosis among emerging economies that they are most vulnerable and thus have to heavily cut down on CO_2 emissions which will impact development. However, it should be kept in mind clearly that climate change is impacting upon both developed and developing countries equally. As the largest developing country, China's economic development was based on extensive utilization of fossil fuels, which leads to a dramatic increase in emissions of both ambient air pollutants and greenhouse gases (GHGs) (Akhtar 2014).

The impact of flooding and the destruction of roads in Colorado in 2013, the Sandy, Harvey and Florence hurricanes in the USA, and flooding in Uttarakhand and Kashmir in India in 2013 and 2014, respectively, are a reflection that disasters affect all regions. Forest fires that ravaged several parts of California and Australia in October–November 2018 will not be underestimated by the US administration which had scrapped an entire section on climate change (Akhtar 2018).

3 IPCC 2018 Special Report, Global Warming of 1.5 °C

IPCC 2018 Special Report entitled, Global Warming of 1.5 °C, asserts that emissions of carbon dioxide (CO_2) caused by human activity must reach "net zero" by 2050 to keep the average rise in global temperatures at 1.5 °C above pre-industrial levels to reduce catastrophic climate-change risk on populations (IPCC 2018). The Report highlights that "Temperature extremes on land are projected to warm more than GMST (high confidence): extreme hot days in mid-latitudes warm by up to about

3 °C at global warming of 1.5 °C and about 4 °C at 2 °C, and extreme cold nights in high latitudes warm by up to about 4.5 °C at 1.5 °C and about 6 °C at 2 °C (high confidence). The number of hot days is projected to increase in most land regions, with highest increases in the tropics (high confidence)." This is relevant as there is likely hood in future of severe heatwave conditions as occurred in 2018 in Japan and in Australia. Europe had its warmest August on record (WMO 2018). IPCC (2018) Special Report also predicts risks from heavy precipitation even sea level will continue to rise well beyond 2100 (high confidence). The Special report also contends that global warming is likely to reach 1.5 °C between 2030 and 2052 if it continues to increase at the current rate (high confidence).

However, the IPCC Special Report put very high confidence projecting lower risks at 1.5 °C than at 2 °C for heat-related morbidity and mortality. Urban heat islands often amplify the impacts of heatwaves in cities (high confidence). Risks from some vector-borne diseases, such as malaria and dengue fever, are projected to increase with warming from 1.5 to 2 °C, including potential shifts in their geographic range (high confidence) (IPCC 2018). It should be noted that several studies have projected altitudinal rise of malaria due to warming as seen in highlands of Africa, mountainous regions in India as well as in some regions in Andes mountains (Akhtar 2007; Mishra et al. 2016; Pinault and Hunter 2012).

4 Human Health

High temperatures and dry gusts resulted in recent wildfires in the USA and in Australia. These quickly spread, threatening hundreds of homes in California. Wildfires, heatwaves and hurricane-induced flooding have resulted in increased mental illnesses in Australia, Spain and California (Taioli et al. 2018; Sarfaty 2018).

According to Andy Haines of London School of Hygiene and Tropical Medicine, warming favors the spread of disease as the events create conditions conducive to disease outbreaks. Agriculture faces warming, more extremes and more diseases. More drought and flooding under the new climate, and accompanying outbreaks of crop pests and diseases, can affect yields, nutrition, food prices and lead to food insecurity. Chemical measures to limit infestations are costly and unhealthy: increased cancer incidence among farmers in Punjab in India is a relevant example (Singh 2015). Extreme high temperature and drought conditions impact water quality, its availability and its accessibility to a vast population.

5 Impact of Heatwaves

Both the intensity and frequency of deadly heatwaves are becoming momentous human health disaster globally encompassing much greater portion of the Earth. Mora et al. (2017), using data from 783 lethal heatwaves in 64 cities and across

36 countries discovered a common threshold where the heatwave becomes deadly. They asserted that currently about 30% of the world's population (and about 13% of the land area) experiences at least 20 days per year on which the deadly threshold is reached. By 2100, this percentage would jump to 74% of the population (47% of the land area) if emissions continued unchecked. It should be noted that WMO confirmed that 2015, 2016, 2017 and 2018 have been confirmed as the four warmest years on record. The year 2016 still holds the global record with an El Nino, while 2017 was the warmest year without an El Niño, which can boost global annual temperatures—an important factor in the occurrence of heatwaves.

The 2018 European drought and heatwave was a period of unusually hot weather that led to record-breaking temperatures and wildfires in many parts of Europe, especially in France, Spain and Portugal during the spring and summer of 2018 (Rubin 2018). Japan suffered an unprecedented two-week heatwave in July 2018 that was declared a natural disaster by the government. The number of deaths rose sharply: at least 65 people died of heatstroke during the last week of July while 22,647 people were hospitalized. In during June–July 2019 Europe has been confounded by another massive heatwave that could break records again.

In South Asia, a region of deep poverty and inequalities, where one-fifth of the world's people live, 73% of the wealth generated went to the richest one per cent in India in 2017 (DTE 2018), new research suggests that by the end of this century climate change could lead to summer heatwaves with levels of heat and humidity that exceed what humans can survive without protection (Chandler 2017).

Forest fires:

With increasing temperatures and levels of drought, Australia and California, besides Spain and Portugal, are suffering from extreme wildfires and intense heatwaves. The 2018 wildfire season was the deadliest and most destructive wildfire season on record in California, with a total of 8527 fires burning an area of 1,893,913 acres (766,439 ha), the largest amount of burned acreage recorded in a fire season, according to the California Department of Forestry and Fire Protection. These high destruction fires are primarily a combination of climate change and severe drought and wind. Climate change is increasing fire season and probability of severe wind and drought. The Camp (northern CA) and Woolsey (Hollywood/Malibu area in southern CA) faced disastrous consequences (Schweizer 2018).

The 2018–19 Australian bushfire season was predicted to be "fairly bleak" in parts of Australia, particularly in the east, by the Bushfire and Natural Hazards Cooperative Research Centre (CRC) in September 2018. Large bushfires had already burned through southern New South Wales during winter. The outlook for Spring was of a higher likelihood of fires with a twice the normal chance of an El Nino for Summer (Wikipedia). With record-breaking temperatures in January 2019 in South Australia and Victoria reaching 49.5 in some parts with night minimums of over 32 °C. Even in Tasmania, there were uncontrolled fires all over the island. Australia has experienced its hottest summer (2018) on record, according to the nation's Bureau of Meteorology. During February–March 2019, forest fires are raging in south-east Australia, particularly Hobart area in Tasmania (Wikipedia 2018–19).

6 Hurricanes, Flooding and Sea-Level Rise

Hurricanes Harvey, Irma and Maria combined with devastating California wildfires and other natural catastrophes to make 2017 the most expensive year on record for disasters in the USA. The disasters caused $306 billion in total damage in 2017, with 16 events that caused more than $1 billion in damage each. The bulk of the damage, at $265 billion, came from hurricanes.

According to Moody Analytics, the impact of Hurricane Florence resulted in the death of 41 people and property damage and disruption and is expected to total at least $17 billion to $20 billion.

It is estimated that property in Sunny Isles alone is now worth more than $10 billion. A number of the wealthiest people in the US have residences in Florida, including 40 billionaires on the Forbes 400 list of richest Americans (BBC 2017). The sea-level rise along the Florida region is eroding home values, and owners might not even know it (Harris 2018).

Rising sea levels are changing the way people think about waterfront real estate. High ground is becoming preferred property. Human health and wellbeing of population is also being impacted by sea-level rise particularly in island countries in the Caribbean and in the Pacific Ocean.

According to oceanographers, among South-Asian countries, Bangladesh is most vulnerable, but India with its vast coastline of nearly 7516 km on the east and west also needs to be proactive, considering the vast numbers of people who are dependent on the oceans for their livelihood.

One research investigation by the National Institute of Oceanography, Goa, concludes that mean sea-level rise trends along with the Indian coast are about 1.30 mm/year and future (global) projections indicate about 0.48 mm (A1B) by the turn of twenty-first century (Unnikrishnan et al. 2010).

A rise in sea levels is certain to impact India badly since the country has a long and densely populated coastline (Akhtar 2017). The projections based on short-precipitation data obtained from Indian cities provides new insights into an adequate stormwater designs in the rapidly and haphazard urbanizing India. The study concludes that 78 Indian cities are vulnerable to flooding (Pandey 2018).

Extreme rainfall in Kashmir in Jammu &Kashmir State of India in 2014 devastated the valley of Kashmir. The devastations in 2013 in both Colorado USA (Taylor 2013; Gochis 2014) and Uttarakhand in India, due to unprecedented flooding. cause terrible destruction of roads and property equally. Uttarakhand flood disaster has been called as the Himalayan Tsunami (Das 2013). With these examples, the argument is strengthened that climate change-induced flooding impacts uniformly to both developed world and developing world.

Nearly 1400 died across India due to rain-related incidents and floods in three months in 2018. The Kerala floods in August 2018 alone accounted for 488 deaths, and Uttar Pradesh in northern India recorded the second-highest number of fatalities (254) due to rain-related incidents.

7 Growing Awareness in the USA

The Climate Museum initiative was conceived of in the aftermath of Superstorm Sandy of 2012, with an objective to create a museum dedicated to climate change and climate solutions in New York City. Thus the contemporary threat of climate change to human existence is being depicted to New Yorkers via a selection of flashing highway signs that have been placed around the city (The Guardian 2018). These "signs sound a warning about the climate crisis and they call us to think and act on the issue," said Miranda Massie, Director of the Climate Museum (The Guardian 2018). Such initiatives should be adopted in other metropolitan cities and towns in both developed and developing countries for better understanding of climate change impacts. As warmer average temperatures will lead to hotter days and more frequent and longer heatwaves, these changes will lead to an increase in heat-related deaths in the USA—reaching as much as thousands to tens of thousands of additional deaths each year by the end of the century during summer months (EPA 2017).

8 Disease Incidence

Changes in climate can affect the breeding of mosquitoes, reduce the availability of food and drinking water and lead to infectious disease outbreaks (Kan 2011). The IPCC also observed that malaria and diarrheal diseases are expected to rise. In Australia, scientists warn that mosquito breeding is likely to change as the ecosystem/rising temperature, changes particularly due to rising temperature and this will increase Australia's susceptibility to outbreaks of vector-borne diseases, like incidence of food-borne diseases like Salmonella. As temperatures rise, more people will spend time indoors to avoid the extreme heat, increasing the risk they will pass on respiratory diseases. This problem will be exacerbated by changes in seasons.

Climate change—temperature rise and climate variability—has already influenced the transmission of a wide range of vector-borne diseases in Europe. Climate change has resulted "in the observed shift of ticks to elevated altitudes and latitudes, notably including the *Ixodes ricinus* tick species that is a vector for Lyme borreliosis and tick-borne encephalitis" (Semenza and Suk 2018). Climate change is also thought to have been a factor in the expansion of other important disease vectors in Europe: *Aedes albopictus* (the Asian tiger mosquito), which transmits diseases such as Zika, dengue and chikungunya, and *Phlebotomus* sandfly species, which transmits diseases including Leishmaniasis (Semenza and Suk 2018).

Warmer weather could make mosquitoes carrying potentially deadly diseases common in the UK within 15 years, warn scientists. Dengue fever, West Nile virus and chikungunya have already reached parts of Europe, and malaria has re-established itself in Greece as tropical mosquitoes spread into new territories (NHS 2015).

Species which transmit dengue fever, West Nile virus and the virus chikungunya could take off in Britain, it is feared (Hope 2015). Due to climate change and other factors, air pollution patterns are changing in several urbanized areas, particularly in metropolitan cities in developing countries of the world, with a significant effect on respiratory health both independently and synergistically with weather conditions; climate scenarios also show Europe as one of the most vulnerable regions in the world. Frequent occurrence of thunderstorms are causing occurrence of asthma in Australia and other countries. Current evidence suggests that inter-annual and inter-decadal climate variability have a direct influence on the epidemiology of vector-borne disease.

In the context of food and nutrition, climate change is impacting global agriculture and is driving up the number of hungry people around the world, according to the United Nations' 2018 State of Food Security and Nutrition in the World report. This report asserts that 821 million people—one in every nine—were malnourished in 2017, up from 815 million in 2016, putting at risk the UN's goal of eradicating hunger in the world by 2030 (Clarke 2018).

9 Conclusion

Extreme weather events, large natural disasters, and failure to curb greenhouse gas emissions and build resilience to climate change are listed as the most prominent global risks. This was in line with the assessment of the Inter-Governmental Panel on Climate Change (IPCC) (2007 and 2014).

Because of such a grim scenario, the late Stephen Hawking has warned that Donald Trump's decision to withdraw from the Paris climate agreement could "could push the Earth over brink" and lead to a point where global warming is "irreversible" (Eleftheriou-Smith 2017). It was in great anguish that Samoa, one of the most vulnerable Pacific islands, has called climate change an "existential threat … for all our Pacific family" and said that any world leader who denied climate change's existence should be taken to a mental hospital (Lyons 2018).

In spite of the US withdrawal from the Paris climate agreement, it should be noted that by the beginning of November 2018, 195 UNFCCC members have signed the agreement and 184 have become party to it. On June 1, 2017, Donald Trump announced the US withdrawal from the Paris agreement. However, despite this the European Union, Russia, China, India, Australia, Japan and all other countries including Israel support the Paris Climate Agreement based on Nationally Determined Contributions (NDC), and all parties report regularly on their emissions and on their process of implementation. Just published, IPCC Report on Global warming of 1.5 °C warns that China, Russia and Canada's current climate policies would steer the world above a catastrophic 5C of warming by the end of 2100, according to an analysis that takes into consideration the climate goals of different countries (Robiou du Pont and Meinshausen 2018; IPCC 2018). Besides IPCC, UNEP 2018 report on emission gaps emphasized that current commitments on NDCs as provided by countries under Paris Climate Agreement are inadequate to bridge the emissions gap in 2030 and suggested the increase in NDC ambitions. The report also highlighted negatively the assessment of actions taken by G20 countries in enhancing their NDC ambitions. The report further elaborated that increased emissions and lagging actions means the gap published in 2018 report is largest than ever (UNEP 2018). The suggestion is that all fossil fuel subsidies must be phased out, and hence the global carbon emissions could be reduced by 10% by 2030. However, it may not be possible for a large number of developing countries who suffer from inequalities and are in the process of economic growth. Slowing down of economic growth means disastrous consequences toward reducing inequalities and poverty alleviation.

References

Akhtar, R. (2007). Health and climate in Kashmir. *Tiempo: A bulletin on climate and development* (Vol. 63, pp. 19–21). London: IIED.

Akhtar, R. (2014, April 14). Hope the West isn't heading towards climate colonialism. *Hindustan Times*. (New Delhi).

Akhtar, R. (2017, July 18). India will be in trouble if breakaway Antarctica iceberg leads to sea level rise. *Hindustan Times*. (e-paper).

Akhtar, R. (2018). COP 21 in Paris: Politics of climate change. In R. Akhtar & C. Palagiano (Eds.), *Climate change and air pollution: The impact on human health in developed and developing countries* (pp 41–48). Springer.

B.B.C. (2017, August 5). *Extreme weather 'could kill up to 152,000 a year in Europe by 2100*, London.

Chandler, D. L. (2017, August 2). Deadly heat waves could hit South Asia this century: Action, climate change could devastate a region home to one-fifth of humanity, study finds. *MIT News Office*.

Clarke, H. (2018, Septemper 11). *Hunger rising with global temperatures, UN report says CNN*, Updated 0902 GMT (1702 HKT).

Das, P. K. (2013). The Himalayan Tsunami'—Cloudburst, flash flood & death toll: A geographical postmortem. *IOSR Journal Of Environmental Science, Toxicology And Food Technology, 7*(2), 33–45.

Davis, N. (2017, August 4). *Extreme weather deaths in Europe 'could increase 50-fold by next century'*. London: The Guardian.

DTE. (2018, September 27). Unequal India: 101 billionaires thrive; 364 million poor struggle to survive. *Down to Earth*.

Eleftheriou-Smith, L.-M. (2017, July 3). Stephen Hawking says Donald Trump could turn Earth into Venus-like planet with 250C and sulphuric acid rain. *The Independent*.

EPA. (2017, January 19). Climate impacts on human health. *United States Environmental Protection Agency*. www.epa.gov.

Gochis, D. (2014, November 4). *The great colorado flood of September 2013*. Colorado: Centre for Atmospheric Research Boulder. https://doi.org/10.1175/BAMS-D-13-00241.1.

Harris, A. (2018, April 24). The risk of sea level rise is chipping away at Miami home values, new research shows. *Miami Herald*.

Hope, J. (2015, March 22). Mosquitoes carrying deadly diseases could take off in Britain, the species could become common in the UK within 15 years. *Daily Mail*.

IPCC. (2018). *Special report: Global warming of 1.5 °C: Summary for policy makers*. Geneva: IPCC.

Kan, H. (2011). Climate change and human health in China. *Environ Health Perspect, 119*(2), A60–A61.

Lyons, K. (2018, Auguset 31). World leaders who deny climate change should go to mental hospital—Samoan PM. *The Guardian*.
Mishra, S.R., Dhimal, M., Guinto, R. R., Adhikari, B., & Chu, C. (2016, August 1). Threats to malaria elimination in the Himalayas. *The Lancet Global Health PE519*.
Mora, C., et al. (2017). Global risk of deadly heat. *Nature Climate Change, 1,* 501–506.
NHS. (2015). Climate change 'might bring rise in UK mosquito-borne diseases'. www.nhs.uk. March 23, 2015.
Pandey, K. (2018, June 29). Warnings from the past: Sea levels will rise, 78 Indian cities to see floods While more than a century-old data projects a warmer world, India still needs a proper regional data on climate vulnerability. *Down To Earth*.
Pinault, L. L., & Hunter, F. F. (2012). Malaria in highlands of Ecuador since 1900. *Emerging Infectious Diseases, 18*(4), 615–622.
Robiou du Pont, Y., & Meinshausen, M. (2018). Warming assessment of the bottom- up Paris Agreement emissions pledges. *Nature Communications, 9*(4810).
Rubin, A. J. (2018, August 4). Scorching summer in Europe signals long-term climate changes. *The New York Times*.
Sarfaty, C. (2018, October 10). Mental health a top priority in Northern California wildfires recovery, providers say. *North Bay Business Journal*.
Schweizer, D. (2018, September 28). Personal communication.
Semenza, J. C., & Suk, J. E. (2018). Vector-borne diseases and climate change: A European perspective. *FEMS Microbiology Letters*, *365*(2), fnx.244.
Singh, J. (2015, July 4). Punjab, cancer capital of India. *Down To Earth*. (New Delhi).
Taioli, E., et al. (2018). Mental health challenges and experiences in displaced populations following Hurricane Sandy and Hurricane Harvey: The need for more comprehensive interventions in temporary shelters. *Journal of Epidemiology & Community Health, 72*(10). http://dx.doi.org/10.1136/jech-2018-210626.
Taylor, A. (2013, September 16). Historic flooding across Colorado. *The Atlantic*.
The Guardian. (2018, September 4). Miranda Massie Climate Museum, New York.
UNEP. (2018, November 2). *The Emissions Gap Report 2018: UN Environment*. Nairobi.
Unnikrishnan, A. S., Manimurali, M., Ramesh Kumar, M. R. (2010). *Sea-level rise along the Indian coast*. Goa: National Institute of Oceanography.
Wikipedia: Bush fires in Australia 2018–19.
WMO. (2018, September 7). Summer sees heat and extreme weather. Geneva: WMO.

Dust Storms and Human Health

Andrew S. Goudie

Abstract Dust storms play a pervasive role in the Earth system. They are events of considerable extent and of frequent occurrence in drylands. The amount of suspended material they contain is an important aspect of dust storms in relationship to health. They transport particulates, pollutants and biological materials for long distances downwind. In some parts of the world, though not all, they appear to be occurring with greater frequency, and some forecasts suggest that this will continue in response to increasing land-use pressures and lower soil moisture contents resulting from climate change. Studies, particularly in east Asia, show associations between dust events and a range of human health issues, including respiratory problems, cardio-vascular complaints, meningococcal meningitis, coccidioidomycosis, conjunctivitis, skin irritation, measles, and transport accidents.

Keywords Dust storms · Health · Respiratory · Cardiovascular · Source areas · Particulates

1 Introduction

This chapter reviews recent studies of dust storms (e.g. their sources areas, frequencies and particulate contents) and considers the evidence that is accumulating to relate these characteristics to health disorders. This issue has already been reviewed by Zhang et al. (2016).

Dust storms play a pervasive role in the Earth system (Goudie and Middleton 2006; UNEP et al. 2016). They result from fronts and convective *haboobs*, which raise dust from desert surfaces. In severe events, dust may reach concentrations in excess of 6000 $\mu g\ m^{-3}$ (Goudie 2014). Much of this load consists of silt. It can be transported over thousands of kilometers and is deposited downwind. The silt comes finer with distance from source. It is dominated by SiO_2 and Al_2O_3, but it may also have a large salt content, an organic component, and crucially from the health point of view, can transport pathogens and anthropogenic pollutants. For example, dust

A. S. Goudie (✉)
The Gatehouse Cottage, Portland, UK
e-mail: andrew.goudie@stx.ox.ac.uk

R. Akhtar (ed.), *Extreme Weather Events and Human Health*,
https://doi.org/10.1007/978-3-030-23773-8_2

from China, collected in Japan, absorbs anthropogenic atmospheric pollutants during transport. Consequently, Asian Dust Events (ADEs) in Japan coincide with increases in hospital admissions and clinical visits for allergic diseases such as asthma, allergic rhinitis and conjunctivitis (Otani et al. 2017). Dust storms can also carry potentially carcinogenic fibrous clay called palygorskite (Rodriguez-Navarro et al. 2018).

1.1 Source Areas

The health impacts of dust storms depend on where humans live with respect to dust sources and the downwind direction of dust transport from them. Satellite sensors have demonstrated the primacy of arid regions and of basins of internal drainage and of alluvial deposits. Estimates of the relative strength of dust emissions demonstrate the importance, firstly of the Sahara (with over half of the global total), secondly of China and Central Asia (with about 20%), thirdly of Arabia and fourthly of Australia. Southern Africa and the Americas are minor sources, together accounting for less than about 5% of the total. Some dust is raised from high-latitude areas.

Saharan dust is moved by the trades toward the Gulf of Guinea, producing Harmattan haze. It is also carried to the Near East, the Arctic, across the Atlantic to the Americas (Prospero et al. 2014), and to the countries bordering the Mediterranean.

In China, dust storms cover immense areas, particularly in the arid west and transport particulates to its more humid parts, Korea, Japan and beyond. As a consequence, some mega-cities, including Beijing, Shanghai, Seoul, Tokyo and Taipei are subjected to ADEs. Dust is also generated in Mongolia, where mining activity has increased its prevalence.

Dust storms in the Middle East occur on the plains of Iraq and Kuwait and have possible health implications (Al-Hemoud et al. 2018). Much dust is lofted off the Tigris-Euphrates alluvial plain, the Jaz Murian depression and the Seistan Basin. In Tehran, the great bulk of particulates are derived from the deserts of Iraq and Syria (Givehchi et al. 2013), while some dust in the United Arab Emirates may be derived from Iran and central Asia. The Tokar Delta in Sudan emits dust over the Red Sea—dust which moves into Arabia and degrades air quality. In India, the greatest number of dust storms occur in the Thar, but eastward-moving dust clouds cross heavily populated North India. In the southern former Soviet Union, dust is blown off the desiccated bed of the Aral Sea and there are many dust storms in Kazakhstan (Issanova et al. 2015).

In southern Africa, Etosha Pan, the Namib river valleys and the Mkgadikgadi Depression (Vickery et al. 2013) are big sources. In South America, the exposed bed of the desiccating Mar Chiquita (Argentina) is a major source (Bucher and Stein, 2016). Dust is also generated from Patagonia, and the Puna and Altiplano. In the USA, dust storms occur in the arid west and material blown from the anthropogenically desiccated Owens Lake has caused health concerns. Australia, though not as dusty as some regions of the world, is the largest source in the Southern Hemisphere. Australian dust causes respiratory problems in New Zealand (Cowie et al. 2010) and impacts upon air quality in Brisbane, Sydney, Newcastle and Melbourne (Leys et al. 2011).

1.2 The Changing Frequency of Dust Storms

Dust storm frequency is important in health terms. Analysis of meteorological data has revealed the changing frequency of events for the last six decades or so. Some areas have shown increasing trends (e.g. the Sahel of Africa and the eastern Gobi of Mongolia). However, others have shown declining trends in the late twentieth century (e.g. China, the Canary Islands, Turkmenistan, Central Asia, Pakistan). In Australia, the declining trend was followed by a spike of activity in the early years of the present century-'the Millennium Drought' (O'Loingsigh et al. 2017). Other areas (e.g. the Kalahari and Seistan) have shown marked fluctuations upwards and downwards in response to such factors as lake flooding and desiccation (Rashki et al. 2015) or climatic fluctuations. In China, both natural and anthropogenic factors are implicated in the observed trends. There is certainly no clear upward trend in dust storm activity in China or Korea (Wang et al. 2018) though the early years of the present millennium saw some severe events (Kurosaki et al. 2011).

Mahowald et al. (2010) have estimated the global picture of changes in dust storm activity for the twentieth century. They suggest that a doubling of atmospheric desert dust loadings took place over much of the globe.

1.3 PM_{10} and $PM_{2.5}$ Levels in Dust Storms

The amount of suspended material they contain is an important aspect of dust storms in relationship to health and in some parts of the world, such as the Negev, dust concentrations are increasing (Krasnov et al. 2016). Concentrations are normally expressed as PM_{10} and $PM_{2.5}$ values. Some storms close to source are associated with very high concentrations, but it is notable that concentrations at large distances downwind can also be elevated, as is evident for some of the big cities in East Asia (Wang et al. 2013) and Australia (Leys et al. 2011). Even in southern Europe and the Caribbean, African dust incursions can produce high particulate levels (Prospero et al. 2014).

According to WHO, the acceptable annual mean value of PM_{10} is 20 $\mu g\ m^{-3}$ and the acceptable 24 h mean value is 50 $\mu g\ m^{-3}$ (de Longueville et al. 2013). The U.S. Environmental Protection Agency regards 24 h $PM_{2.5}$ levels between 35.5 and 55.4 $\mu g\ m^{-3}$ as being unhealthy for sensitive groups, 55.5–150.4 as being unhealthy, 150.5–250.4 as being very unhealthy, and 250.4–500 as hazardous. These values are regularly exceeded in dust storms (Goudie 2014; Middleton 2017). PM_{10} values may remain high for extended periods, particularly in some Iranian sites (Goudarzi et al. 2017). Engelbrecht et al. (2009) showed that over large parts of the Middle East people were exposed to average annual PM_{10} values ranging from 72 to 303 $\mu g\ m^{-3}$ and $PM_{2.5}$ values ranging from 35 to 111 $\mu g\ m^{-3}$.

1.4 Biomaterials

Dust storms pick up and transport bacteria, pollen spores, fungi and viruses (Weil et al. 2017). These have implications for disease incidence, including meningococcal meningitis and coccidiomycosis. Desert dust may contain dangerous cyanotoxins (Richer et al. 2015). In Japan, Watanabe et al. (2011) found that asthma in adults was augmented on days when Asian Dust carried pollen in comparison with dusty days without pollen. El-Askary et al. (2017) suggested that Kawasaki Disease (KD), a rare vascular disease that can result in cardiac damage in children, could be associated with a fungus known as *Candida*, which is transported in dust from China to Japan.

1.5 Miscellaneous Pollutants

Dust storms pick up and transport anthropogenic materials such as pesticides, herbicides, heavy metals, dioxins and radioactive isotopes. Those that have moved across heavily industrialized and polluted areas, such as northern China, may therefore pose a greater risk than those storms that have passed over less heavily developed areas (Almeida-Silva et al. 2013). In addition, dust storms remove toxic materials, including arsenic, from desiccated lake beds and some dune areas (Soukup et al. 2012; Morman and Plumlee 2013). There has been especial concern about the deflation of pesticides, including Lindane, from the exposed bed of the Aral Sea (Balmagambetova et al. 2017). Radionuclides may be blown from sites of past nuclear tests and from uranium mines tailings (Csavina et al. 2012).

2 Human Health Disorders

As Sandstrom and Forsberg (2008) have pointed out, particle size is a main determinant of where in the respiratory tract a dust particle will come to rest when inhaled. Because of their small size, particles on the order of ~10 μ or less (PM_{10}) can penetrate the deepest part of the lungs such as the bronchioles or alveoli. Larger particles are generally filtered in the nose and throat via cilia and mucus. Particles which are <2.5 μ penetrate into the gas exchange regions of the lung, and ultra-fine particles (<100 nm) may pass through the lungs to affect other organs, with possible cardiovascular consequences (Martinelli et al. 2013).

2.1 Respiratory Disorders

The pathogenic effect of dust inhalation on respiratory tissues can be attributed to the direct physical action of dust particles on the epithelium of the human airways and may be exacerbated by the toxic effects of both trace elements (including arsenic) and of biologically active compounds (Schweitzer et al. 2018). Studies in East Asia have related ADEs to asthma, pneumonia and tracheitis (Yang 2013). Tao et al. (2012), working in Lanzhou, China, found that dust storms led to increased respiratory hospitalizations, particularly for those aged >65. Tam et al. (2012), working in Hong Kong, found significant increases in emergency hospital admissions due to respiratory problems, with a 1.05% increase per 10 $\mu g\ m^{-3}$. Chien et al. (2012), working in Taipei, found that compared with weeks before ADEs respiratory clinic visits during the weeks after an event increased by 2.5% for children younger than 6, and by 5.0% for children aged 7–14. It poses a particular risk to the elderly (Lin et al. 2016). Studies in Seoul, Korea, by Lee and Lee (2013) showed that ADEs led to a 22% increase in the rate of asthma treatments with a 6-day lag. Pneumonia and allergic rhinitis are respiratory complaints that appears to be influenced by ADEs in Taiwan (Kang et al. 2012).

In Australia, dust blowing from the interior is significantly associated with asthma severity. A 15% increase in non-accidental mortality at a lag of 3 days from a dust event was found for Sydney (Johnston et al. 2011), where Merrifield et al. (2013) also found large increases in asthma emergency department visits.

The effects of African dust outbreaks on asthma and other respiratory and cardiac afflictions in southern Europe has been noted (Stafoggia et al. 2016), and there is now a large corpus of work on the particulate loads associated with African dust incursions, and the effects that these have on respiratory diseases (see, e.g., Trianti et al. (2017) on Greece and Díaz et al. (2017) on Spain).

Other locations where a link between dust events and respiratory problems has been established include Seistan and other parts of Iran (Khaniabadi et al. 2017). There are also high levels of asthma in Arabia, with, for example, 24% of the population of Saudi Arabia being asthmatic, and dust storms are one possible factor (along with lifestyle, smoking and indoor dust levels) (Al-Ghazawy 2013). On the other hand, a significant association between dust levels and respiratory mortality has not always been found (Zhang et al. 2016).

Much dust storm material is silt-sized quartz, and if this is inhaled over a sustained period, it causes non-occupational silicosis in some areas of High Asia. Silicosis may be a factor in the prevalence of tuberculosis in the Thar (Mathur and Choudhary 1997). In the 'dust belt' of North Africa to China, a substantial proportion of lung cancers may be caused by exposure to the $PM_{2.5}$ particles in desert dust (Giannadaki et al. 2014), especially in countries like Mauritania and Mali.

2.2 Cardiovascular Disorders

High particulate levels in the air cause cardiovascular disease, including myocardial infarction, stroke, heart failure, arrhythmias, and venous thromboembolism (Martinelli et al. 2013). Kang et al. (2013), working in Taiwan, found that ADEs were associated with an acute increase in stroke hospitalization, while Liu et al. (2017) presented data which indicated that $PM_{2.5}$ concentrations from ADEs in Taiwan were highly correlated with emergency visits for cardiovascular diseases. However, also working in Taiwan, Yang et al. (2009) were unable to find a statistically significant association with congestive heart failure. In Nagasaki, Japan, Ueda et al. (2012) reported that heavy ADEs caused emergency ambulance dispatches for cardiovascular diseases to rise by 20.8%. Likewise, in Cyprus, Middleton et al. (2008) found that cardiovascular admissions increased after dust episodes, and Neophytou et al. (2013) found there was a 2.4% increase in daily cardiovascular associated with each 10 ug m^{-3} increase in PM_{10} concentrations on African Dust days. In Spain, Perez et al. (2012) found an increased cardiovascular mortality associated with $PM_{10\text{-}2.5}$ levels, as did Alessandrini et al. (2013) in Rome, Italy. In Israel, Vodonos et al. (2014) found a strong correlation between PM_{10} concentrations and chronic obstructive pulmonary disease (COPD). In the USA, Crooks et al. (2016) found a positive association between dust storms and non-accidental, cardiovascular, and other non-accidental mortality, particularly in the two states with the highest dust storm incidence, Arizona and California.

2.3 Measles

Although little work has been undertaken, Ma et al. (2017) found an association between dust events and measles incidence in the Gansu region of China. The Dust Bowl years of the 1930s were a time of great measles epidemics in Kansas, USA (Brown et al. 1935). The causes of this association need further investigation.

2.4 Miscellaneous Other Diseases and Health Issues

2.4.1 Coccidioidomycosis

Coccidioidomycosis, an infection caused by inhalation of airborne spores of soil-dwelling fungi found in the south-west USA and parts of Mexico and central and South America, can affect the lungs, cause extreme disfigurement, and be fatal (Gorris et al. 2018). In the USA, it is known as Valley Fever. Around 150,000 cases are reported annually (Anderson 2013). There is a strong bimodal seasonality of the disease in Arizona, with peaks in the drier and dusty months of June–July and

August–November. Tong et al. (2017) have found that in the American south-west, the frequency of windblown dust storms has increased 240% from 1990s to 2000s. This trend is associated with variations of sea surface temperature caused by the Pacific Decadal Oscillation. They found that the frequency of dust storms is correlated with Valley Fever incidences, and that the infection rate went up more than 800% from 2000 to 2011. Disturbance of desert surfaces in California is exacerbating the problem (Gorris et al. 2018) by increasing the probability of aerosolizing *Coccidioides* spp. spores.

2.4.2 Meningococcal Meningitis

Meningococcal meningitis is a serious problem in West Africa (Thomson et al. 2009). For instance, during the 2017 epidemic in northern Nigeria, 14,473 cases (including 1158 deaths) were reported, of which over 50% involved children under the age of 16. There is a clear temporal correlation between times of great atmospheric dust contents (the Harmattan) and cases of meningitis (Deroubaix et al. 2013). One theory is that dust, combined with extremely dry air, damages the pharyngeal mucosa, thereby easing bacterial invasion. Attempts at vaccination are a major priority.

2.4.3 Conjunctivitis

Prolonged exposure to dusty air can lead to conjunctivitis. Chien et al. (2014) found elevated conjunctivitis clinic visits during dust storm periods, particularly for children. Ko et al. (2016) also found a link between ADEs and conjunctivitis in Japan.

2.4.4 Dermatological Disorders

The presence of heavy metals such as nickel in dust may cause skin irritation (Onishi et al. 2015). In addition, it has been suggested that ADE particles can 'exert toxicological effects on human skin through the activation of the cellular detoxification system, the production of pro-inflammatory and immunomodulatory cytokines and changes in the expression of proteins essential in normal epidermal differentiation' (Choi et al. 2011, p. 92).

2.4.5 Algal Blooms

The nutrients carried in dust storms, including iron (Mahowald et al. 2009), can indirectly impact human health by stimulating chlorophyll production and toxic algal blooms in coastal environments (Gallisai et al. 2016). These have a range of effects including shellfish poisoning and respiratory complaints (Fleming et al. 2011).

2.4.6 Transport Accidents

Dust storms cause death through their role in causing road and air accidents (Baddock et al. 2013). Some fatal commercial air crashes have been attributed to visibility reduction or to the adverse mechanical effects of dust storms.

3 Conclusions

Dust storms are events of considerable extent and of frequent occurrence in deserts and on their margins. They transport particulates, pollutants and biological materials for long distances downwind. In some parts of the world, though not all, they appear to be occurring with greater frequency and some forecasts suggest that this will continue in response to increasing land-use pressures and lower soil moisture contents resulting from climate change. An increasing corpus of studies, particularly in east Asia, show associations between dust events and a range of human health issues, including respiratory problems, cardiovascular complaints, meningococcal meningitis, conjunctivitis, skin irritation, measles and transport accidents. It is probable that some groups of people may be especially susceptible to these effects, including the elderly, those who are socio-economically disadvantaged (Ho et al. 2018) and the very young (Foreman 2018). Plainly, dust storm control and mitigation are of immense importance (Middleton and Kang 2017).

References

Alessandrini, G. R., Stafoggi, M., Faustini, A. (2013). Saharan dust and the association between particulate matter and daily hospitalisations in Rome, Italy. *Occupational and Environmental Medicine, 70*(6), 432–434.

Al-Ghazawy, O. (2013). The rising danger of asthma. *Nature Middle East*. https://doi.org/10.1038/middleeast.2013.79.

Al-Hemoud, A., Al-Dousari, A., Al-Shatti, A., Al-Khayat, A., Behbehani, W., Malak, M. (2018). Health impact assessment associated with exposure to PM10 and dust storms in Kuwait. *Atmosphere, 9*(1). https://doi.org/10.3390/atmos9010006.

Almeida-Silva, M., Almeida, S. M., Freitas, M. C., Pio, C. A., Nunes, T., & Cardoso, J. (2013). Impact of Sahara dust transport on Cape Verde atmospheric element particles. *Journal of Toxicology and Environmental Health, Part A, 76,* 240–251.

Anderson, C. (2013). Agent-based modelling of coccidioidomycosis. *Doctoral dissertation*. USA: University of Pittsburg.

Baddock, M. C., Strong, C. L., Murray, P. S., & McTainsh, G. H. (2013). Aeolian dust as transport hazard. *Atmospheric Environment, 71,* 7–14.

Balmagambetova, A., Abdelazim, I. A., Zhurabekova, G., Rakhmanov, S., Bekmukhambetov, Y., & Ismagulova, E. K. (2017). Reproductive and health-related hazards of Lindane exposure in Aral Sea area. *Environmental Disease, 2*(3), 70–75.

Brown, E. G., Gottlieb, S., & Laybourn, R. L. (1935). Dust storms and their possible effect on health: With special reference to the dust storms in Kansas in 1935. *Public Health Reports (1896-1970),* 1369–1383.

Bucher, E. H., & Stein, A. F. (2016). Large salt dust storms follow a 30-year rainfall cycle in the Mar Chiquita Lake (Córdoba, Argentina). *PLoS One, 11*(6), e0156672.

Chien, L. C., Lien, Y. J., Yang, C. H., & Yu, H. L. (2014). Acute increase of children's conjunctivitis clinic visits by Asian Dust Storms exposure—A spatiotemporal study in Taipei, Taiwan. *PloS One, 9*(10), e109175.

Chien, L. C., Yang, C. H., & Yu, H. L. (2012). Estimated effects of Asian Dust Storms on spatiotemporal distributions of clinic visits for respiratory diseases in Taipei children (Taiwan). *Environmental Health Perspectives, 120,* 1215–1220.

Choi, H., Shin, D. W., Kim, W., Doh, S. J., Lee, S. H., & Noh, M. (2011). Asian dust storm particles induce a broad toxicological transcriptional program in human epidermal keratinocytes. *Toxicology Letters, 200,* 92–99.

Cowie, G., Lawson, W., & Kim, N. (2010). Australian dust causing respiratory disease admissions in some North Island, New Zealand hospitals. *New Zealand Medical Journal, 123,* 87–88.

Crooks, J. L., Cascio, W. E., Percy, M. S., Reyes, J., Neas, L. M., & Hilborn, E. D. (2016). The association between dust storms and daily non-accidental mortality in the United States, 1993–2005. *Environmental Health Perspectives, 124*(11), 1735–1743.

Csavina, J., Field, J., Taylor, M. P., Gao, S., Landázuri, A., Betterton, E. A., et al. (2012). A review on the importance of metals and metalloids in atmospheric dust and aerosol from mining operations. *Science of the Total Environment, 433,* 58–73.

De Longueville, F., Ozer, P., Doumbia, S., & Henry, S. (2013). Desert dust impacts on human health: An alarming worldwide reality and a need for studies in West Africa. *International Journal of Biometeorology, 57,* 1–19.

Deroubaix, A., Martiny, N., Chiapello, I., & Marticoréna, B. (2013). Suitability of OMI aerosol index to reflect mineral dust surface conditions: Preliminary application for studying the link with meningitis epidemics in the Sahel. *Remote Sensing of Environment, 133,* 116–127.

Díaz, J., Linares, C., Carmona, R., Russo, A., Ortiz, C., Salvador, P., et al. (2017). Saharan dust intrusions in Spain: Health impacts and associated synoptic conditions. *Environmental Research, 156,* 455–467.

El-Askary, H., LaHaye, N., Linstead, E., Sprigg, W. A., & Yacoub, M. (2017). Remote sensing observation of annual dust cycles and possible causality of Kawasaki disease outbreaks in Japan. *Global Cardiology Science & Practice, 22.* https://doi.org/10.21542/gcsp.2017.22.

Engelbrecht, J. P., McDonald, E. V., Gillies, J. A., Jayanty, R. K. M., Casuccio, G., & Gertler, A. W. (2009). Characterising mineral dusts and other aerosols from the Middle East—Part 1: Ambient sampling. *Inhalation Toxicology, 21,* 297–326.

Fleming, E., Kirkpatrick, B., Backer, L. C., Walsh, C. J., Nierenberg, K., Clark, J., et al. (2011). Review of Florida red tide and human health effects. *Harmful Algae, 10,* 224–233.

Foreman, T. (2018). *The effect of dust storms on child health in West Africa.* Columbia University, CDEP-CGEG Working Paper No. 47.

Gallisai, R., Volpe, G., & Peters, F. (2016). Large Saharan dust storms: Implications for chlorophyll dynamics in the Mediterranean Sea. *Global Biogeochemical Cycles, 30*(11), 1725–1737.

Giannadaki, D., Pozzer, A., & Lelieveld, J. (2014). Modeled global effects of airborne desert dust on air quality and premature mortality. *Atmospheric Chemistry and Physics, 14*(2), 957–968.

Givehchi, R., Arhami, M., & Tajrishy, M. (2013). Contribution of the Middle Eastern dust source areas to PM_{10} levels in urban receptors: case study of Tehran, Iran. *Atmospheric Environment, 75,* 287–295.

Goudarzi, G., Daryanoosh, S. M., Godini, H., Hopke, P. K., Sicard, P., De Marco, A., et al. (2017). Health risk assessment of exposure to the Middle-Eastern dust storms in the Iranian megacity of Kermanshah. *Public Health, 148,* 109–116.

Goudie, A. S. (2014). Desert dust and human health disorders. *Environment International, 63,* 101–113.

Goudie, A. S., & Middleton, N. J. (2006). *Desert dust in the global system*. Heidelberg: Springer.

Gorris, M. E., Cat, L. A., Zender, C. S., Treseder, K., & Randerson, J. (2018). Coccidioidomycosis dynamics in relation to climate in the southwestern United States. *GeoHealth, 2*(1), 6–24.

Ho, H. C., Wong, M. S., Yang, L., Chan, T. C., & Bilal, M. (2018). Influences of socioeconomic vulnerability and intra-urban air pollution exposure on short-term mortality during extreme dust events. *Environmental Pollution, 235,* 155–162.

Issanova, G., Abuduwaili, J., Kaldybayev, A., Semenov, O., & Dedova, T. (2015). Dust storms in Kazakhstan: frequency and division. *Journal Geological Society of India, 85*(3), 348–358.

Johnston, F., Hanigan, I., Henderson, S., Morgan, G., & Bowman, D. (2011). Extreme air pollution events from bushfires and dust storms and their association with mortality in Sydney, Australia 1994–2007. *Environmental Research, 111,* 811–816.

Kang, J. H., Keller, J. J., Chen, S. C., & Lin, H. C. (2012). Asian dust storm events are associated with an acute increase in pneumonia hospitalization. *Annals of Epidemiology, 22,* 257–263.

Kang, J. H., Liu, T. C., Keller, J., & Lin, H. C. (2013). Asian storm events are associated with an acute increase in stroke hospitalisation. *Annals of Epidemiology, 67,* 125–131.

Khaniabadi, Y. O., Daryanoosh, S. M., Amrane, A., Polosa, R., Hopke, P. K., Goudarzi, G., et al. (2017). Impact of Middle Eastern dust storms on human health. *Atmospheric Pollution Research, 8*(4), 606–613.

Ko, R., Hayashi, M., Hayashi, H., Hayashi, K., Kato, H., Kurata, Y., et al. (2016). Correlation between acute conjunctivitis and Asian dust on ocular surfaces. *Journal of Toxicology and Environmental Health, Part A, 79*(8), 367–375.

Krasnov, H., Katra, I., & Friger, M. (2016). Increase in dust storm related PM10 concentrations: A time series analysis of 2001–2015. *Environmental Pollution, 213,* 36–42.

Kurosaki, Y., Shinoda, M., & Mikami, M. (2011). What caused a recent increase in dust outbreaks over East Asia? *Geophysical Research Letters, 38,* L11702. https://doi.org/10.1029/2011GL047494.

Lee, J. W., & Lee, K. K. (2013). Effects of Asian dust events on daily asthma patients in Seoul, Korea. *Meteorological Applications*. https://doi.org/10.1002/met.1351.

Leys, J. F., Heidenreicj, S. K., Strong, C. L., McTainsh, G. H., & Quigley, S. (2011). PM_{10} concentrations and mass transport during "Red Dawn"—Sydney 23 September 2009. *Aeolian Research, 3,* 327–342.

Lin, Y. K., Chen, C. F., Yeh, H. C., & Wang, Y. C. (2016). Emergency room visits associated with particulate concentration and Asian dust storms in metropolitan Taipei. *Journal of Exposure Science & Environmental Epidemiology, 26*(2), 189–196.

Liu, S. T., Liao, C. Y., Kuo, C. Y., & Kuo, H. W. (2017). The effects of $PM_{2.5}$ from Asian dust storms on emergency room visits for cardiovascular and respiratory diseases. *International Journal of Environmental Research and Public Health, 14*(4). https://doi.org/10.3390/ijerph14040428.

Ma, Y., Zhou, J., Yang, S., Zhao, Y., & Zheng, X. (2017). Assessment for the impact of dust events on measles incidence in western China. *Atmospheric Environment, 157,* 1–9.

Mahowald, N. M., Engelstaedter, S., Luo, C., Sealy, A., Artaxo, P., Benitez-Nelson, C., et al. (2009). Atmospheric iron deposition: global distribution, variability, and human perturbations. *Annual Review of Marine Science, 1,* 245–278.

Mahowald, N. M., Kloster, S., Engelstaedter, S., Moore, J. K., Mukhopadhyay, S., McConnell, J. R., et al. (2010). Observed 20th century desert dust variability: Impact on climate and biogeochemistry. *Atmospheric Chemistry and Physics, 10,* 10875–10893.

Martinelli, N., Olivieri, O., & Girelli, D. (2013). Air particulate matter and cardiovascular disease: A narrative review. *European Journal of Internal Medicine, 24,* 295–302.

Mathur, M. L., & Choudhary, R. C. (1997). Desert lung syndrome in rural dwellers of the Thar Desert, India. *Journal of Arid Environments, 35,* 559–562.

Merrifield, A., Schindeler, S., Jalaludin, B., & Smith, W. (2013). Health effects of the September 2009 dust storm in Sydney, Australia: Did emergency department visits and hospital admissions increase? *Environmental Health*. https://doi.org/10.1186/1476-069x-12-32.

Middleton, N., Yiallouros, P., Kleanthous, S., Kolokotroni, O., Schwartz, J., Dockery, D.W., et al. (2008). A 10-year time-series analysis of respiratory and cardiovascular morbidity in Nicosia, Cyprus: The effect of short-term changes in air pollution and dust storms. *Environmental Health*. https://doi.org/10.1186/1476-069x-7-39.

Middleton, N. J. (2017). Desert dust hazards: A global review. *Aeolian Research, 24,* 53–63.

Middleton, N. J., & Kang, U. (2017). Sand and dust storms: Impact mitigation. *Sustainability, 9*(6). https://doi.org/10.3390/su9061053.

Morman, S. A., & Plumlee, G. S. (2013). The role of airborne dusts in human disease. *Aeolian Research, 9,* 203–212.

Neophytou, A., Yiallouros, P., Coull, B. A., Kleanthous, S., Pavlou, P., Pashiardis, S., et al. (2013). Particulate matter concentrations during desert dust outbreaks and daily mortality in Nicosia, Cyprus. *Journal of Exposure Science & Environmental Epidemiology, 23,* 275–280.

O'Loingsigh, T., Chubb, T., Baddock, M., Kelly, T., Tapper, N. J., De Deckker, P., et al. (2017). Sources and pathways of dust during the Australian "Millennium Drought" decade. *Journal of Geophysical Research: Atmospheres, 122*(2), 1246–1260.

Otani, S., Kurosaki, Y., Kurozawa, Y., & Shinoda, M. (2017). Dust storms from degraded drylands of Asia: Dynamics and health impacts. *Land, 6*(4). https://doi.org/10.3390/land6040083.

Onishi, K., Otani, S., Yoshida, A., Mu, H., & Kurozawa, Y. (2015). Adverse health effects of Asian dust particles and heavy metals in Japan. *Asia Pacific Journal of Public Health, 27*(2), 1719–1726.

Perez, L., Tobías, A., Querol, X., Pey, J., Alastuey, A., Díaz, J., et al. (2012). Saharan dust, particulate matter and cause-specific mortality: A case-crossover study in Barcelona (Spain). *Environment International, 48,* 150–155.

Prospero, J. M., Collard, F. X., Molinié, J., & Jeannot, A. (2014). Characterizing the annual cycle of African dust transport to the Caribbean Basin and South America and its impact on the environment and air quality. *Global Biogeochemical Cycles, 28*(7), 757–773.

Rashki, A., Kaskaoutis, D., Francois, P., Kosmopoulos, P. G., & Legrand, M. (2015). Dust-storm dynamics over Sistan region, Iran: Seasonality, transport characteristics and affected areas. *Aeolian Research, 16,* 35–48.

Richer, R., Banack, S. A., Metcalf, J. S., & Cox, P. A. (2015). The persistence of cyanobacterial toxins in desert soils. *Journal of Arid Environments, 112,* 134–139.

Rodriguez-Navarro, C., di Lorenzo, F., & Elert, K. (2018). Mineralogy and physicochemical features of Saharan dust wet deposited in the Iberian Peninsula during an extreme red rain event. *Atmospheric Chemistry & Physics*. https://doi.org/10.5194/acp-2018-211.

Sandstrom, T., & Forsberg, B. (2008). Desert dust. An unrecognized source of dangerous pollution? *Epidemiology, 19,* 808–809.

Schweitzer, M. D., Calzadilla, A. S., Salamo, O., Sharifi, A., Kumar, N., Holt, G., et al. (2018). Lung health in era of climate change and dust storms. *Environmental Research, 163,* 36–42.

Soukup, D., Buck, B., Goossens, D., Ulery, A., McLaurin, B. T., Baron, D., et al. (2012). Arsenic concentrations in dust emissions from wind erosion and off-road vehicles in the Nellis Dunes Recreational Area, Nevada, USA. *Aeolian Research, 5,* 77–89.

Stafoggia, M., Zauli-Sajani, S., Pey, J., Samoli, E., Alessandrini, E., Basagaña, X., et al. (2016). Desert dust outbreaks in Southern Europe: Contribution to daily PM_{10} concentrations and short-term associations with mortality and hospital admissions. *Environmental Health Perspectives, 124*(4), 413–419.

Tam, W. W. S., Wong, T. W., Wong, A. H. S., & Hui, D. S. C. (2012). Effect of dust storm events on daily emergency admissions for respiratory diseases. *Respirology, 17,* 143–148.

Tao, Y., An, Z., Sun, Z., Hou, Q., & Wang, Y. (2012). Association between dust weather and number of admissions for patients with respiratory diseases in spring in Lanzhou. *Science of the Total Environment, 423,* 8–21.

Thomson, M. C., Jeanne, I., & Djingarey, M. (2009). Dust and epidemic meningitis in the Sahel: A public health and operational research perspective. In *IOP Conference Series: Earth and Environmental Science* (Vol. 7). https://doi.org/10.1088/1755-1307/7/1/012017.

Tong, D. Q., Wang, J. X., Gill, T. E., Lei, H., & Wang, B. (2017). Intensified dust storm activity and Valley fever infection in the southwestern United States. *Geophysical Research Letters, 44*(9), 4304–4312.

Trianti, S. M., Samolim E., Rodopoulou, S., Katsouyanni, K., Papiris, S. A., & Karakatsani, A. (2017). Desert dust outbreaks and respiratory morbidity in Athens, Greece. *Environmental Health, 16*(1). https://doi.org/10.1186/s12940-017-0281-x.

Ueda, K., Shimizu, A., Nitta, H., & Inoue, K. (2012). Long-range transported Asian Dust and emergency ambulance dispatches. *Inhalation Toxicology, 24,* 858–867.

UNEP, WMO, UNCCD. (2016). *Global assessment of sand and dust storms*. Nairobi: UNEP.

Vickery, K. J., Eckardt, F. D., & Bryant, R. G. (2013). A sub-basin scale dust plume source frequency inventory for southern Africa, 2005–2008. *Geophysical Research Letters, 40*(19), 5274–5279.

Vodonos, A., Friger, M., Katra, I., Avnon, L., Krasnov, H., Koutrakis, P., et al. (2014). The impact of desert dust exposures on hospitalizations due to exacerbation of chronic obstructive pulmonary disease. *Air Quality, Atmosphere and Health, 7*(4), 433–439.

Wang, L., Du, H., Chen, J., Zhang, M., Huang, X., Tan, H., et al. (2013). Consecutive transport of anthropogenic air masses and dust storm plume: Two case events at Shanghai, China. *Atmospheric Research, 127,* 22–33.

Wang, X., Liu, J., Che, H., Ji, F., & Liu, J. (2018). Spatial and temporal evolution of natural and anthropogenic dust events over northern China. *Scientific Reports, 8*(1). https://doi.org/10.1038/s41598-018-20382-5.

Watanabe, M., Igishi, T., Burioka, N., Yamasaki, A., Kurai, J., Takeuchi, H., et al. (2011). Pollen augments the influence of desert dust on symptoms of adult asthma patients. *Allergology International, 60*, 517–524.

Weil, T., De Filippo, C., Albanese, D., Donati, C., Pindo, M., Pavarini, L., et al. (2017). Legal immigrants: Invasion of alien microbial communities during winter occurring desert dust storms. *Microbiome, 5*(1). https://doi.org/10.1186/s40168-017-0249-7.

Yang, C. Y., Cheng, M. H., & Chen, C. C. (2009). Effects of Asian dust storm events on hospital admissions for congestive heart failure in Taipei, Taiwan. *Journal of Toxicology and Environmental Health, Part A, 72,* 324–328.

Yang, Z. H. (2013). Association of inhalable particles in dust events with daily outpatient visits for Tracheitis. *Journal of Environmental & Occupational Medicine, 30,* 88–92.

Zhang, X., Zhao, L., Tong, D. Q., Wu, G., Dan, M., Teng, B. (2016). A systematic review of global desert dust and associated human health effects. *Atmosphere, 7*(12). https://doi.org/10.3390/atmos7120158.

Developed Countries

The Impacts of Climate Change on Health and Development in Canadian Arctic and Sub-arctic Communities in the Twenty-First Century: A Systematic Review

Fox Underwood and Stefania Bertazzon

Abstract Climate change in the arctic affects the movement and health of animals and the health and development of Inuit communities in Canada. We investigated such associations by performing a systematic review to find studies where community members reported changes in their communities attributed to climate change. In the 22 studies examined, we discovered that increased rainfall, storms, and erosion are likely causing greater infrastructure damage. As well, unpredictable and unstable weather has made travel difficult and dangerous along with thinning ice and changes in the timing of ice break-up and ice freeze-up. Lastly, a declining perceived health in wildlife, coupled with various obstacles in hunting, is a threat to the health and livelihoods of community members.

Keywords Canada · Arctic · Sub-arctic · Spatial variability · Climate change · Health

1 Introduction

A polar bear on a small piece of floating sea ice is a common image in people's minds when considering climate change in the arctic. That loss of ice will change, and has changed, the behaviour of that polar bear by forcing it to search for food in-land. Caribou, fish, whales, and seals will all have to migrate and adapt. So, too, will the Inuit communities that rely greatly on such country food for subsistence and livelihood be forced to adapt as well. Our aim was to examine changes in climate and weather on the health and development in primarily Inuit communities in Canada and do so by using systematically gathered information reported at the community level by members of each community.

F. Underwood (✉) · S. Bertazzon
Department of Geography, University of Calgary, Calgary, Canada
e-mail: feunderw@ucalgary.ca

S. Bertazzon
Department of History, Archaeology, Geography, Fine & Performing Arts, University of Florence, Florence, Italy

R. Akhtar (ed.), *Extreme Weather Events and Human Health*,
https://doi.org/10.1007/978-3-030-23773-8_3

2 Methods

Our area of interest was all arctic and sub-arctic communities in or near the four Inuit regions of Inuvialuit, Nunavut, Nunavik, and Nunatsiavut. To this end, we searched Scopus and Web of Science using the query "Canada AND arctic AND 'climate change'" from the beginning of the databases to 5 March 2018. We would additionally search reference lists of papers found in the database search for papers potentially not retrieved from the databases. Papers would be excluded if they reported data from before the year 2000, did not report data on people, did not report a link between climate change and health or development, did not report data for any specific Canadian community (e.g. values were only reported at the region level), did not report on arctic or sub-arctic communities in or near the four Inuit regions (Inuvialuit, Nunavut, Nunavik, and Nunatsiavut), or reported data already reported by other studies we had included.

3 Results

From our search of Scopus and Web of Science, 2388 records were found for abstract review, and 403 of those were selected for full-text review, with 22 papers suitable for data extraction. The four regions and 29 communities represented are shown in Fig. 1 and Table 1, along with community locations, study citations, years of study, and data type. The majority of studies included reported data between 2000 and 2010, with only three studies reporting data after 2010. All except one study had interview, workshop, or focus-group components where community members discussed climate changes and its impacts with researchers.

Geographically, we had recorded data for the majority of communities in and around Inuvialuit and Nunatsiavut, yet closer to half of communities in and around Nunavut and Nunavik. Most climate and weather variables were presence or absence values (Table 2). Some variables were increased or decreased (e.g. less snow or more

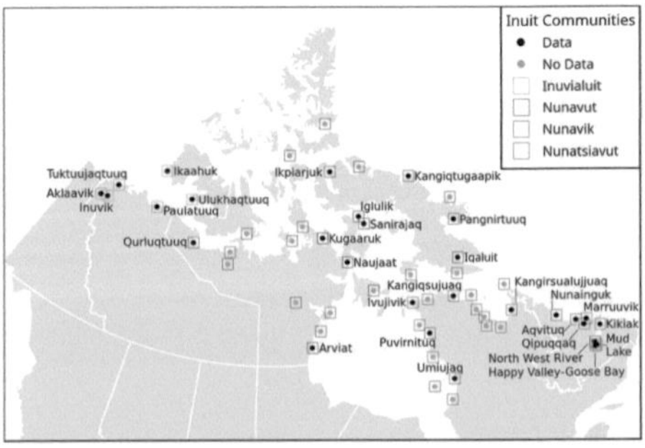

Fig. 1 Inuit communities in Canada

Table 1 Community and study list

Region	Community	Location	Study	Year of study	Data type
Inuvialuit Settlement Region (ISR) and area	Aklaavik (Aklavik)	68° 13′ 13″N 135° 0′ 42″W	Nickels et al. (2005, pp. 67–69, 90–93)	2002–2005	Workshops
			Wesche and Chan (2010, pp. 365–367)	2002–2005	Workshops
	Ikaahuk (Sachs Harbour)	71° 59′ 8″N 125° 14′ 53″W	Berkes and Jolly (2001, p. 5)	1999–2001	Interviews, workshops, focus groups
			Nichols et al. (2004, pp. 71–77)	1999–2001	Interviews
	Inuvik	68° 21′ 42″N 133° 43′ 50″W	Ford et al. (2013, p. 10)	2009–2012	Interviews, food bank usage statistics
			Nickels et al. (2005, pp. 67–69, 90–93)	2002–2005	Workshops
	Paulatuuq (Paulatuk)	69° 21′ 5″N 124° 4′ 10″W	Nickels et al. (2005, pp. 67–69, 90–93)	2002–2005	Workshops
			Wesche and Chan (2010, pp. 365–367)	2002–2005	Workshops
	Tuktuujaqtuuq (Tuktoyaktuk)	69° 26′ 34″N 133° 1′ 52″W	Andrachuk and Smit (2012, pp. 872–875)	not stated	Interviews
			Nickels et al. (2005, pp. 67–69, 90–93)	2002–2005	Workshops
			Wesche and Chan (2010, pp. 365–367)	2002–2005	Workshops
	Ulukhaqtuuq (previously Holman Island)	70° 44′ 11″N 117° 46′ 5″W	Nickels et al. (2005, pp. 67–69, 90–93)	2002–2005	Workshops

(continued)

Table 1 (continued)

Region	Community	Location	Study	Year of study	Data type
			Wesche and Chan (2010, pp. 365–367)	2002–2005	Workshops
Nunavut and area	Arviat	61° 6′ 29″N 94° 3′ 25″W	Beaumier et al. (2015, pp. 555–556)	not stated	Interviews
	Iglulik (Igloolik)	69° 22′ 34″N 81° 47′ 58″W	Ford et al. (2006a, pp. 131–133)	2004–2006	Interviews
			Ford et al. (2008, pp. 50–53)	2004–2006	Interviews
			Ford et al. (2009, pp. 142–149)	2006	Interviews, measurements of physical environment
			Laidler et al. (2009, pp. 376–383)	2004–2006	Interviews, measurements of physical environment
	Ikpiarjuk (Arctic Bay)	73° 2′ 11″N 85° 9′ 9″W	Ford et al. (2006b, pp. 149–150)	2004–2006	Interviews
			Ford et al. (2008, pp. 50–53)	2004–2006	Interviews
			Nickels et al. (2005, pp. 67–69, 90–93)	2002–2005	Workshops
	Iqaluit	63° 44′ 55″N 68° 31′ 11″W	Bunce et al. (2016, pp. 1426–1433)	2015	Interviews
			Hatcher and Forbes (2015, p. 463)	2009–2011	Measurements of physical environment
	Kangiqtugaapik (Clyde River)	70° 28′ 26″N 68° 35′ 10″W	Gearheard et al. (2010, pp. 273–279)	2000–2004	Interviews

(continued)

Table 1 (continued)

Region	Community	Location	Study	Year of study	Data type
	Kugaaruk or Arvilikjuaq (previously Pelly Bay)	68° 31′ 59″N 89° 49′ 36″W	Nancarrow and Chan (2010, pp. 5–9)	2005	Focus groups
			Nickels et al. (2005, pp. 67–69, 90–93)	2002–2005	Workshops
	Naujaat (previously Repulse Bay)	66° 31′ 19″N 86° 14′ 6″W	Nancarrow and Chan (2010, pp. 5–9)	2005	Focus groups
			Nickels et al. (2005, pp. 67–69, 90–93)	2002–2005	Workshops
	Pangnirtuuq (Pangnirtung)	66° 8′ 52″N 65° 41′ 58″W	Giles et al. (2013, pp. 209–210)	2008–2009	Interviews, focus groups
	Qurluqtuuq or Kugluktuk (previously Coppermine)	67° 49′ 32″N 115° 5′ 42″W	Prno et al. (2011, pp. 6–11)	2007	Interviews
	Sanirajaq (Hall Beach)	68° 46′ 38″N 81° 13′ 27″W	Ford et al. (2009, pp. 142–149)	2006	Interviews, measurements of physical environment
Nunavik and area	Ivujivik	62° 25′ 0″N 77° 54′ 30″W	Martin et al. (2007, pp. 197–199)	2003–2004	Interviews
			Nickels et al. (2005, pp. 67–69, 90–93)	2002–2005	Workshops
	Kangiqsujuaq	61° 36′ 0″N 71° 58′ 0″W	Cuerrier et al. (2015, pp. 384–389)	2007–2009	Interviews
			Martin et al. (2007, pp. 197–199)	2003–2004	Interviews

(continued)

Table 1 (continued)

Region	Community	Location	Study	Year of study	Data type
			Nickels et al. (2005, pp. 67–69, 90–93)	2002–2005	Workshops
	Kangirsualujjuaq	58° 41′ 0″N 65° 57′ 0″W	Cuerrier et al. (2015, pp. 384–389)	2007–2009	Interviews
	Puvirnituq	60° 1′ 48″N 77° 16′ 48″W	Martin et al. (2007, pp. 197–199)	2003–2004	Interviews
			Nickels et al. (2005, pp. 67–69, 90–93)	2002–2005	Workshops
	Umiujaq	56° 32′ 0″N 76° 33′ 0″W	Cuerrier et al. (2015, pp. 384–389)	2007–2009	Interviews
			Martin et al. (2007, pp. 197–199)	2003–2004	Interviews
Nunatsiavut and area	Aqvituq (Hopedale)	55° 29′ 2.54″N 60° 12′ 11.48″W	Crate (2012, pp. 62, 71–72)	2008–2011	Interviews or surveys
	Happy Valley-Goose Bay	53° 18′ 7″N 60° 25′ 0″W	Nickels et al. (2005, pp. 67–69, 90–93)	2002–2005	Workshops
	Kikiak (Rigolet)	54° 10′ 47″N 58° 25′ 44″W	Goldhar et al. (2014, pp. 76–78)	2009	Interviews
			Nickels et al. (2005, pp. 67–69, 90–93)	2002–2005	Workshops
	Marruuvik (Makkovik)	55° 4′ 38″N 59° 11′ 16″W	Crate (2012, pp. 62, 71–72)	2008–2011	Interviews or surveys
			Nickels et al. (2005, pp. 67–69, 90–93)	2002–2005	Workshops

(continued)

Table 1 (continued)

Region	Community	Location	Study	Year of study	Data type
	Mud Lake	53° 18′ 20″N 60° 10′ 1″W	Crate (2012, pp. 62, 71–72)	2008–2011	Interviews or surveys
	North West River	53° 31′ 31.32″N 60° 8′ 41.8″W	Nickels et al. (2005, pp. 67–69, 90–93)	2002–2005	Workshops
	Nunainguk (Nain)	56° 32′ 32″N 61° 41′ 34″W	Nickels et al. (2005, pp. 67–69, 90–93)	2002–2005	Workshops
	Qipuqqaq (Postville)	54° 54′ 37″N 59° 48′ 8″W	Nickels et al. (2005, pp. 67–69, 90–93)	2002–2005	Workshops

Table 2 Heat Map key: climate and weather

Key	Event occurring	Red	Blue	Grey
Ew	Extreme weather	Yes		
Wp	Unpredictable weather	Yes		
Wd	Unstable wind direction	Yes		
Wt	Warmer temperatures	Yes		
Sc	Season length change	Yes		
Wc	Prevailing wind direction change	Yes		
Ib	Earlier ice break-up	Yes	No	Yes and no
If	Later ice freeze-up	Yes	No	
Wl	Water level changes	Lower	Higher	
Ti	Ice thickness changes	Thinning		
Sf	Snow frequency	Less snow	More snow	Less and more
Rf	Rain frequency	More rain	Less rain	
Fr	Freezing rain frequency	More freezing rain		
S	Storm frequency	More storms		
E	Erosion frequency	Increased		

snow) with a couple having opposite values reported in the same community (e.g. less snow and more snow).

Overwhelmingly, communities in the four Inuit regions reported unpredictable and unstable weather (89%), later ice freeze-up (89%), thinning ice (86%), and less snow (72%) (Tables 3 and 4). Changes in timing and length of seasons (62%) and higher temperatures (51%) were reported throughout all regions except Inuvialuit. Increased erosion (58%), earlier ice break-up (58%), and falling water levels (58%) were reported in all regions, although two communities (Ikpiarjuk and Iqaluit) reported higher sea levels. Wind direction instability (24%) was reported only by community members in Nunavut and Nunavik; however, change in prevailing winds (58%) was reported in all regions. Rain was reported as more frequent in 48% of communities (except those in Nunavik) and less frequent in 13% of communities (and only those in Nunavik). Freezing rain (27% overall) occurred in communities in all regions except Nunatsiavut. More frequent storms (48%) were reported in all regions, although four communities in Nunavik reported less frequent storms. And while just over half of all communities reported extreme weather events (55%), the majority occurred in the eastern regions of Nunavik and Nunatsiavut. Extreme weather in Nunatsiavut was water drying-up (Goldhar et al. 2014, p. 76) and extremely high summer temperatures (Nickels et al. 2005, p. 67); in Nunavik, drought (Martin et al. 2007, p. 198) was reported along with a lightning event that caused fire in a dried-out area (Cuerrier et al. 2015, p. 387); in Nunavut were extremely high summer temperatures (Nickels et al. 2005, p. 67) and mudslides (Nickels et al. 2005, p. 69); and in Inuvialuit were extremely high summer temperatures (Nickels et al. 2005, p. 67), mudslides (Nickels et al. 2005, p. 69), and thunderstorms where none occurred before (Nichols et al. 2004, p. 77).

The health and development variables were predominantly yes or no questions (Table 5). Bear sightings were classified as more or less. Members from a few communities reported both the presence and the absence of a few variables.

Difficulties faced by communities from climate change varied by region (Tables 6 and 7). Although 20% of Inuit region communities reported more dangerous hunting, 44% reported more difficult hunting, more expensive hunting, and shorter or delayed hunting seasons. Nunavik reported only more difficult hunting and not more dangerous hunting, more expensive hunting, or shorter or delayed hunting seasons. A decline in wildlife health was reported in 62% of communities, along with a decline of or increased inaccessibility of caribou (34%), fish (27%), and berries (37%). Half of communities (48%) reported resorting to store-bought food, although none were communities in Nunatsiavut. Communities in all areas reported infrastructure damage (58%) from weather. Difficult travel was reported in 65% of communities, while more dangerous travel was reported in the majority (82%) of communities. More polar bears were sighted in Nunavut (Beaumier et al. 2015, p. 555; Bunce et al. 2016, p. 1427; Nancarrow and Chan 2010, p. 6) along with more grizzly bears around Arviat (Beaumier et al. 2015, p. 556). Members of Nunavik communities reported seeing more black bears (Cuerrier et al. 2015, p. 389). Fewer polar bears were seen around Ikaahuk (Berkes and Jolly 2001, p. 5; Nichols et al. 2004, p. 73).

Table 3 Heat Map: climate and weather by community

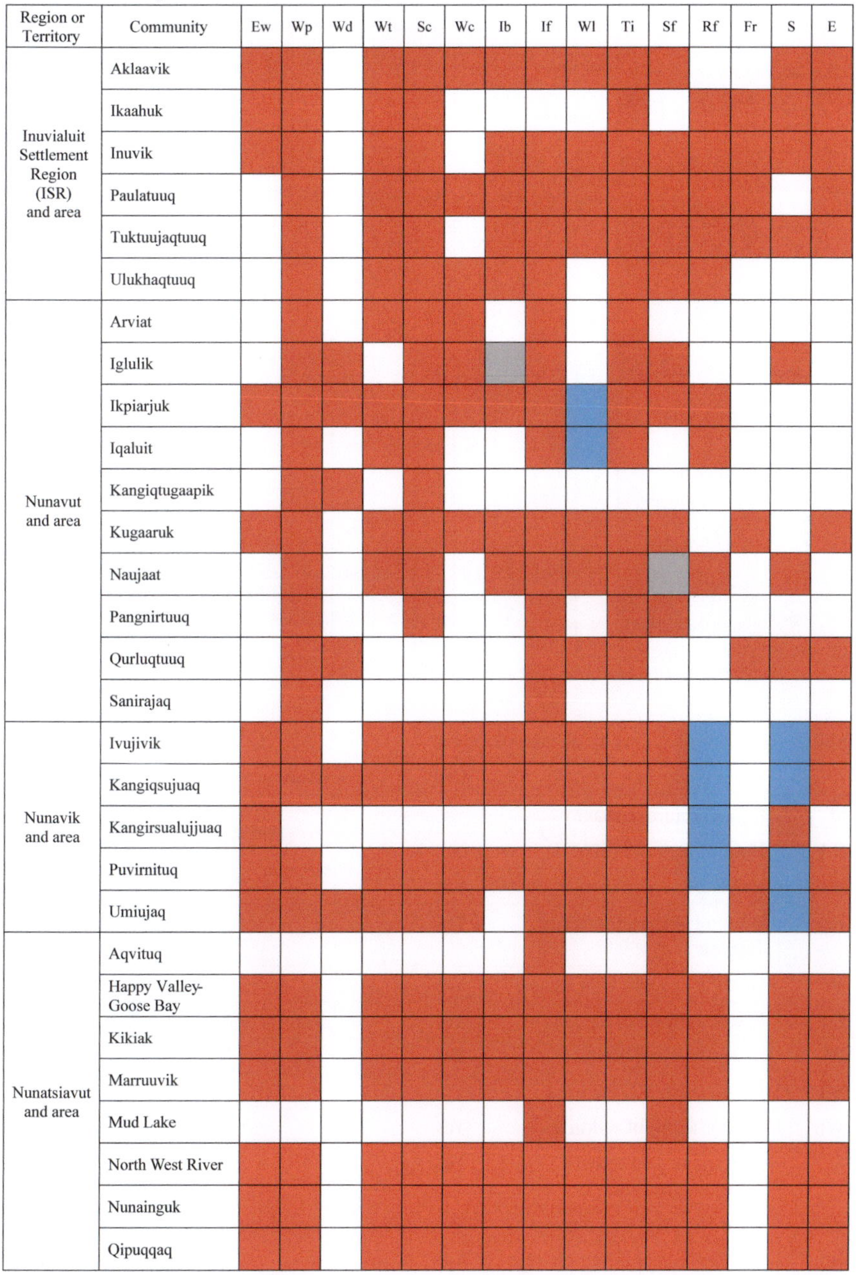

Table 4 Heat Map: climateClimate and weather by Inuit region

Region	Variable Direction	Ew	Wp	Wd	Wt	Sc	Wc	Ib	If	Wl	Ti	Sf	Rf	Fr	S	E
Inuvialuit Settlement Region (ISR) and area	% of 6 communities	50	100	0	0	0	50	83	83	66	100	83	83	66	66	83
	% of 6 communities									0			0		0	
	% of 6 communities							0				0				
Nunavut and area	% of 10 communities	20	100	40	50	80	40	30	90	30	80	40	30	20	30	20
	% of 10 communities									20			0		0	
	% of 10 communities							10				10				
Nunavik and area	% of 5 communities	100	80	40	80	80	80	60	80	80	100	80	0	40	20	80
	% of 5 communities									0			80		80	
	% of 5 communities							0				0				
Nunatsiavut and area	% of 8 communities	75	75	0	75	75	75	75	100	75	75	100	75	0	75	75
	% of 8 communities									0			0		0	
	% of 8 communities							0				0				
All Regions	% of 29 communities	55	89	24	51	62	58	58	89	58	86	72	48	27	48	58
	% of 29 communities									6			13		13	
	% of 29 communities							3				3				

Table 5 Heat Map key: health and development

Key	Event occurring	Red	Blue	Grey
Id	Infrastructure damage	Yes		
Hd	Hunting is more difficult	Yes		
Hn	Hunting is more dangerous	Yes		
Hs	Hunting season shorter or delayed	Yes		
He	Hunting is more expensive	Yes		
Cd	Fewer caribou in the area	Yes	No: more caribou	
Fd	Fewer fish in the area	Yes	No: more fish	Yes and no
Bd	Fewer berries in the area	Yes		Yes and no
Wh	Lower health of wildlife	Yes		
Si	Increase in purchase of store-bought food	Yes		
Bp	Polar bear sighting change	More bears	Less bears	
Bg	Grizzly bear sighting change	More bears		
Bb	Black bear sighting change	More bears		
Td	Travel is more difficult	Yes		Yes and no
Tn	Travel is more dangerous	Yes		

Table 6 Heat Map: health and development by community

Region or Territory	Community	Id	Hd	Hn	Hs	He	Cd	Fd	Bd	Wh	Si	Bp	Bg	Bb	Td	Tn
Inuvialuit Settlement Region (ISR) and area	Aklaavik	Red	Red		Red	Red	Red	Red	Red	Red	Red				Red	Red
	Ikaahuk		Red	Red			Red	Grey				Blue			Grey	
	Inuvik	Red	Red			Red	Red		Red	Red	Red				Red	Red
	Paulatuuq	Red	Red		Red	Red	Red	Grey	Red	Red	Red				Red	Red
	Tuktuujaqtuuq	Red	Red		Red	Red	Red	Grey	Red	Red	Red				Red	Red
	Ulukhaqtuuq	Red	Red		Red	Red	Red	Blue	Red	Red	Red				Red	Red
Nunavut and area	Arviat	Red				Red	Red				Red	Red	Red			
	Iglulik		Red	Red	Red	Red					Red				Grey	Red
	Ikpiarjuk		Red	Red		Red				Red					Red	Red
	Iqaluit					Red	Red					Red				Red
	Kangiqtugaapik		Red	Red		Red									Red	Red
	Kugaaruk		Red				Blue		Red	Red	Red	Red			Red	Red
	Naujaat	Red	Red				Blue		Red	Red	Red	Red			Red	Red
	Pangnirtuuq		Red	Red											Red	Red
	Qurluqtuuq	Red	Red			Red		Red		Red						Red
	Sanirajaq		Red	Red		Red	Red				Red					
Nunavik and area	Ivujivik	Red	Red						Red	Red	Red				Red	Red
	Kangiqsujuaq	Red	Red						Red	Red	Red			Red	Red	Red
	Kangirsualujjuaq		Red				Red		Grey					Red		
	Puvirnituq	Red	Red						Red	Red	Red				Red	Red
	Umiujaq								Red					Red		
Nunatsiavut and area	Aqvituq				Red											Red
	Happy Valley-Goose Bay	Red			Red			Red		Red					Red	Red
	Kikiak	Red	Red		Red	Red		Red		Red					Red	Red
	Marruuvik	Red			Red			Red		Red					Red	Red
	Mud Lake				Red											Red
	North West River	Red			Red			Red		Red					Red	Red
	Nunainguk	Red			Red			Red		Red					Red	Red
	Qipuqqaq	Red			Red			Red		Red					Red	Red

Table 7 Heat Map: health and development by Inuit region

Region	Variable Direction	Id	Hd	Hn	Hs	He	Cd	Fd	Bd	Wh	Si	Bp	Bg	Bb	Td	Tn
Inuvialuit Settlement Region (ISR) and area	% of 6 communities	83	100	16	66	83	100	16	83	83	83	0	0	0	83	83
	% of 6 communities						0	16				16				
	% of 6 communities							50	0						16	
Nunavut and area	% of 10 communities	30	80	50	10	70	30	10	20	40	50	40	10	0	50	80
	% of 10 communities						20	0				0				
	% of 10 communities							0	0						10	
Nunavik and area	% of 5 communities	60	80	0	0	0	20	0	80	60	60	0	0	60	60	60
	% of 5 communities						0	0				0				
	% of 5 communities							0	20						0	
Nunatsiavut and area	% of 8 communities	75	12	0	100	12	0	75	0	75	0	0	0	0	75	100
	% of 8 communities						0	0				0				
	% of 8 communities							0	0						0	
All Regions	% of 29 communities	58	44	20	44	44	34	27	37	62	48	13	3	10	65	82
	% of 29 communities						6	3				3				
	% of 29 communities							10	3						6	

4 Discussion

Far more similarities were seen with changes in weather and climate among the regions than with health and development. Unpredictable weather, later ice freeze-up, and changes in timing and length of seasons no doubt contributed to difficult travelling and danger travelling as reported by the vast majority of communities. A rise in extreme weather events, rain, storms, and erosion has also likely lead to greater infrastructure damage, all of which were reported in communities of all regions. Additionally, the substantial changes in weather and climate may be lowering the health of wildlife, which was a common observation throughout all regions along with various difficulties of hunting. Most communities reporting a rise in store-bought food also reported a decline in the health of wildlife, which is an understandable correlation. Communities in Nunatsiavut were the only ones to report declining wildlife health while at the same time not reporting resorting to store-bought food; however, the region may have better prospects for hunting that has allowed its members to avoid store-bought food in favour of country food. Although reports of polar bears, with less ice to hunt on, moving in-land, along with more common sightings of grizzly bears and black bears, did come up, relatively few communities reported this or, perhaps, omitted such events for the lack of concern (e.g. bears can be targets in sport hunting, so hunters may not be concerned if more bears are nearby). We also suspect that, while changes in climate and weather may be easier to see and measure, there may be a lag between environmental change and the change and reporting of health and development in arctic communities.

5 Conclusion

In this systematic review, we identified studies that included observations of community members of communities in Canada's arctic and sub-arctic regions. Changing weather and climate in the twenty-first century has adversely affected the health of people in Inuit communities by causing damage to infrastructure and increasing risk of injury when travelling in unpredictable weather and unsafe conditions. Changing climate, too, has lowered the health of much wildlife and caused caribou, fish, and bear to move to different territories, sometimes close to or further away from communities, which has affected the ability for hunters to catch game for nutrition and employment. And although no single observation about hunting was present across all regions, different regions reported at least one detrimental fact of hunting life, whether it was more difficult hunting, more dangerous hunting, more expensive hunting, shorter or delayed hunting seasons, or a combination of them. As an activity that is a large part of subsidence and employment in many communities, the increasing difficulty, and danger, of hunting cannot be understated as just as damaging to the health and development of these communities as damaging thunderstorms. More in-depth analyses and more discussion with the Inuit who live in northern communities are most assuredly desired and absolutely necessary to understand the effects of climate change on their lives.

References

Andrachuk, M., & Smit, B. (2012). Community-based vulnerability assessment of Tuktoyaktuk, NWT, Canada to environmental and socio-economic changes. *Regional Environmental Change, 12,* 867–885.

Beaumier, M. C., Ford, J. D., & Tagalik, S. (2015). The food security of Inuit women in Arviat, Nunavut: the role of socio-economic factors and climate change. *Polar Record, 51,* 550–559.

Berkes, F., & Jolly, D. (2001). Adapting to climate change: Social-ecological resilience in a Canadian Western Arctic community. *Conservation Ecology, 5,* 1–15.

Bunce, A., Ford, J., Harper, S., Edge, V., & Team, I. R. (2016). Vulnerability and adaptive capacity of Inuit women to climate change: A case study from Iqaluit, Nunavut. *Natural Hazards, 83,* 1419–1441.

Crate, S. A. (2012). Climate change and ice dependent communities: Perspectives from Siberia and Labrador. *Polar Journal, 2,* 61–75.

Cuerrier, A., Brunet, N. D., Gerin-Lajoie, J., Downing, A., & Levesque, E. (2015). The study of Inuit knowledge of climate change in Nunavik, Quebec: A mixed methods approach. *Human Ecology, 43,* 379–394.

Ford, J. D., Gough, W. A., Laidler, G. J., MacDonald, J., Irngaut, C., & Qrunnut, K. (2009). Sea ice, climate change, and community vulnerability in northern Foxe Basin, Canada. *Climate Research, 38,* 137–154.

Ford, J. D., Lardeau, M. P., Blackett, H., Chatwood, S., & Kurszewski, D. (2013). Community food program use in Inuvik, Northwest Territories. *BMC Public Health, 13,* 1–15.

Ford, J. D., Smit, B., Wandel, J., Allurut, M., Shappa, K., Ittusarjuat, H., et al. (2008). Climate change in the Arctic: Current and future vulnerability in two Inuit communities in Canada. *Geographical Journal, 174,* 45–62.

Ford, J. D., Smit, B., Wandel, J., & MacDonald, J. (2006a). Vulnerability to climate change in Igloolik, Nunavut: What we can learn from the past and present. *Polar Record, 42,* 127–138.

Ford, J. D., Smit, B., & Wandel, J. (2006b). Vulnerability to climate change in the Arctic: A case study from Arctic Bay, Canada. *Global Environmental Change-Human and Policy Dimensions, 16,* 145–160.

Gearheard, S., Pocernich, M., Stewart, R., Sanguya, J., & Huntington, H. P. (2010). Linking Inuit knowledge and meteorological station observations to understand changing wind patterns at Clyde River, Nunavut. *Climatic Change, 100,* 267–294.

Giles, A. R., Strachan, S. M., Doucette, M., & Stadig, G. S. (2013). Adaptation to aquatic risks due to climate change in Pangnirtung, Nunavut. *Arctic, 66,* 207–217.

Goldhar, C., Bell, T., & Wolf, J. (2014). Vulnerability to freshwater changes in the Inuit Settlement Region of Nunatsiavut, Labrador: A case study from Rigolet. *Arctic, 67,* 71–83.

Hatcher, S. V., & Forbes, D. L. (2015). Exposure to coastal hazards in a rapidly expanding Northern Urban Centre, Iqaluit, Nunavut. *Arctic, 68,* 453–471.

Laidler, G. J., Ford, J. D., Gough, W. A., Ikummaq, T., Gagnon, A. S., Kowal, S., et al. (2009). Travelling and hunting in a changing Arctic: Assessing Inuit vulnerability to sea ice change in Igloolik, Nunavut. *Climatic Change, 94,* 363–397.

Martin, D., Bélanger, D., Gosselin, P., Brazeau, J., Furgal, C., & Déry, S. (2007). Drinking water and potential threats to human health in nunavik: Adaptation strategies under climate change conditions. *Arctic, 60,* 195–202.

Nancarrow, T. L., & Chan, H. M. (2010). Observations of environmental changes and potential dietary impacts in two communities in Nunavut, Canada. *Rural and Remote Health, 10,* 1–12.

Nichols, T., Berkes, F., Jolly, D., & Snow, N. B. (2004). Climate change and sea ice: Local observations from the Canadian western Arctic. *Arctic, 57,* 68–79.

Nickels, S., Furgal, C., Buell, M., & Moquin, H. (2005). *Unikkaaqatigiit-putting the human face on climate change: Perspectives from Inuit in Canada* (pp. 1–124). Joint Publication of Inuit Tapiriit Kanatami, Nasivvik Centre for Inuit Health and Changing Environments at Universite Laval and the Ajunnginiq Centre, National Aboriginal Health Organization.

Prno, J., Bradshaw, B., Wandel, J., Pearce, T., Smit, B., & Tozer, L. (2011). Community vulnerability to climate change in the context of other exposure-sensitivities in Kugluktuk, Nunavut. *Polar Research, 30,* 1–21.

Wesche, S. D., & Chan, H. M. (2010). Adapting to the Impacts of Climate Change on Food Security among Inuit in the Western Canadian Arctic. *EcoHealth, 7,* 361–373.

Wildland Fire, Extreme Weather and Society: Implications of a History of Fire Suppression in California, USA

Donald Schweizer, Tom Nichols, Ricardo Cisneros, Kathleen Navarro and Trent Procter

Abstract Wildland fire is a natural process integral to the formation and health of forest ecosystems globally. California, USA, is case study where large areas of wildland have a recent 100 plus year history of human suppression of fire that with extreme weather is combining to create large high intensity burns changing both species composition and increasing threats to life, health and property. The cool wet winters and hot dry summers in California produce a climate where fire is common and many environmental systems have evolved to rely on frequent fire for reproduction and health. Fire has been systemically removed creating a backlog of fuels as vegetation normally burned accumulates. Extreme weather enhanced by climate change is increasing the duration of the fire season and occurrence of extreme fire weather and events. The abundance of fuels and increase in probability of fire, primarily due to human-caused ignitions in the wildland–urban interface, are creating an increase of large catastrophic fires not typical of these ecosystems. These large high-intensity fires are an immediate threat to life and property, produce large amounts of smoke impacting human health far from the fire, and leave behind a burn area then susceptible to extreme rainfall events that create landslides and mudslides. Returning fire to the historic role it has played in sustaining these systems reduces the probability of catastrophic fire and the conditions where extreme rainfall can additionally cause further threats to life and property from debris flows. Exposure of the public to smoke from wildfire increases when high intensity burns occur. Wildland fire typical

D. Schweizer (✉)
Pacific Southwest Region, USDA Forest Service, Bishop, CA 93514, USA
e-mail: dschweizer@ucmerced.edu

D. Schweizer · R. Cisneros
Public Health Graduate Group, Health Sciences Research Institute, University of California, Merced, CA 95343, USA

T. Nichols
Eagle, ID 83616, USA

K. Navarro
USDA Forest Service, Manti-La Sal National Forest, Price, UT 84501, USA

T. Procter
USDA Forest Service, Pacific Southwest Region, Vallejo, CA 94592, USA

R. Akhtar (ed.), *Extreme Weather Events and Human Health*,
https://doi.org/10.1007/978-3-030-23773-8_4

of this ecosystem which occurred before suppression limited the extent and amount of such exposure. There are current attempts to effect positive change to policy and give a voice to the role of fire in the ecosystem. Long-standing policy based on the unsustainable complete exclusion of fire and public pressure on air quality inhibits functional change to smoke and fire management. The collision of what current fire and smoke science advises as the appropriate action (inclusion of fire as a land management tool), and public opinion driving implementation of fire management decisions in California (the exclusion of fire) illustrates a global problem where climate change and policy driven by belief are synergistically worsening environmental and human health.

Keywords Wildland fire · Extreme weather · Smoke management policy

1 Introduction

Disturbance from fires is essential to the ecological health of forests and natural areas worldwide (Stephens et al. 2014, pp. 116–117; Thonicke et al. 2008, p. 670). Fuel loading was relatively light in areas where fire was frequent before the era of suppression which has caused multiple fire cycles to be missed and fuels to accumulate. Increases in temperature and drought in others areas, such as the northern Rocky Mountains, USA or the boreal forest biome, with longer intervals between fires are also causing concern because fires are expected to increase in occurrence, area and severity (Gauthier et al. 2015, p. 820).

Fire has an important role to play in the forest ecosystem; however, increased human development of forested areas has reduced the role of fire in the environmental system. Fire is widely suppressed across the world. Suppression is integral to good fire management in a modern world but may undervalue the long-term benefits of this natural disturbance, as well as the social and ecological consequences of the attempted suppression of all fires. Large destructive fires are occurring globally from Mediterranean region including Greece and Portugal to the arctic boreal forests of Canada, Sweden and Russia to the Australian bushfires. Media coverage tells of the destruction of life and property and the devastation of fire when suppression fails. This cycle of using suppression to postpone the occurrence of fire and then reacting to the devastation when suppression does not work with even greater suppression efforts appears to be a policy failure of modern land use because it largely ignores the importance of natural process in forest health and as a mitigation to the effects of climate change.

Extreme weather events and their frequency are a concern globally. How often and where extreme events have been occurring is not uniform spatially (Easterling et al. 2000, p. 417). Where extreme weather combines with an increased fire size and intensity, large high intensity “megafires” can occur. Anthropogenic-driven climatic stressors are contributing to an increase in megafires (Flannigan et al. 2013, pp. 59–60). The negative effects of these fires are heightened by policies and human behavior which lead to increased development susceptible to burning in fire-prone ecosystems (Calkin et al. 2015, pp. 5–6) and extreme levels of fuel loading from active suppression of fire (Steel et al. 2015, pp. 19–20).

Megafire is increasing largely from increased fuel loading and continuity, rising temperatures and longer fire seasons, and increased development within and around fire-prone forests. The 2017 fire season in California, USA, is an excellent example of the conflict between natural process and modern land use and the consequences when suppression fails. The 2017 wildfire season in California was foreseeable with 45 plus years warning of an out of balance fire-prone ecosystem (Kilgore 1973, p. 511) and anthropogenic emissions altering climate. Despite scientific evidence, fire is rarely allowed to burn as it did for millennia. This is particularly true in wilderness areas far from development where the short-term benefits from suppression are largely elimination of smoke.

The use of naturally occurring fire as a management tool along with both suppression and prescribed fire (planned fire with an arranged set of weather and safety conditions to meet burn objectives) each have a role in comprehensive fire management. A question to consider is whether we are accurately valuing the role fire has in the ecosystem for long-term human and ecological health or are we unduly biasing our actions to human intervention. In this chapter, we explore the consequences when suppression policy, or the implementation of the policy, fails and suggest that current air and fire policy and management need substantive change to limit catastrophic, unwanted fire. We discuss the importance of fire as a natural process for forest and human health and how air management policy must recognize that while smoke may be undesirable to the public, it is inevitable and that current air and land management policy actions are contributing to the problem through institutionalized lack of foresight.

2 Fuel Loading and Extreme Weather Influence on Wildland Fire in California, USA

Wildland fire is a natural process with widespread occurrence throughout much of California, affecting natural and cultural resources, human health, infrastructure and public safety. It has a wide range of behavior and effects, which are due to combinations of topography, climate, weather patterns, vegetation types and both natural and human sources of ignition.

The general relationship among fire cycle (years between fires), fire behavior (flame lengths, rate of spread) and vegetation type is called a fire regime. Federally protected land managed for resource objectives include much of the landscape of the American West. Although federal land management agencies (e.g. US Forest Service, National Park Service, Bureau of Land Management, Bureau of Indian Affairs, and the Fish and Wildlife Service) have different founding legislation and missions from resource extraction to conservation, they all are required to minimize the impact of development and use of the land they manage. The agencies additionally have large areas of federal Class I Wilderness Areas (Fig. 1), protected and managed so as to preserve its natural conditions, which are protected by the 1964 Wilderness Act.

This large area of little developed land and wilderness in the western USA embodies the scope of the issue of smoke impacts to populated areas adjacent and dependent on fire-prone areas with fire-adapted species dependent on an active fire regime to remain healthy and unimpaired for future generations. Prior to European settlement

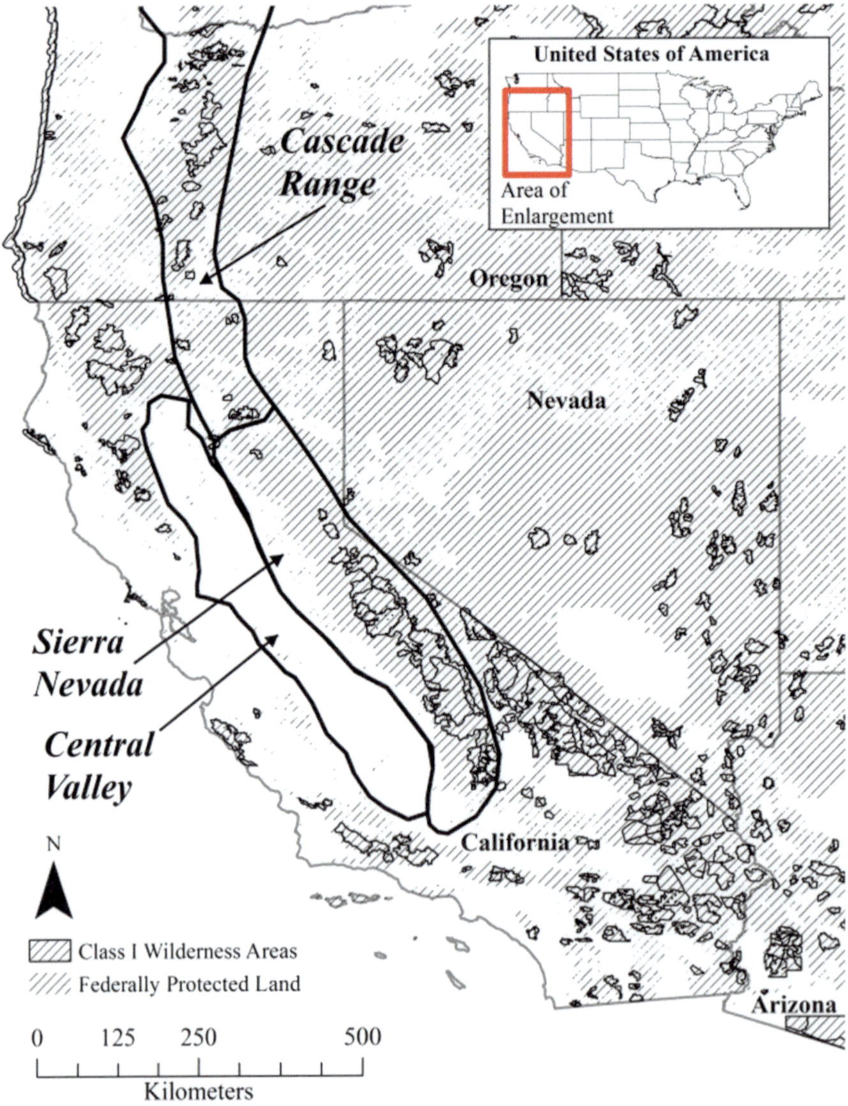

Fig. 1 Wilderness and other federal land management areas in California area, USA

(~1800), stands of mixed conifer forests on the western Sierra Nevada of California (Fig. 1) burned frequently. This fire regime is characterized by low-intensity fires which occurred historically on a cycle of twenty years or less (Sugihara et al. 2006, p. 270). Other vegetation types, such as subalpine forests or coastal chaparral, have fire regimes with longer cycles but can burn more intensely.

Ecosystems tend to be adapted to a specific fire regime. If fire occurs too infrequently, fire regimes with short cycles will be impaired by increasing tree density, an accumulation of fuel, and unnaturally intense fires which often kill mature trees. If fire is too frequent, such as the result of increased human-caused ignitions from arson or carelessness, fire regimes with longer fire cycles will be impaired and over-burn. For example, chaparral can be replaced by invasive grasses if burned too frequently.

Beginning in the early twentieth century, wildfires ignited by human and natural sources begin to be suppressed by local, state and federal fire agencies. While wildfire suppression had occurred before this period, the 1910 wildfires in the Northern Rockies influenced the US Forest Service to approach fire suppression more systematically, and as a cornerstone of its land management policy, which in turn influenced other agencies. The use of fire, such as prescribed fires ignited to produce various post-fire benefits, was also discouraged, although the practice continued in the southeastern USA.

The suppression of wildfires which threaten public safety continues to be an important objective. Protection of life and property is an important component of sound wildland fire management. However, the suppression of all natural fires has significant consequences to forest health and resiliency, especially in areas of California which are characterized by fire regimes with shorter fire cycles, such as the Sierra Nevada and southern Cascades.

Prior to the onset of fire suppression, recently burned areas had reduced quantities of wildland fuels and new, green growth which acted as barriers to fire spread from subsequent ignitions, producing a mosaic of different aged burned areas which could affect and limit the spread of wildfires.

Due to fire suppression, the historic fire mosaic has been lost in many forested ecosystems in the western and southwestern USA. This has been replaced by a layer of continuous wildland fuels which supports the rapid growth of wildfires, especially during wind events. As noted by the California Department of Forestry and Fire Protection (Cal Fire), since accurate records started to be kept in 1932, the top five and seven of the ten largest wildfires in the state have occurred since 2000 (Cal Fire 2018). California fire statistics throughout this chapter only include the last full year of record (2017). It should be noted that 2018, without a complete annual record, already has the largest California wildfire (Mendocino Complex) and the at least 13th deadliest (Carr Fire), suggesting the large destructive wildfires of 2017 were not an anomaly (Fig. 2).

Fire is important to ecological health. Past policies removing fire from the landscape have made these systems susceptible to extreme weather and delayed smoke impacts to the future (Cisneros et al. 2018b, p. 123). Additionally, the changing climate increases fire potential (Liu et al. 2010, pp. 688–691). Extreme weather resulting from climate change is increasing and wildland fire is being affected. Anthropogenic

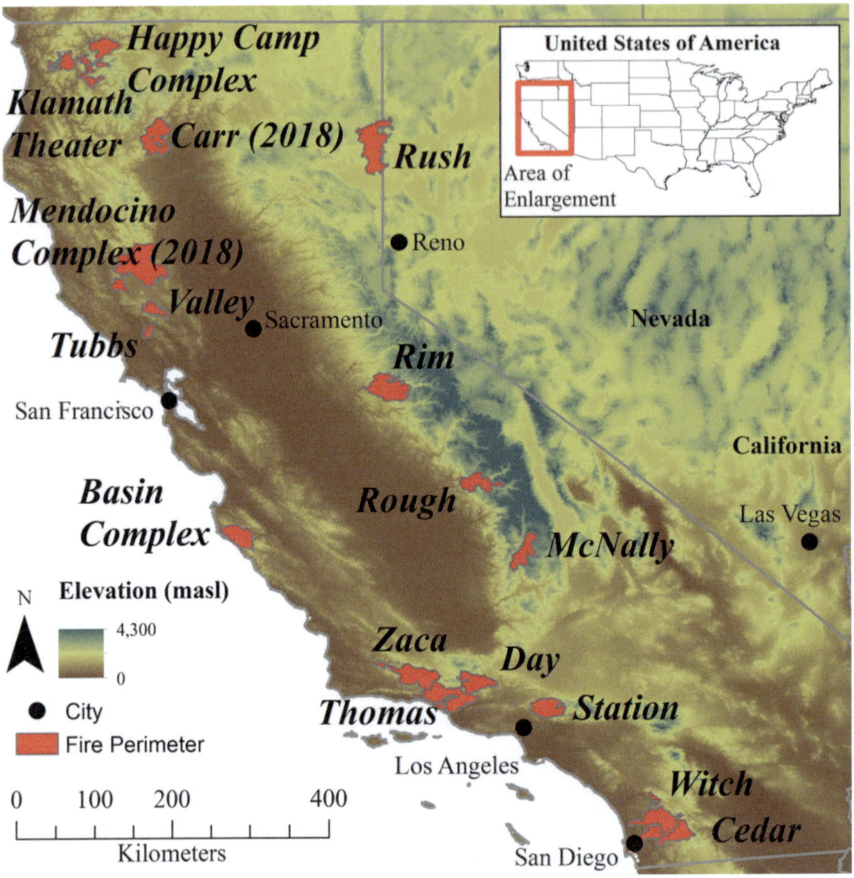

Fig. 2 Significant fires in California since 2002 including the 2018 Carr and Mendocino Complex Fires

emissions are enhancing extreme fire conditions, in which the risk of fire ignition and spread is increasing and fire seasons are being extended (Kirchmeier-Young et al. 2017, pp. 374–376).

More extreme weather and fire conditions are a reality for fire management (Dowdy 2018, p. 233), but fire management strategies and policy are slow to change (North et al. 2015, pp. 1280–1281). Climate and weather impacts to forest fuels provide one metric for early warning and planning of fire management resources (Boer et al. 2017, p. 1201; Littell et al. 2009, pp. 1017–1019). Suppression, a reactive management strategy, is usually the primary tool used by management because the decision process tends to favor the elimination of immediate wildfire risk, with little to no incentive to consider consequences of such decisions that can occur a year or many years in the future. This risk avoidance combined with other factors such as the availability of fire management resources such as engines and aircraft, air quality and public tolerance of smoke, other wildfire activity, etc. creates an environment where there is always a reason not to use fire as a management tool today.

While reactive measures to fire are integral to protect life and property, a healthy forest using proactive fire management for resilience and adaptation is integral to long-term prevention of this loss (Stephens et al. 2018, p. 85). There is an ecological necessity of fire as a natural process for forest health (Hutto et al. 2016, pp. 4–6; Kilgore 1973, pp. 510–511). Species which have evolved in the presence of fire have developed many responses to it, often integrating it into their life cycles. Chaparral species may re-sprout after fire, while some forest species such as knobcone pine (*Pinus attenuata*) and giant sequoia (*Sequoiadendron giganteum*) possess cones which are opened by the heat of a fire, dropping their seeds on nutrient-rich ash. Native grasses such as deergrass (*Muhlenbergia rigens*) are stimulated by fire.

3 US Fire Policy History

Pre-European settlement of the USA, fire was not actively suppressed and was used for resource enhancement. Native Americans were aware of the post-fire responses of many species, and ignited fires to encourage these benefits (Sugihara et al. 2006, p. 272). As settlement increased, suppression became the dominant means to manage fire in the early twentieth century particularly in the western USA. As a result of wildland fire research in the mid-twentieth century, it became apparent that fire suppression as the only management strategy produced long-term negative consequences such as the accumulation of wildland fuels which supported more intense and damaging fires than had occurred during the pre-settlement period.

In the 1970s, federal fire policy changed from one of fire control, in which all wildfires were aggressively suppressed, to one of fire management, which involved a combination of aggressive wildfire suppression and the use of fire for various purposes. This concept of "fire use" could include allowing wildfires to burn for multiple objectives, as well as the execution of prescribed fires ignited by agency staff, for the restoration or maintenance of fire-dependent ecosystems. Fire could also be used to reduce hazardous quantities of wildland fuel, especially near the wildland–urban interface.

The change in federal fire policy has presented land management agencies with a broad spectrum of methods by which public and firefighter safety, property protection and resource management objectives may be pursued. However, wildfire suppression remains the dominant approach, with the protection of safety and property of greatest concern. The use of prescribed and natural fire to restore fire dependent ecosystems which have become impaired by fire suppression, or to meet other management objectives such as wildfire hazard fuel reduction, occurs much less often.

4 Current Conditions: The Backlog of Fire on the Landscape Scale

Nationally, over 4 million hectares of wildfire occurred in 2017 with a federal firefighting cost of 2.9 billion dollars (NIFC 2018). Wildfires during 2015 and 2017 covered the largest area of any years since 1983 with recent prescribed years (2015, 2016 and 2017) also having the largest areas treated (Table 1).

Except for the Bureau of Land Management, the majority of prescribed fire acres occurred in the southeastern USA; the exception was the southwestern USA for the Bureau of Land Management. Generally, wildland fuels in these areas are grassy or brushy, burn more quickly, and produce less smoke than do the heavier forest fuels found in the western USA.

There have been 187 large fires (over 100,000 acres or ~40,000 ha) in the USA since 1997 (NIFC 2018). The largest to burn land in California (32nd largest in the nation 1997–2017) was the 2012 Rush Fire (127,710 ha). The Rush Fire burned in both California (110,039 ha) and Nevada (17,671 ha) and thus the third largest California wildfire ((1) 2017 Thomas 114,078 ha; (2) 2003 Cedar 110,579 ha) to date (Cal Fire 2018). While California has not experienced the largest fires in the USA, the state is fire-prone and densely populated. The year 2017 was a destructive year for fire in California. In 2017, 5 fires in California destroyed 9386 structures and directly caused 40 deaths (Cal Fire 2018).

Table 1 Area burned by wildfire in the USA

Year	Wildfire (ha)	Prescribed (ha)	Total (ha)
2017	4,057,417	2,601,819	6,659,236
2016	2,229,818	1,625,021	3,854,839
2015	4,097,506	1,197,166	5,294,672
2014	1,455,094	967,118	2,422,212
2013	1,748,060	809,388	2,557,448
2012	3,774,198	797,974	4,572,172
2011	3,525,368	855,025	4,380,393
2010	1,385,128	980,903	2,366,032
2009	2,396,464	1,024,306	3,420,770
2008	2,141,788	783,068	2,924,856
2007	3,774,929	1,274,383	5,049,313

From the National Interagency Fire Center fire information statistics (NIFC 2018)

5 Impacts to Life and Property

There are also indirect impacts to life, health and property from wildfire. Table 2 summarizes California fires with the largest losses to life and property (Cal Fire 2018). Loss of life and property can be expected to increase even with advances in firefighting technology and tactics as population increases and fires get larger and more intense. Large loss of life and property can occur on both large and small wildfires. Through 2017, the 19 wildfires covering the largest area in California are all since 1999 and correspond to a period of significant growth of communities into California wildlands. Fire managers are increasingly challenged to suppress larger and more intense wildfires while safeguarding their employees and the public while protecting ever more structures in and adjacent to wildlands.

Burned areas are more vulnerable to erosion and endanger water supplies (Bladon et al. 2014, p. 8939). In addition to ongoing threats to drinking water, extreme rainfall can induce mudslides that directly threaten life and property. For example, the 2017 Thomas Fire, which was the largest recorded California wildfire (until the 2018 Mendocino Complex Fire), was responsible for 1 death (Cal Fire 2018) while the subsequent mudslides after high intensity rains were responsible for over 20 deaths. The Thomas Fire burned north of Los Angeles in Ventura and Santa Barbara Counties in December of 2017 (Fig. 2). Extreme weather including the prolonged Santa Ana winds (strong, dry down-slope winds that flow from inland areas to coastal California) helped spread the fire and, along with the lack of significant recent precipitation, contributed to the large area burned. The fire forced over a hundred thousand residents to be evacuated. Windy, warm and dry weather conditions were then followed by heavy rains starting in early January, 2018, which caused flash floods and mudflows. High-intensity fires increase the potential for debris flows because they remove vegetation that helps hold soil in place and creates a hydrophobic top layer of burned soil (Ren et al. 2011, p. 336). Multiple heavy rain events occurred in this area of southern California from January through July of 2018 with thousands of residents being told to evacuate some multiple times. The extreme weather altering the seasonal timing, extent and severity of the fire followed by extreme rainfall immediately after the fire compounded and extended the disruption to residents of southern California.

Table 2 Selected California wildfires with large loss of life (firefighter and public) and property

Fire	Size (ha)	Year	Deaths	Structures lost
Griffith Park	19	1933	29	0
Tunnel	647	1991	25	2900
Tubbs	14,895	2017	22	5643
Cedar	110,579	2003	15	2820
Rattlesnake	542	1953	15	0
Valley	30,783	2015	4	1955
Witch	80,124	2007	2	1650

It is difficult to capture the impacts of extreme weather and wildfire as it is a complex interaction of the environment with far-reaching effects that can come with the wildfire, the rainy season after, or many years later to ecosystem services such as municipal water supplies. However, increases in extreme weather events amplify the direct impacts to life and property from wildfire.

In order to mitigate the increasing threat of wildfires, the western USA has enormous wildland fuel treatment needs; for example, 3.8 million hectares of forests in Oregon and Washington are in need of prescribed fire and thinning for restoration of conditions which are more resilient to wildfire (Haugo et al. 2015, p. 37). Not all of this area must be treated; as noted by (Finney et al. 2006, p. 125), prescribed fire and other methods of wildland fuels reduction which treat only 10% to 20% of the landscape each decade are effective in disrupting the growth of large wildfires.

Even so, the area treated annually with prescribed fire and wildfire managed for multiple objectives falls far short of the need for such treatments such as for wildland fuel hazard reduction, even as wildfires are becoming larger, more damaging, and expensive. This continuing pattern indicates the discrepancy between the flexible fire policy which supports the use of fire, and the results of the implementation of this policy.

6 Impediments for Beneficial Fire

Air quality regulations, and their application to smoke management, are commonly listed as an impediment to the use of prescribed fire or of wildfire managed for multiple objectives. However, the fire management community also contributes to the treatment shortfall; even if air quality regulators identify conditions under which prescribed fires may be ignited, fire managers may decline the opportunity. For example, while spring weather may have excellent smoke dispersal conditions, fire managers are reluctant to ignite prescribed fires which may continue to burn into the warming and drying summer season, because of a risk that the fire may eventually escape the intended burn unit. Consequently, fire managers prefer to conduct prescribed fires in the fall, when conditions are cooler; however, smoke dispersal is also poorer, and burning permits are more difficult to obtain.

Added to these factors are the frequent occurrence of large wildfires in California in the summer and fall, which cause significant amounts of smoke. Even when such wildfires are not present, there is at least a concern that such fires may occur during this time and will require massive numbers of firefighters to fight, and so there is a reluctance by fire mangers to commit fire staff to prescribed fires and wildfires allowed to burn for multiple objectives and so be unavailable for firefighting if the need arises.

The use of fire will continue to be impaired because the same fire staff charged with short-term fire suppression emergencies is also charged with long-term planning, staffing and management of prescribed fires and wildfires managed for multiple objectives which may cover thousands of hectares and last for several weeks. Area treated will continue to be a fraction of what is needed, and this pattern is unlikely to change unless fire staff is established, dedicated and funded specifically to conduct proactive fire management strategies throughout the fire season.

7 Why not Burn More?

Some ecosystems can benefit from fire suppression, such as chaparral. If burned too often, it can be converted to grassland. However, many California ecosystems are adapted to short fire cycles, and the suppression of fire has caused them to degrade and be more vulnerable to fire. Such forests are dependent on frequent low-intensity fire where smoke will be visible. At the least, wilderness users and rural communities near the fire will be impacted. Smoke will inevitably become a reason to slow down, speed up, or suppress the burn.

Wildfire smoke is arguably the most reviled ecosystem process. While the role wildland fire has in forest and ecosystem health is well studied, the ecosystem role of smoke is not. Limited research has indicated smoke has a role to play in the fire-prone ecosystems of California (Keeley et al. 2005, p. 2320; Klocke et al. 2011, p. 11; Parmeter and Uhrenholdt 1975, p. 31; Thomas et al. 2007, p. 605). Estimations of emissions and area burned in California suggest years that are currently considered extreme fire years are actually more typical of the pre-European settlement period (Stephens et al. 2007, p. 213). The lack of the historic smoke cycle in California may have unintended consequences unrelated to the immediate impacts to human health and comfort. More thoroughly understanding the role smoke plays in the ecosystem will help to further quantify the impacts of fire suppression.

8 Biases and Shortfalls of Smoke Management

Wildfire is inevitable and there are multiple reasons not to burn. But, is smoke perhaps the easiest to remedy? Smoke is in no means the only reason not to burn but it is an important consideration in the anthropogenically polluted airsheds of California. Even though wildland fire smoke has little impact on attainment of air pollution standards (Preisler et al. 2015, p. 347), it is still unpleasant and, for a population that has grown up and through the era of suppression, can be intolerable. In addition, there have been documented adverse health impacts from smoke to the public. Past studies have reported evidence of wildfire smoke being associated with respiratory morbidity, which included exacerbations of symptoms of chronic obstructive pulmonary disease and asthma, and possibly increased mortality from all causes and respiratory infections (Reid et al. 2016, p. 1341).

The policy driving smoke management is a clash of unattainable public wants of a smoke-free environment versus the inevitable smoke from fire. Restricting prescribed and ecologically beneficial fire puts the burden of increased fire risk and smoke exposure onto future generations. Under current policy, air regulators can pass impacts to the future with little or no consequences. The fire next year is rarely attributed to the fire put off today. In addition, fire managers often live in the most smoke impacted areas and often take the brunt of local dissatisfaction. It is easy and convenient to discount the long-term ecological and fire risk impacts of their decisions with regard to the management of natural or prescribed fire where their neighbors are incensed over smoke.

Event management, or documenting the amount and location of smoke and trying to help the public reduce their exposure, is now common on most fires assessing management decisions and is not designed to assess long-term degradation of the environmental system from suppression actions (Schweizer and Cisneros 2017, p. 34). Wildfire smoke, a natural process, is not managed like other naturally occurring emissions such as pollen or volcanic gasses. Culturally, we are biased to believe we can control wildland fire smoke and it is therefore different. Emissions are calculated but unlike anthropogenic emissions have no economic pressure to bypass emission controls and calculations. Wildfire emissions additionally are a much more convenient source to explain air quality. Unlike automobile, agricultural, or industrial emissions which have large financial incentives to err to the lowest amounts or avoid emission controls, naturally occurring wildfire helps these sources most when smoke emissions can be overestimated thereby providing another source for air regulators to focus upon.

Fire management smoke policy is largely dictated by small vocal groups who have an unrealistic belief that all fires can be suppressed (Schweizer and Cisneros 2017, p. 34; Cisneros et al. 2018a, p. 1). But, that is an underestimate of public perception and support. Californians living in and around smoke-prone areas with high levels of anthropogenic pollution have a good understanding of the relative importance of smoke to their overall air quality (Cisneros et al. 2017, p. 5). Nuisance complaints can be an excellent input for fire and air managers but should not be used to determine smoke management (Cisneros and Schweizer 2018, pp. 428–429).

9 A Path to the Future: Using Area Monitoring and Regulatory Thresholds to Assess Smoke Impacts?

While impossible, attempting to eliminate smoke through suppression can help garner support from smoke sensitive groups but ignores both the increased threat to health and safety when suppression fails and the unknown effects to forest ecosystem health in a naturally smoky area. However, current event-driven smoke management is not the solution (Schweizer and Cisneros 2017, p. 34). Model estimates of emissions and atmospheric transport, while important to getting information to the public to reduce smoke exposure, do not accurately reflect the ground-based conditions and without empirical data for exposure are contributing to a false narrative of impacts over time (Cisneros and Schweizer 2018, p. 423; Schweizer and Cisneros 2014, p. 276).

For this reason, there needs to be two distinct independent strategies when managing smoke—(1) event management and (2) proactive smoke management. Event management is needed for a given fire where priority is given to early warning of worse case scenarios. Proactive smoke management needs to consider all potential fires and past impacts and look comparatively at management decisions critically. While event management is widely in place in the USA, proactive smoke management is not. Currently, smoke management is biased to single event analysis or how bad is the smoke in a given location and how adamant is public disapproval.

Smoke from fires burning at historic normal levels of size and intensity can reduce the smoke impacts from other fire types and remain largely below attainment standards (Schweizer and Cisneros 2014, p. 276). Prescribed fire can also limit the impacts through reducing the order of magnitude in size from a large high-intensity fire (Navarro et al. 2018, p. 1). However, any manipulation of a fire that is burning within the historic normal, including the addition of fire during times of good dispersal to allay public smoke exposure concerns, increases smoke exposure (Schweizer et al. 2017, p. 354). Using established air quality standards and thresholds at a landscape level is important for smoke management to be effective (Schweizer et al. 2017, pp. 351–353).

Net smoke exposure seems to be tied fundamentally to the extent of fire management ground actions. Smoke impacts from planned prescribed fire can be beneficial in that the total emissions per day are reduced and meteorological conditions can be chosen while unplanned high intensity forest fires are not. But, for fire in areas such as wilderness where no life or property is at stake, burning at historic timing, size and intensity reduces smoke exposure by limiting dispersal primarily to the less populated area near the fire. This is in contrast with the high plume heights associated with high intensity burns that can impact large urban centers long distances from the fire. Sound smoke management to reduce net exposure needs to favor wildland fires that need little to no fire management action wherever it is safe and possible (e.g. wilderness). Natural process dominated ecosystems will provide the best health outcomes. Favoring a return to natural process fire is integral to sound smoke management policy particularly in the era of increased extreme weather events from climate change. This ideal is not immediately possible because of fuel build-up, increased wildland–urban development and climate change but should be the goal. Particularly in remote wilderness locations, fire that can be left alone should be and fire and air managers along with the public need patience. There is a time and a place for suppression. There is a time and place for prescribed fire. But, there is also a time and place for naturally occurring wildfire to function in the environmental system.

10 Summary and Conclusion

Although many people and organizations are attempting to effect positive change to policy and its implementation and giving a voice to the role of fire in the ecosystem, long-standing policy based on what the public wants, particularly in regards to clean air, inhibits functional change to smoke and fire management. The short-term benefits, including putting off smoke to future generations, belie the narrative that there are no consequences and man's dominion over environmental systems is infallibly beneficial. Management actions largely suppress wildfire while prescribed fire is currently the advocated action by policymakers and implementers. While there is a role for prescribed fire in fire-prone ecosystem restoration, attempting to replace suppression with prescribed fire is largely dealing with the symptoms of fuel buildup and development in the wildland–urban interface. This strategy may allow air and land managers and the public to feel good about reintroducing fire to the system but assumes our process is superior to natural process. Wildland fire smoke from natural process fire takes patience and assumes some of the liability now when it can often be put off.

Extreme weather, climate change, and a history of suppression are combining to limit our ability to defer fire in many areas. Increased wildland fire will come with increased smoke. Smoke can be at the highest exposure levels from megafires when suppression fails (Cisneros et al. 2018b, p. 122). Alternatively, restoring the fire cycle to forests and wilderness far from developed areas may prove the best to reduce adverse health outcomes from smoke exposure and additionally provide forest resiliency to help mitigate loss of life and property from large high-intensity fires more likely under extreme weather events. However, wildland fire smoke in a community comes with a myriad of short-term consequences to public comfort and health. The long-term benefits can easily be disregarded today when we think we can put off the burn to some indefinite future. Suppression also created an artificial expectation of smoke free skies in the American West. Compounded with this unrealistic expectation is the reality that a century of fire suppression has led to an unnatural accumulation of fuels that ultimately will burn producing levels of smoke emissions orders of magnitude larger than occurred before suppression. This "smoke debt" is difficult for a community to accept now because it is more palatable to put it off to some undetermined date in the future. The burden of debt is significant for the current generation, which will have to accept, address and live through the smoke debt when fire occurs and the debt comes due. Deferring both the fuel buildup and smoke debt is a convenient and well-learned behavior developed over many generations in the American West. However, treating these fuels and reducing smoke exposure will not be easier tomorrow.

References

Bladon, K. D., Emelko, M. B., Silins, U., & Stone, M. (2014). Wildfire and the future of water supply. *Environmental Science & Technology, 48,* 8936–8943. https://doi.org/10.1021/es500130g.

Boer, M. M., Nolan, R. H., Resco De Dios, V., Clarke, H., Price, O. F., & Bradstock, R. A. (2017). Changing weather extremes call for early warning of potential for catastrophic fire. *Earth's Future, 5,* 1196–1202. https://doi.org/10.1002/2017EF000657.

Cal Fire. (2018). California Department of Forestry and Fire Protection Statistics and Events [WWW Document]. http://cdfdata.fire.ca.gov/incidents/incidents_statsevents. Accessed July 9, 2018.

Calkin, D. E., Thompson, M. P., & Finney, M. A. (2015). Negative consequences of positive feedbacks in US wildfire management. *Forest Ecosystems, 2,* 9. https://doi.org/10.1186/s40663-015-0033-8.

Cisneros, R., & Schweizer, D. W. (2018). The efficacy of news releases, news reports, and public nuisance complaints for determining smoke impacts to air quality from wildland fire. *Air Quality, Atmosphere & Health, 11,* 423–429. https://doi.org/10.1007/s11869-018-0545-9.

Cisneros, R., Brown, P., Cameron, L., Gaab, E., Gonzalez, M., Ramondt, S., et al. (2017). Understanding public views about air quality and air pollution sources in the San Joaquin Valley, California. *Journal of Environmental and Public Health, 2017,* 1–7. https://doi.org/10.1155/2017/4535142.

Cisneros, R., Alcala, E., Schweizer, D., & Burke, N. (2018a). Smoke complaints caused by wildland fire in the southern Sierra Nevada region, California. *International Journal of Wildland Fire.* https://doi.org/10.1071/wf18060.

Cisneros, R., Schweizer, D., Tarnay, L., Navarro, K., Veloz, D., Trent Procter, C., et al. (2018b). Climate change, forest fires, and health in California. In R. Akhtar & C. Palagiano (Eds.), *Climate change and air pollution* (pp. 99–130). Springer International Publishing. https://doi.org/10.1007/978-3-319-61346-8_8.

Dowdy, A. J. (2018). Climatological variability of fire weather in Australia. *Journal of Applied Meteorology and Climatology, 57,* 221–234. https://doi.org/10.1175/JAMC-D-17-0167.1.

Easterling, D. R., Evans, J. L., Groisman, P. Y., Karl, T. R., Kunkel, K. E., & Ambenje, P. (2000). Observed variability and trends in extreme climate events: A brief review. *Bulletin of the American Meteorological Society, 81,* 417–425. https://doi.org/10.1175/1520-0477(2000)081%3c0417:OVATIE%3e2.3.CO;2.

Finney, M. A., Seli, R. C., Mchugh, C. W., Ager, A. A., Bahro, B., & Agee, J. K. (2006). Simulation of long-term landscape-level fuel treatment effects on large wildfires. In P. L. Andrews & B. W. Butler (comps) *Fuels Management*: *How to Measure Success*: *Conference Proceedings*. 28–30 March 2006, Portland, OR. *Proceedings of RMRS-P-41*. Fort Collins, CO: U.S. Department *of* Agriculture, *Forest* Service, Rocky Mountain.

Flannigan, M., Cantin, A. S., De Groot, W. J., Wotton, M., Newbery, A., & Gowman, L. M. (2013). Global wildland fire season severity in the 21st century. *Forest Ecology and Management, 294,* 54–61. https://doi.org/10.1016/j.foreco.2012.10.022.

Gauthier, S., Bernier, P., Kuuluvainen, T., Shvidenko, A. Z., & Schepaschenko, D. G. (2015). Boreal forest health and global change. *Science, 349,* 819–822. https://doi.org/10.1126/science.aaa9092.

Haugo, R., Zanger, C., DeMeo, T., Ringo, C., Shlisky, A., Blankenship, K., et al. (2015). A new approach to evaluate forest structure restoration needs across Oregon and Washington. *Forest Ecology and Management, 335,* 37–50. https://doi.org/10.1016/j.foreco.2014.09.014.

Hutto, R. L., Keane, R. E., Sherriff, R. L., Rota, C. T., Eby, L. A., & Saab, V. A. (2016). Toward a more ecologically informed view of severe forest fires. *Ecosphere, 7,* 1–13. https://doi.org/10.1002/ecs2.1255.

Keeley, J. E., McGinnis, T. W., & Bollens, K. A. (2005). Seed germination of Sierra Nevada postfire chaparral species. *Madrono, 52,* 175–181.

Kilgore, B. M. (1973). The ecological role of fire in Sierran conifer forests. Its application to National Park management. *Quaternary Research, 3,* 496–513. https://doi.org/10.1016/0033-5894(73)90010-0.

Kirchmeier-Young, M. C., Zwiers, F. W., Gillett, N. P., & Cannon, A. J. (2017). Attributing extreme fire risk in Western Canada to human emissions. *Climatic Change, 144,* 365–379. https://doi.org/10.1007/s10584-017-2030-0.

Klocke, D., Schmitz, A., & Schmitz, H. (2011). Fire-adaptation in Hypocerides nearcticus Borgmeier and Anabarhynchus hyalipennis hyalipennis Marquart and New Notes about the Australian "Smoke Fly" Microsania australis Collart (Diptera: Phoridae, Therevidae and Platypezidae). *The Open Entomology Journal, 5,* 10–14. https://doi.org/10.2174/1874407901105010010.

Littell, J. S., Mckenzie, D., Peterson, D. L., & Westerling, A. L. (2009). Climate and wildfire area burned in western U.S. Climate and wildfire. *Ecological Applications, 19,* 1003–1021.

Liu, Y., Stanturf, J., & Goodrick, S. (2010). Trends in global wildfire potential in a changing climate. *Forest Ecology and Management, 259,* 685–697. https://doi.org/10.1016/j.foreco.2009.09.002.

Navarro, K. M., Schweizer, D., Balmes, J. R., & Cisneros, R. (2018). A review of community smoke exposure from wildfire compared to prescribed fire in the United States. *Atmosphere (Basel), 9.* https://doi.org/10.3390/atmos9050185.

NIFC. (2018). National Interagency Fire Center [WWW Document]. http://www.nifc.gov/fireInfo/fireInfo_main.html. Accessed March 4, 2016.

North, M. P., Stephens, S. L., Collins, B. M., Agee, J. K., Aplet, G., Franklin, J. F., et al. (2015). Reform forest fire management. *Science, 349,* 1280–1281. https://doi.org/10.1126/science.aab2356.

Parmeter, J. R., Jr., & Uhrenholdt, B. (1975). Some Effects of Pine-Needle or Grass Smoke on Fungi. *Phytopathology, 65*, 28–31. https://doi.org/10.1094/Phyto-65-28.

Preisler, H., Schweizer, D., Cisneros, R., Procter, T., Ruminski, M., & Tarnay, L. (2015). A statistical model for determining impact of wildland fires on Particulate Matter (PM 2.5) in Central California aided by satellite imagery of smoke. *Environmental Pollution, 205,* 340–349. https://doi.org/10.1016/j.envpol.2015.06.018.

Reid, C. E., Brauer, M., Johnston, F. H., Jerrett, M., Balmes, J. R., & Elliott, C. T. (2016). Critical review of health impacts of wildfire smoke exposure. *Environmental Health Perspectives, 124,* 1334–1343. https://doi.org/10.1289/ehp.1409277.

Ren, D., Fu, R., Leslie, L. M., & Dickinson, R. E. (2011). Modeling the mudslide aftermath of the 2007 Southern California wildfires. *Natural Hazards, 57,* 327–343. https://doi.org/10.1007/s11069-010-9615-5.

Schweizer, D., & Cisneros, R. (2014). Wildland fire management and air quality in the southern Sierra Nevada: Using the Lion Fire as a case study with a multi-year perspective on PM2.5 impacts and fire policy. *Journal of Environmental Management, 144,* 265–278. https://doi.org/10.1016/j.jenvman.2014.06.007.

Schweizer, D. W., & Cisneros, R. (2017). Forest fire policy: Change conventional thinking of smoke management to prioritize long-term air quality and public health. *Air Quality, Atmosphere & Health, 10,* 33–36. https://doi.org/10.1007/s11869-016-0405-4.

Schweizer, D., Cisneros, R., Traina, S., Ghezzehei, T. A., & Shaw, G. (2017). Using National Ambient Air Quality Standards for fine particulate matter to assess regional wildland fire smoke and air quality management. *Journal of Environmental Management, 201,* 345–356. https://doi.org/10.1016/j.jenvman.2017.07.004.

Steel, Z. L., Safford, H. D., & Viers, J. H. (2015). The fire frequency-severity relationship and the legacy of fire suppression in California forests. *Ecosphere, 6,* 8. https://doi.org/10.1890/ES14-00224.1.

Stephens, S. L., Martin, R. E., & Clinton, N. E. (2007). Prehistoric fire area and emissions from California's forests, woodlands, shrublands, and grasslands. *Forest Ecology and Management, 251,* 205–216. https://doi.org/10.1016/j.foreco.2007.06.005.

Stephens, S. L., Burrows, N., Buyantuyev, A., Gray, R. W., Keane, R. E., Kubian, R., et al. (2014). Temperate and boreal forest mega-fires: Characteristics and challenges. *Frontiers in Ecology and the Environment, 12,* 115–122. https://doi.org/10.1890/120332.

Stephens, S. L., Collins, B. M., Fettig, C. J., Finney, M. A., Hoffman, C. M., Knapp, E. E., et al. (2018). Drought, tree mortality, and wildfire in forests adapted to frequent fire. *Bioscience, 68,* 77–88. https://doi.org/10.1093/biosci/bix146.
Sugihara, N. G., Van Wagtendonk, J. W., Fites-Kaufman, J., Shaffer, K. E., & Thode, A. E. (2006). *Fire in California's ecosystems*, First ed. University of California Press.
Thomas, P. B., Morris, E. C., & Auld, T. D. (2007). Response surfaces for the combined effects of heat shock and smoke on germination of 16 species forming soil seed banks in South-east Australia. *Austral Ecology, 32,* 605–616. https://doi.org/10.1111/j.1442-9993.2007.01730.x.
Thonicke, K., Venevsky, S., Sitch, S., & Cramer, W. (2008). The role of fire disturbance for global vegetation dynamics: Coupling fire into a Dynamic Global Vegetation Model. *Global Ecology and Biogeography, 10,* 661–677. https://doi.org/10.1046/j.1466-822X.2001.00175.x.

Extreme Weather Events and Air Pollution Peaks in the Light of Climate Change: The Limits of the Notion of Risk

Isabelle Roussel

Abstract With the Anthropocene, the global risks induced by a deep degradation of the planet and a disturbance of the great biochemical cycles cannot be compensated or indemnified any more. Climate Change requires rethinking the notion of risk and hazard through the resilience of territories and their way of adapting to the environment. The management of climate control and air quality are closely linked and question the daily governance of cities that must integrate air, climate and energy into innovative and appropriate policies by all residents.

Keywords Climate change · Air pollution · Sustainable cities · Energy system · Adaptation · Resilience

The influence of humanity on the biosphere is such that the present era can be called Anthropocene[1] because, as D. Bourg points out (Bourg 2018): "*it is no longer a question of scenery, nor environment to which we would print a static mark, but of a sequence of which we are only a link and whose general movement escapes us and carries us away, without us ever being able to Stop.*" The fate of humanity seems more and more linked to that of its environment. It is interesting, in this respect, to resume the beginning of the book of M. Serres (1990) showing two men fighting while moving-sands, in the image of the great disturbances to which the planet is subject, are engulfing them. Human activities are responsible to these dysfunctions; they question the urgency of the management methods to be implemented. Given such perspectives, the notion of risk linked to extreme events is no longer operative.

[1]Qualifier proposed in 2002 by Paul Crutzen, Nobel Prize in Chemistry. Its final adoption by geologists is still controversial.

I. Roussel (✉)
University of Lille, Lille, France
e-mail: appa.irou@gmail.com

R. Akhtar (ed.), *Extreme Weather Events and Human Health*,
https://doi.org/10.1007/978-3-030-23773-8_5

Fight with cudgiels by Goya (1748–1826)

With the Anthropocene, global risks from a deep degradation of the planet and disrupted major cycles biochimic cannot be compensated or indemnified. Given the magnitude of the phenomenon of climate change and announced unpredictability of disaster, the notion of risk in insurance sense of the term is no longer appropriate. Henri de Castries, CEO of Axa,[2] makes the link between the one-off risk and the longer-term risk linked to Climate Change, by declaring: "*We do not have the choice: a world at +2 °C would still be insurable, a world at + 4 °C would certainly not be.*"

Even though IPCC (Intergovernmental Panel on the evolution of the climate) climate scientists are still reluctant to consider extreme events as markers of Climate Change, they are alert to public opinion as to what the reality of everyday life will be in a sustainable Climate Change. The lessons learned from these alerts help to better support a prevention policy that has been made possible by the entire natural risk management policy. Alerts and the concept of health risk have also been used for the management of air pollution, which at the present time can no longer be considered separately from climate risk, as is shown by experiments in many cities. The only possible attitude is that advocated by U. Beck in "the risk society" (Beck 2001) which is a pragmatic adaptation to the materiality of the world. The perspective of global climate change has upsetted the apprehension of extreme weather events as they anticipate the reality of future climate. How climate change management and the quality of the air they take into account these forms of adaptation to natural resources, the requirements of human health and that of the planet, too neglected at the time of the industrial revolution?

[2]AXA, a major French insurance company.

It is this question that is the subject of this contribution which is based both on the example of Climate Change and on that of air pollution whose evolutions, very different, tend to converge today to show little operational nature of the concept of risk when it comes to ensuring a better livability of cities that are, at present places that concentrate the vulnerabilities and challenges. Projection of these issues toward a certain future must not be seen as an inescapable threat that calls for passivity but rather as an incentive to better control the environment and to integrate it into the different planetary and local policies. However, the obstacles to overcome are numerous.

1 From the Extreme Climate Events to Climate Change Adaptation

Are heatwaves and extreme events are part of the chaotic and inherently variable nature of the climate, or are their occurrence and frequency increasing with global warming? However, the logic of insurance risk and that of Climate Change are different. The question of the cohabitation of these two different approaches has been analyzed in the SREX report, in 2012.[3] The confrontation with extreme events in the present, nevertheless, makes it possible to set up adaptation policies, as shown by the lessons learned from the 2003 heat wave.

1.1 Disaster and Risk Management Exist Before the Emergence of Climate Change

The management of risks, considered as the crossroads between a hazard and the vulnerability of territories, exists before the emergence of Climate Change and the first reports of the IPCC (Intergovernmental Panel on Climate Change). In his thesis (2015a, p. 154), Buffet (2015a) showed how the theme of disaster risk reduction has been gradually developed since the 1970s, and more particularly in the context of the International Decade for the Reduction of Disaster Risk Reduction, (1990–1999).[4]

The crossover between the two approaches has been difficult to achieve and Climate Change has been classified as emerging risk regardless of recorded disasters. On average, from 2000 to 2016, more than 220 million people were affected by natural disasters and more than 92,000 died. To USA (Valantin 2017), the estimated average cost for damage caused by so-called natural disasters was $3 billions a year during the 1980s and then $20 billions during the first decade of this century and $40

[3]https://www.ipcc.ch/pdf/special-reports/srex/IPCC_SREX_FR_web.pdf.

[4]It then experienced a further acceleration following the tsunami in South Asia in 2004. In 1990, a United Nations-specific body (IDNDR, which became UNISDR208 in 2000) is responsible for coordinating a work plan aimed at strengthening disaster risk.

billions from 2011 to 2012. In 2017, with hurricane Harvey and other disasters, the cost reaches 200 billions.

1.1.1 Disasters Have Given Rise to Different Types of Damage Mutualisation

Climate Change, integrated into the concept of the Anthropocene, reports a new type of hazard which challenges human responsibility. The early IPCC reports have kept a very climate-centric approach and therefore very (too) theoretical in dealing with climate risk control by separating natural phenomena from their social aspects (Kelman et al. 2015, p. 28). However, climate risk is an essential economic challenge because of the key role that can be played by insurance because it generates worldwide economic losses every year that the reinsurance industry estimates at 100 billions euros, of which only 30 billion are insured. Hence, the end of the insurance system is declared by Henri de Castries (cf. above). The second report from December 1995 confirms the responsibility of human activities on Climate Change. The publication of the third report from January 2001 concludes that there is **a probable link** between human activities and the increase in temperature since 1950. Following the IPCC's fourth report in 2007, the theme of natural disasters has been increasingly associated with Climate Change: "*Climate change contributes to a wide variety of extreme conditions*: *heat waves, torrential rains, fires, droughts, melting snow and ice*[5]." In France, the IFOP survey, conducted in September 2015,[6] shows that 31% of respondents state that humanitarian crises due to meteorological phenomena (floods, storms or droughts) are the consequences of Climate Change. The 5th report of the IPCC is published in September 2013, which states that the link between human activities and the increase in temperatures observed since 1950 is **extremely likely**. It would seem that the sixth report, from the introduction, would affirm: "Greenhouse gas emissions from human activities are the main causes of global warming, that has been occurring at an average rate of 0.17 °C (±0.07 °C) per decade since 1950."

It is thus the man who can prevent the climatic risk either by controlling the emissions of GHG (greenhouse gas) or by limiting the vulnerability of the territories. These adaptation actions are very diverse as shown by the example of heat waves (cf. below).

Risk management policies and those of controlling Climate Change belong to two different logics, however, and tend to subside within the "risk society" (Beck 2001). These two logics are differentiated by the longer time scale for Climate Change than for the emergency of the disaster. Thus, are two management modes that can even be contradictory: the actions brought to avoid the catastrophe, can neglect more radical

[5]Our Common Future in the Face of Climate Change—Final Declaration of the CFCC Scientific Committee 15, chaired by Chris Field and organized by UNESCO, Future Earth and ICSU.

[6]Ifop survey for Cap 21 Solutions conducted in September 2015 with a sample of 1000 people representative of the French population aged 18 and over.

solutions but more effective in the long term. The classic example is the construction of a dike to preserve a building when it would be better to leave or to move it.

The risks of increasing the frequency and intensity of natural disasters activate a discourse presenting each disaster as a sign of the impacts of Climate Change. This discourse is more about fear or threat than about integrating risk into everyday life (Charles and Kalaora 2017).

1.2 The Example of the Heat Wave of 2003

1.2.1 The Heat Wave of 2003 as a Precursor of Future Heat Waves

From 7 to 14 August 2003, in several cities in southern Europe, minimum temperatures were almost continuously above 20 °C (Fig. 1). Paris and Lisbon were particularly affected. Between 4 and 12 August, the maximum temperatures recorded at Paris-Montsouris remained well above 36 °C.

Studies on the modalities of this heatwave and its effects are still in progress (European project CANICULE) to draw the lessons of this episode whose frequency, according to the scientists of the IPCC is called to be multiplied. According to GHGs, emission scenarios beyond 2050, heat wave will be more intense, more durable and will happen early (from May) and/or later (until October). Thus, under the most pessimistic scenario envisaged by the IPCC (RCP 8.5), heat waves could become two to three times more numerous by the middle of the twenty-first century. We could observe on average on the horizon 2021–2050: nearly a heat wave a year. Studies

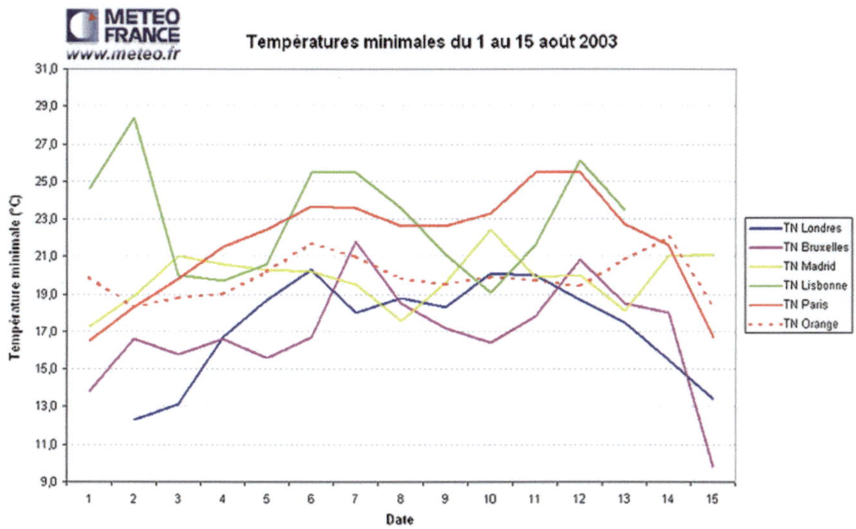

Fig. 1 2003, Mean temperatures of European cities, between August 1 and 15 (Météo France)

show that the most vulnerable regions to heat waves (South of USA) are also the most exposed to Climate Change.

1.2.2 The Damaging Effects of This Climatic Phenomenon Have Mainly Been Felt in Cities

According to the C40,[7] «*The number of cities exposed to extreme temperature will nearly triple over the next decades. By 2050 more than 970 cities will experience average summertime temperature highs of 35 °C (95 °F). Today, only 354 cities are so hot. The urban population exposed to these high temperatures will increase by 800 percent to reach 1.6 billions by mid-century*».

The city of Paris, where 1067 excess deaths were recorded during the episode, contributed by itself to 7.2% of the excess mortality while the Parisian population represents only 3. 7% of the French population (BEH 2018; Cadot and Spira 2006).

The geographical distribution of Parisian mortality during this episode highlights the importance of certain individual socio-demographic characteristics, specially the age and social isolation of people (Canouï-Poitrine et al. 2006). O'Neill et al. (2005) had highlighted the excess mortality of the most disadvantaged populations during the heat waves and attributed this excess to the unequal access to air conditioning which, according to him, should be generalized.

1.2.3 Although Age and Isolation Were the Main Factors in This Summer Mortality, the Characteristics of Housing and Urban Planning also Played a Role in Paris

Mortality was higher in dwellings located on the upper floors of buildings as the highest flats are more sensitive to the sun and less spacious. In Paris in 2003, living under the roof increased the risk of mortality by four. The density of urbanization favors the existence of an urban heat Island, sensitive specially at night when the walls restore the heat stored at the heart of the day. However, it is the remaining of high night temperatures that are more difficult to bear, from a health point of view. On the other hand, the proximity of sea, river or green spaces reduces the temperature of neighboring districts.

1.2.4 These Observations Make It Possible to Envisage What Measures Should Be Taken to Avoid the Harmful Consequences of Such Phenomena

This fine analysis of the characteristics of the Paris heat wave made it possible to outline solutions, or bundles of solutions, likely to mitigate the effects of the future

---

[7] https://www.c40.org/other/the-future-we-don-t-want-for-cities-the-heat-is-on.

heat waves. Indeed, sustainable development avoids the use of long-term harmful expedients such as air conditioning (Benmarhnia and Beaudeau 2018).

Thermal insulation of the building can affect both winter energy consumption and summer comfort by dividing the mortality risk by five, if natural ventilation is maintained. The orientation of the dwelling, the presence of shutters (occultation of the windows from the outside) and the opening of opposite facades result in significant differences temperature.

The introduction of green spaces, real islands of freshness, the greening of the walls can also refresh the city. The decrease in car traffic (or its ban as in Athens) in the centers can contribute to the same result.

The French health authorities have set up a national heat wave plan, aimed at improving forecasting by anticipating warnings from weather forecasts that are sometimes difficult to understand. The plan also has a social component encouraging the reporting of isolated seniors, while institutions have to provide a refreshed room to residents. It seems that the implementation of these devices results as a reduction in mortality during episodes of extreme heat that occurred in 2006 and 2017, but weather conditions were not comparable.

These considerations highlight the difficulties of adaptation policies that are impossible to manage by the only point of view of regulations because they correspond to technical and social innovations that cannot emerge from a strongly normative framework.

1.3 Some Reflections on Adaptation to Climate Change

The heat wave of 2003 helped to raise awareness of how badly the local climate has been integrated into planning policies. For example, the architecture, promoting the bay windows or neglecting the topoclimatic conditions of the exhibition, contributed to the unbridled consumption of energy. The first step in the actions to be taken is to put an end to bad adaptation. The last century and the beginning of this century have been marked in the Western world by a process of vulnerability and artificialization of environments.

Adaptation presents paradoxes since modern society has progressively freed itself from the climate to ensure better food and sanitary security. For example, in many countries the summer heat, thanks to the cold chain and air conditioning, is no longer reflected in a resurgence of dysentery and infant mortality. Adaptation, paradoxically, is not directly anchored in an approach of integration with nature but rather in protection approach which, to overcome the fluctuation of climate, tends to resort to sophisticated technical processes precisely to take refuge in a better-controlled world. The obvious health benefit of this type of adaptation raises ethical questions, as access to security is not available to everyone. What is the right mode of resilience to develop the territory: integrate the climate or protect it?

In fact, unlike climate models, adaptation must start from an analysis of the territory and its capacity for resilience. It must go beyond the analysis of risks and possible disasters to create new opportunities, not just to fight against phenomenon with undesirable effects. Adaptation, according to the writings of the Stockholm Environment Institute,[8] affects behaviors, practices and policies. According to Buffet (2015b), "*In about fifteen years, adaptation has gone from a question of impact modeling to a human experience.*" He pointed to the example of Typhoon Haiyan, which devastated the Philippines and caused over 7000 deaths at the opening of COP-19 in Warsaw. The COP bulletin of 19 November 2013 does not hesitate to report on this event to encourage participants to combine actions to prevent Climate Change with those of extreme events: "*the world has entered an era of loss and damage devastating because of climate change. The collective failure of reducing our emissions and the necessary support for adaptation actions means that vulnerable communities, ecosystems and countries are increasingly faced with irreversible loss and damage. Typhoon Haiyan, which may have affected more than 9 millions people in the Philippines, is a stark reminder of the damage caused by the most severe storms. This position implies integrating not only emergency response but also prevention.*"

Adaptation should not be based on a threat but on the desire to improve the quality of life for all integrating health concerns and the notion of environmental health into policies. But the modern welfare state that wants to be protective uses the registers of fear and threat. The answers given are often partial and unsuitable because adaptation supposes the flexibility and dynamics of individuals likely to build their most resilient environment and the most favorable to their health (Charles and Kalaora 2017).

This type of adaptation, favorable to the health of individuals, is inseparable from the maintenance of breathable air quality. Indeed, the management of air quality, after having followed logic of the alert related to adverse weather situations must be thought today in a dynamic integrated to the control of Climate Change. Here again, the example of heat waves illustrates this requirement of an integrated approach since the health risks associated with heat are combined with those caused by an increase in ozone levels that accompanies, especially in the city, high temperatures. It is especially in cities that Climate Change and air quality create a new dynamic favorable to the implementation of innovations or experiments.

2 The Atmospheric Pollution Between Peak Management and Better-Integrated Air/Climate/Energy Management from Risk Management to Resilient Territories

The management of air pollution is much older and has had a trajectory quite different from that of Climate Change since air pollution has never been considered as a natural hazard. It has been regulated in the modern era on the respect of the standards to avoid the peaks of pollution than to control the chronic health risk. As with extreme

[8]Climate Change and Disaster Risk Reduction, 2014, http://www.preventionweb.net/english/hyogo/gar/2015/en/bgdocs/SEI,%202014.pdf.

weather events, air quality management has shifted from controlling the exceptional phenomenon (the peak of pollution), linked to unfavorable weather conditions to controlling daily chronic pollution. Currently, the vision of pollution is much more global and is a more radical transformation of the economy and society, while relying on a local scale of quality of life and justice. This necessary integrated vision no longer focuses on exceptional events, but introduces daily health concerns in the construction of urban governance.

2.1 A Brief History of Air Pollution from Meteorological Risk to Health Risk

The concept of risk in the field of air pollution has been used in different ways. The risk in the sense of the occurrence of a climatic hazard made it possible to understand the peak phenomenon and to trigger warnings but the risk has also become more sanitary and chronic (Roussel 2015).

For centuries, air pollution has been more a matter of impurity and waste, and thus of an anthropogenic origin. In the eighteenth century, it was taken into account by the miasma theory and the aerism and then dropped out, due to advances in chemistry and after discovery Pasteur microbes.

In Europe, the murderous London smog episode of 1952 marked a turning point in the management of air pollution, the control of which became imperative to accompany industrial development. The political agenda of atmospheric pollution is contemporary with the development of metrology, which has generalized the measurement and allowed to enact health standards at the European level. Thus, thanks to the objectivities of pollution, its management has gone from a subjective complaint based on the notion of inconvenience or nuisance to compliance with emission or concentrations standards. This orientation accompanies the massive industrialization of Europe after the Second World War. The development of electronics on the devices used revealed peaks of pollution while their reliability was lower to measure the background noise."

2.1.1 Industrial Alerts Guided by Weather Conditions

Thanks to advances in metrology, the authorities were able to set up warning systems when, adverse weather conditions were forecasted, so pollution levels might exceed the authorized standards.

These devices were located in industrial areas to encourage companies to change their fuel in case of weather risk. Their effectiveness reflected the linearity between sulfur dioxide concentrations in the atmosphere and industrial emissions and, with little media coverage, the alerts were part of an effort to control industrial waste by

Fig. 2 Air Pur n°56 proceedings of Seminar presenting results of the CNRS study called "*does the alert in the field of the air quality correspond to a sanitary, physical or media risk?*" June 1999

the state. Over the years, the number has decreased due to greater control of polluting industrial emissions.

The development of reliable measurement networks has made it possible to undertake more precise epidemiological studies (Fig. 2).

2.1.2 The Health Risk Related to Air Pollution Becomes Clearer

The results of the Parisian measurement network AIRPARIF allowed some epidemiologists, trained at the Canadian school, to be interested in the effects of Parisian pollution. The ERPURS (Urban Health Hazard Risk Assessment) study carried out by Ile-de-France health observatory between 1985 and 2000 is based on the temporal ecological epidemiological method. It highlighted a strong link between pollution levels and a number of health indicators such as the number of days of hospitalization. This work was supported by the results of many other studies built on the same pattern and covering several agglomerations, such as the NMMAPS projects in the USA (Samet et al. 2000a, b, c) and APHEA in Europe (Katsouyanni et al. 1996). These studies put air pollution at the heart of public health and health risks that question the responsibility of the state in the same way as other health scandals that occur at the same time about AIDS and asbestos or even dioxin. It is in this context that the EU is developing strong regulation and many standards[9] which the Member States undertake to respect.

Epidemiological studies have shown the existence of a relationship without threshold between pollution and health risk. References to annual averages emphasize chronicity risk that is not only focused on the goal of alert thresholds.

Yet in France, the air law in 1996 Air opened the way for quality management of air by alerts issued from the state and highly publicized.

2.1.3 Highly Publicized French Alerts

The persistence of the management by the alerts is a French specificity which is explained by the weight of the State which, under the pretext of its responsibility for health risk, manages the alerts instituted by the law and specified by procedures which rely on the AASQA (Associations approved for the monitoring of air quality). The authorities have had to manage three types of alerts by thinking of imposing restrictions on emissions to improve air quality: alerts related to an exceedance of the threshold observed on nitrogen oxides, and therefore in connection with traffic, alerts related to ozone as a result of the heat wave of 2003 and the Directive "ozone" in 2002. The link between traffic restrictions and the health risks associated with this pollutant in power irritatingly strong was more difficult to admit to population because ozone is a secondary pollutant spreading over vast areas.

According to the French High Public Health Council (HCSP 2012), the effectiveness of these alerts is limited only the fight against chronic pollution has a health effect. In addition, the peaks of pollution have little mobilized public opinion despite intense media coverage. The lack of results rather discouraged inhabitants who felt pollution as a threat related to a meteorological hazard that was difficult to control (Charles and Kalaora 2017). In the French context of management by the peaks considered as communication tools, the absence of alert for the particles had contributed to neglect this pollutant yet very sensitive (Fig. 3). Particle alerts, introduced in 2007, often corresponding to foggy days, have hit people's minds even more, since

[9]Directive 96/62/EC, as amended in 2008, distinguishes alert thresholds: "level beyond which short-term exposure poses a risk to human health and from which Member States take appropriate and immediate action" and limit values: "a level established on the basis of scientific knowledge, with the aim of avoiding, preventing or reducing the harmful effects on human health and/or the environment as a whole, to be achieved within a given period and not to exceed once reached."

Fig. 3 Particles that darken the horizon strike the minds and impose means of prevention such as the wearing of masks during the Beijing Marathon (APPA)

the extent of the harmful effects of this pollutant has recently been discovered and has helped to raise public awareness.

2.2 *The Importance of the Deleterious Effects of Particles in the Mobilization of Public Opinion*

Scientific research on the effects of pollution has progressively recognized the role of particles and their massive health impact; this led to a reassessment, worldwide, of air pollution balance estimated at between seven million (WHO 2012) and nine million annual deaths (Lancet Commission on Pollution and Health 2018) estimated[10] outdoor and indoor air combined. In 2013, after ten years of work on this complex issue, the WHO recognized carcinogenic to humans, classified into group 1, the outdoor air pollution.[11]

The publication of the number of pathologies affected by high levels of particles has struck the minds because the finest particles, crossing the boundaries of the organs, can contribute to the emergence of different cancers that the population dreads. In addition, the visibility of the dust that veil the atmosphere reinforces the impression of malaise (even if the ultra-fine particles, the most harmful, are invisible) and gives a reality to the figures of 48,000 deaths attributable to the pollution by PM announced by Public Health France (Fig. 3).

In Europe, particles come mainly from three sources that play a strong cultural role: car traffic, agriculture and the inside the houses so practically all inhabitants become emitters therefore actors of the air quality that has become "the business of all." Yet survey conducted at the level of Europe[12] shows that only 30% of women and 26% of men want better health information while the majority of EU citizens are opting for a regulation very institutional for reducing emissions at European level.

2.3 *The Links Between Atmospheric Pollution and Climate Change*

Now, ozone and particles (called short-lived climate forcers[13]) have effects on the climate, despite their shorter lifetime than greenhouse gases (GHGs) that can persist for several decades in the atmosphere. In addition, all these gases are mainly from combustion phenomena whose control is beneficial in many ways.

The challenges of climate change have emerged recently in spirits, it was able in drawing attention to CO_2, to hide the question of atmospheric pollution introducing some confusion with GHG emissions as shown studies on the perception of climate

[10]http://www.who.int/mediacentre/news/releases/2014/air-pollution/fr/. With the order of 1.2 millions annual deaths in China related to outdoor air, about 500,000 annual deaths in Europe, 240,000 in the USA and 48,000 in France.

[11]"The Working Group found that there is sufficient evidence in humans and in experimental animals for the carcinogenicity of outdoor air pollution in general and of PM in outdoor air pollution more specifically. These findings are supported by strong mechanistic evidence in exposed humans, including studies showing increased frequencies of micronuclei and chromosomal aberrations in individuals occupationally or residentially exposed to polluted air, as well as by studies showing genetic and related effects in animals and various experimental systems. At wide range of other effects related to carcinogenesis, including oxidative stress, inflammation and epigenetic alterations, have been observed in exposed humans and animals and in various experimental systems." (IARC monograph 109, p. 34).

[12]Eurobarometer 468, November 2017. http://ec.europa.eu/commfrontoffice/publicopinion/index.cfm/Survey/getSurveyDetail/yearFrom/1974/yearTo/2017/surveyKy/2156.

[13]Their reduction begins to interest international bodies, http://ccacoalition.org/en/news/declaration-short-lived-climate-pollutants-ratified-members-parliamentary-confederation.

change (Boy 2013). Nevertheless, air pollution and climate change represent the aspects of the environment which, according to various surveys, concern between 30 and 40% of French and Europeans. These two topics are, rightly, very mixed. The emergence of the fear of a large-scale climate change has contributed to broadening the dimension of air pollution toward all environmental issues impacted and to better highlight the link between pollution and the energy system. The end of fossil fuels is very beneficial for air quality, which benefits from all the investments made in favor of energy innovation. On the other hand, the fight against Climate Change has been confined for a long time to the technical sphere in a top/down movement, while moving toward adaptation, and highlights the health benefits of the measures to be taken. China is a good example of the identification of co-benefit air/climate/energy/health.

Despite these conclusions, institutions are still very sectored and separate management of the air quality of the climate change that could lead to contradictions such as heating wood encouraged to reduce fossil fuels but may, without additional precautions, contribute to the increase of particulate levels in indoor and outdoor air.

On the other hand, at the city level, the two issues are better integrated and the benefits or collateral damage better controlled.

3 Cities and the Implementation of an Integrated Air/Climate Policy/Energy/ Health

More than 50% of the world's population lives in cities (77% in Europe) and 70% in 2050. Cities currently emit about 80% of global greenhouse gases. In France, 19 agglomerations exceed the European standards and show the insufficiency of the solutions used until now to clean the air of the cities.

The Paris Climate agreement in 2015 recognized the need for the involvement of "stakeholders" in the implementation of transformations related to Climate Change in a context where the mobilization of states is far below the stakes. In particular, cities and local communities have shown a real dynamism to engage, through innovative initiatives, the transformations needed to build a carbon-free world. They are particularly active in the exchange of their experiments within networks[14] that have been organized at different levels since the 1990s. The articulation between different sectors of urban life is a major challenge at the local level, precisely between the air quality, energy and climate. The latter, built according to distinct logics and constituting independent intervention sectors, must henceforth be the subject, in France, of a joint planning approach through the elaboration of the Territorial Climate-Air-

[14]UCCRN is a network of cities dedicated to providing good the Information That city leaders—from government, the private sector, non-Governmental organizations and the community—need in order to for Assessment current and future Risks, make choices That improve resilience to Climate Change and climate extremes, and take actions to reduce greenhouse gas emissions.

Energy Plans.[15] (PCAET) also meet difficulties, since some actions that favor the reduction of greenhouse gases are not necessarily beneficial for the reduction of atmospheric pollution, and vice versa.

3.1 Promotional Urban Policies for Air and Climate

The change in the energy system and the remains of fossil fuels can only be beneficial for air quality, which encourages cities to promote all innovations that can result in a reduction or optimization of energy consumption. Thus, climate change by accelerating the reduction of the most polluting energies is favorable to the quality of the air. New technologies promoted by "smart cities" can support this transition as long as they are open to everyone and not only to the technophile population. Innovations are developing in many sectors both technical and social: search for energy-efficient buildings, urban densification, reflections on modes of transport, health benefits, etc.

The services of the city, connected to international networks (C40 for example[16]), are important for partners to network and support local initiatives. This new context is fueling the creativity of managers tempted by the implementation of interesting experiments for urban life and creating local jobs not only in large industrial groups but also in small and medium-size enterprises that have seen the profit they could take from working with the city to create endogenous economic development.

These environmental imperatives in cities can only be imagined by considering, in the long term, a set of solutions involving the housing, city planning and mobility possibilities it offers.

In the field of urban transport, car pooling and its various variants, transport on demand, car-sharing systems and self-service bikes for pedibus and bikebus, the range of services to the mobility has grown steadily in recent years by relying on the development of digital systems.

The promotion of public transport and electric vehicles has the advantage of giving way to the smallest polluting diesel vehicles with quieter vehicles, which can reduce the noise level of cities, which is a major nuisance for city dwellers.

European cities seek to implement solutions to reduce vehicle emissions in the city. Thus, 232 LEZ "Low Emission Zones" have been established in 12 European countries[17] (Pouponneau 2018). In France, the principle of ZCR (restricted traffic zones) is supported by the energy transition law, the device operates according to a

[15]The procedures for drawing up and updating these new plans (compulsory for inter-municipal cooperation institutions (EPCI) with own taxation of more than 20 000 inhabitants) are specified in Law No. 2015-993 of 17 August 2015 on the energy transition for green growth, the decree n° 2016-849 of June 28, 2016. They will have to be realized at the latest on December 31, 2018.

[16]https://www.c40.org/.

[17]In September 2017.

selection of authorized vehicles[18] to enter the city center according to their seniority and therefore their membership in a Euro standard level. The presence of a LEZ, like that of an urban toll, should, in theory, accelerate the transformation of the car fleet and encourage the disappearance of the engine and contribute to increase the adherence to the soft modes or even the decrease of the frequentation of the city centers. This good evolution for the environment can only be beneficial for the health of the population.

3.2 In the Area of Housing and Urban Planning, Some Points of Vigilance Are Needed

The residential and tertiary housing sectors have a significant margin of GHG reduction since this sector contributes nearly a quarter of the final energy consumption responsible for greenhouse gas emissions. Housing insulation is beneficial in the event of heat waves (see above), but to reconcile the reduction of GHG emissions with the requirements of air quality, it is important that energy savings do not deteriorate the indoor air quality or thermal comfort.

The urban heat islands (ICU) are the object of growing concern in the urban spaces whose sensitivity to the heat waves is recognized (cf. above) The night temperatures recorded in the city center show 4 even 5° than in neighboring peripheries. The dense city, which is particularly important from the point of view of energy restrictions, is not without its health disadvantages, especially when it is composed of "canyons" streets, poorly ventilated, which trap pollutants in addition to heat. The main solutions considered for reducing the heat island are the reduction of energy consumption in the city and the creation of islands of freshness through the planting of vegetation and the development of urban agriculture.

Wood heating should be recommended with caution because if toxic emissions can be easily controlled in large boiler rooms, only efficient domestic heating methods should be encouraged. This vigilance focuses on open fireplaces and old wood stoves that not only generate indoor pollution but also contribute to increased particulate concentrations in the ambient air.

3.3 But Above All, Vigilance Must Concern Urban Governance

The challenges are considerable because they concern the future of the planet; the implementation of these challenges is based on profound cultural changes as it is

---

[18]The vignette The Crit'Air sticker is a macaron visibly placed on cars, motorbikes and light vehicles, indicating their level of pollutant emission according to their date of registration and to their type of carburant.

nothing less than revisiting the notion of inhabiting the land or the city by leaving more room for climate and the environment.

While international agreements grant a great deal of flexibility to local initiatives, the responsibility of the elected representatives, in partnership with the various stakeholders, is strong because they must ensure that the progress made contribute to a greater justice, as well in countries than in cities that need to become more inclusive.

3.3.1 The Inclusive City

The new tools with great flexibility of form allow all the actors of the city to imagine the city, with dwelling and mobility of the future remain inclusive. However, the innovative aspects tend to favor the city center compared to the periphery and the inhabitants most able to invest in new technologies. However, tools for assessing well-being and fair distribution are still modest because it is easier to define quantitative rather than qualitative objectives, while knowing that these question concern many sectors of urban management (Wheel-Legall et al. 2015).

It is necessary not to consider the air quality or energy savings are only profits or collateral damage of climate policies. Integrating management transgresses silo management logic of different urban projects, it is difficult to apply because health issues are still often considered a curative and non-preventive logic. Environmental health leaves the strictly sanitary world to question all sectors of urban life.

These topics question not only horizontal and transectorial fluidity but also vertical fluidity since the issues to be taken into account range from the proximity of the interior environments to the global dimension.

3.3.2 Management of Scales from Local to Global

Local initiatives by actors convinced of the need for action face the heterogeneity of situations and lifestyles that delay the implementation of coherent policies to train the entire population. About 81% of Europeans are aware of their environment as a factor in their quality of life and health, and Climate Change (Fig. 4) and air pollution are high on the agenda. On the other hand, we see that the importance given to these two factors depends to a large extent on the sensitivity of individuals to the environment and therefore to their level of information, which shows how much raising the awareness level remains an important goal for associations, because it is the inhabitants who, through their motivation, can advance the policies.

According to the 2017 Eurobarometer, 58% of knowledgeable individuals place a lot of importance on Climate Change and 49% on air quality, and this percentage is reduced to 24% in the general population for Climate Change and to 30% in air pollution that enters more into the sphere of proximity.

However, as demonstrated ozone alert, the solutions to be implemented don't necessarily lie locally and expected results can be shifted in time and space, which invites the political to remain consistent at different spatio-temporal scales.

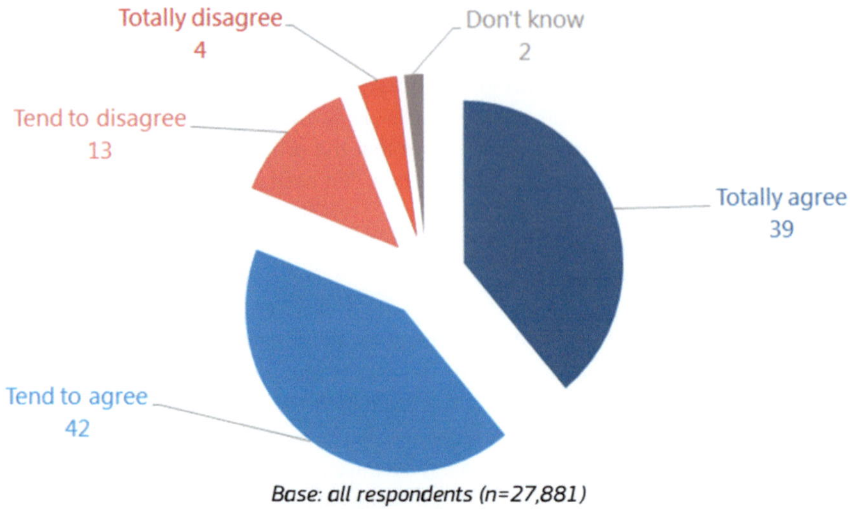

Fig. 4 Percentage of Europeans who consider the environment to have a direct impact on their daily lives and their health in response of this question: «Please tell me to what extent you agree or disagree to the following statement: environmental issues have a direct effect on your daily life and health». *Source* Eurobarometer Special Eurobarometer 468 2017

4 Conclusion

It seems that public opinion, carried by a number of NGOs, is beginning to realize that the control of Climate Change and air pollution is a condition for the survival of humanity in the medium term and even in the short term, since the recommended solutions are beneficial for health. The extreme events that will become more and more frequent illustrate what can become the daily reality of the future climate. In the context of the Anthropocene, the notion of hazard has evolved since the anthropic responsibility has become evident; thus, conventional risk and alerts responses are inoperative as well as state policies based on injunctions or regulations. It is a true transformation of the relationship between humanity and the planet that imposes itself; on the other hand, the actions undertaken are not at the scale of these stakes even if, especially in the cities, a dynamic, still confined to a few actions, begins to appear. The cultural change required is deep and institutions have not always taken the measure. Only the economic and financial world, supported by the consumer power and the sensitivity of the people can act in favor of the transformation of the energy system on a planetary scale and the reduction of the weight of the human activities on the Earth system.

References

Beck, U. (2001). *The society of risk: On the road to another modernity* (521 p.). Alto Aubier.

BEH. (2018). Understanding and preventing the health impacts of heat in a context of climate change. *Bulletin Epidémiologique Hebdomadaire* (16–17), 354–357. http://invs.santepubliquefrance.fr/beh/2018/16-17/2018_16-17_6.html.

Benmarhnia, T., & Beaudeau, P. (2018). Cities and the heatwave: Preparing for the future and preventing the health effects of urban heat islands. *Bulletin Epidémiologique Hebdomadaire* (16–17), 354–7. http://invs.santepubliquefrance.fr/beh/2018/16-17/2018_16-17_6.html.

Bourg, D. (2018). *A new land* (240 p.). Desclée de Brouwer.

Boy, D. (2013). Social representations of Climate Change. *Pollution Atmosphérique*, pp. 190–201 http://www.appa.asso.fr/_docs/1/fckeditor/file/Revues/PollutionAtmospherique/Hors-serie-climat-juin-2013/Boy.pdf.

Buffet, C. (2015a). *Adaptation to climate change: construction, framing and actors, global negotiation arenas for vulnerable populations in Bangladesh*. Doctoral thesis, Ecole des Hautes Etudes en Sciences Sociales.

Buffet, C. (2015b). Towards a new geography of climate expertise? BCAS and ICCCAD, expertise "local" In Bangladesh. *Review Air Pollution*. http://lodel.irevues.inist.fr/pollution-atmospherique/index.php?id=5279.

C40. (2018). *The future we do not want to*. https://c40-production-images.s3.amazonaws.com/other_uploads/images/1789_Future_We_Don't_Want_Report_1.4_hi-res_120618.original.pdf.

Cadot, E., & Spira, A. (2006). Heat wave and excess mortality in Paris in August 2003. *Espace-Populations, Society*, 239–249. https://journals.openedition.org/eps/1383.

Canouï-Poitrine, F., Cadot, E., & Spira, A. (2006). Excess deaths during the August 2003 heat wave in Paris, France. *Review of Epidemiology and Public Health, 54*(2), 127–135. https://doi.org/10.1016/S0398-7620(06)76706-2.

Charles, L., & Kalaora, B. (2017). Borderless environment, risk society and threat potentiation: a challenge for the social sciences. Proceedings of a Seminar, Paris, January 2017 (forthcoming).

HCSP. (2012). Particulate matter in outdoor air pollution. https://www.hcsp.fr/explore.cgi/avisrapportsdomaine?ae=avisrapportsdomaine&clefr=273.

HCSP. (2013). Notice on health messages to be broadcast during episodes of ambient air pollution by particles, ozone, nitrogen dioxide and/or sulfur dioxide, November [Online]. http://www.airparif.asso.fr/_pdf/avis-.

Katsouyanni, K., Schwartz, J., Spix, C., et al. (1996). Short term effects of air pollution on health: An European approach using epidemiologic time series data: The APHEA protocol. *Journal of Epidemiology & Community Health, 50*(Suppl 1), S12–S18.

Kelman, I., Gaillard, J., & Mercer, J. (2015). Climate change's role in disaster risk reduction's future beyond vulnerability and resilience. *International Journal of Disaster Risk Science, 6,* 21–27.

Lancet Commision on Pollution and Health (2018) Lancet. 391:462–512. Published Online 19 Oct 2017. http://dx.doi.org/10.1016/S0140-6736(17)32345-0

O'Neill, M. S., Zanobetti, A., & Schwartz, J. (2005). Disparities by race in heat-related mortality in four US cities: The role of air conditioning prevalence. *Journal of Urban Health, 82*(2), 191–197.

Pouponneau, M. (2018). Beyond circulation restrictions: Accompanying and informing cities. *Atmospheric Pollution,* No. 237–238. Living in the City.

Roussel, I. (2015). Episodes of pollution and warning devices in France: a historical and socio-political vision. *Atmospheric Pollution* (special issue—March 2015), 40–66. http://www.appa.asso.fr/_docs/1/fckeditor/file/Revues/PollutionAtmospherique/HS_pointes/Roussel_Historique.pdf.

Samet, J.-M., Dominici, F., Zeger, S. L., et al. (2000a). The national morbidity, mortality, and air pollution study. Part I: Methods and methodologic issues. *Research report (Health Effects Institute)*, 5–14, discussion pp. 75–84.

Samet, J. M., Zeger, S. L., Dominici, F., et al. (2000b). The national morbidity, mortality, and air pollution study. Part II: Morbidity and mortality from air pollution in the United States. *Research report (Health Effects Institute), 94*, 5–70.

Samet, J. M., Dominici, F., Curriero, F. C., et al. (2000c). Fine particulate air pollution and mortality in 20 US cities, 1987–1994. *New England Journal of Medicine, 343*(24), 1742–1749.

Serres, M. (1990). *The natural contract* (p. 191). Paris: Bourin.

Valantin, J. M. (2017). *Geopolitics of a disordered planet*. Paris: Le Seuil.

Wheel-Legall, A., Gall, J., Potelon, J.-L., & Cuzin, Y. (2015). Guide "Acting for healthy urban planning, concepts and tools". In *National Guide Town Planning and Health. EHESP/DGS* (192 p.).

World health statistics (2012) 177 p. http://www.who.int/gho/indicatorregistry

Extreme Weather and Human Health in Italy

Cosimo Palagiano

Abstract Italy is a very fragile country due to its recent geological formation. Frequent earthquakes and volcanic phenomena create situations of danger with a large number of victims. Moreover, earthquakes are sudden and unpredictable, and many structures and houses have not been built with earthquake-proof constructions. The situation is getting worse and worse due to climate change, which leads to irregularities in precipitation and temperatures. Some consequences of this situation are landslides and floods, which, on the other hand, involve the whole country. Therefore, the watercourses break their riverbeds causing floods and landslides, which the lack of vegetation cover makes frequent and dangerous.

Keywords Italy · Climate change · Earthquakes · Floods · Landslides · Health

1 Introduction

Climate change has a significant impact on Italy for the following reasons: (i) the extent of the country in latitude, (ii) the geological and morphological constitution, with relatively recent ridges, (iii) immersion in the Mediterranean, of which receives strong and changing influences. The consequences of these peculiarities are strong variations in temperature and precipitation in the same season and year by year, which produce floods and landslides throughout the territory. Many buildings are destroyed and many lives are lost. According to ISTAT (Italian Statistical Institute 2010) in the decade 2000–2009, the average annual temperature, equal to 13.3 °C, was higher by 0.8 °C compared to the climatic period 1971–2000.[1] Also the maximum temperature (18.0 °C) and the minimum temperature (8.5 °C) were higher than their respective climatic values of 0.9 and 0.6 °C, respectively. In all the years of the decade, with the exception of 2005, the average, maximum and minimum temperatures recorded climatic values always higher than the reference ones. In the decade 2000–2009,

[1] The data of period considered are the more recent available.

C. Palagiano (✉)
Sapienza Università di Roma, Rome, Italy
e-mail: cosimo.palagiano@uniroma1.it

R. Akhtar (ed.), *Extreme Weather Events and Human Health*,
https://doi.org/10.1007/978-3-030-23773-8_6

the average annual precipitation was 763 mm, with 30 mm of rain less than the climatic value of the period 1971–2000. The least rainy year was 2001 with 189 mm of deviation from the climatic value, while the wettest one 2002 with 88 mm more. Precipitation not only they differ between the north and south of the country, but above all present oscillations in the amount of annual rainfall fallen to soil, which is the most worrying aspect of the phenomenon, because in every territory it is necessary to manage years of heavy rainfall and years of strong water shortage, resulting in increased risk of landslides and floods in the first case and water shortages and droughts in the second. InValle d'Aosta, for example, has gone from 1894 mm of precipitation in 2000 to 520 in 2005, while in Sardinia from 668 mm in 2004 to 351 of 2001. It should be noted that due to climate change not only temperature and rainfall differ from previous periods. In addition, the alternation of high and low temperatures, and high and low rainfall is rapid. Therefore, it is important to consider the short duration of the so-called good and bad weather (Fig. 1).

Exploitation of the territory, lack of interventions and climate change put more and more people in a dangerous situation. To date, seven million citizens live in areas at risk—a number that could increase in the future. In Italy, there are seven million people living under the threat of hydro-geological instability with the nightmare of landslides and floods. A constant threat is caused by years of wild building speculation, land consolidation and interventions that remain stationary and blocked by bureaucracy and appeals. And the future risks being already marked, with an exploitation of the territory that continues inexorable, while meteorological events are becoming increasingly violent and sudden.

1. **Heat waves**

Heat waves are another risk to human health due to high temperatures. In the summer of 2015, the statistics found in 21 Italian cities, 13% of deaths of people over 65 years of age are attributable to the heat. The deaths of individuals between 65 and 74 years

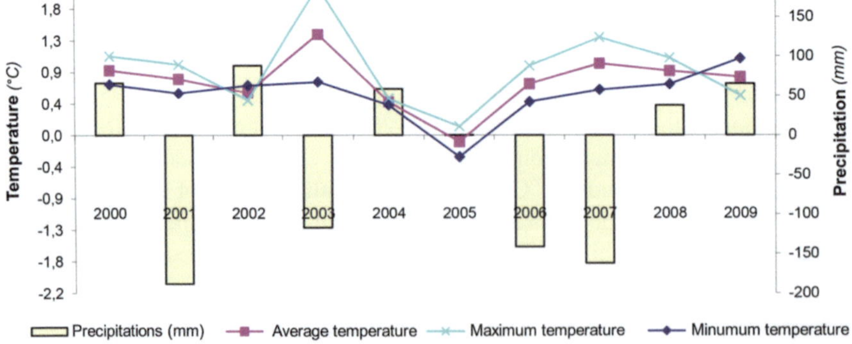

Fig. 1 Deviation of average, maximum and minimum temperature and total precipitation from the climatic value in Italy. Years 2000–2009 (temperature in degrees Celsius and precipitation in millimeters)

of age were 305 (9%), of individuals between 75 and 84 years of age were 791 (11%) and individuals of okltre 85 years of age there were 1559 (15%) (http://www.salute.gov.it/caldo).

2. **Floods occurred in Italy between 2000 and 2017**

Floods are a rather common phenomenon in Italy, with the loss of buildings and human lives. This phenomenon is due to the irregularity of the rains, the hydrographic regimes and the construction of houses too close to the waterways. Many homes are illegally built in areas subject to flooding (Table 1).

In total, 189 were dead and 12 were missing. But we have to consider that in many cases houses and constructions (roads, railways, bridges, etc.) were destroyed. Many families were evacuated and lost their properties. The most severe damages of constructions and infrastructures were suffered by the population of Genoa, Liguria, Province of Alessandria and Milan on November 15, 2014: in Province of Alessandria, 238 mm of rain had fallen in a day. On May 3, in the commune of Senigallia, the waters and the mud of the River Misa covered the streets up to 3 meters up to the first floors of the buildings. On November 18, 2013, in Sardinia the rivers Cedrino and Posada flooded the village of Torpè with over 3000 cubic meters per second. On October 25, 2011, 520 mm of rain had fallen in 6 h in Lunigiana. In the same year 2011, on 11 June, the damages amounted to 7,200,000 Euros for individuals and businesses and about 450,000 Euros for public damages. About 185 families and 50 productive slabs were affected. Other extraordinary events have to be registered in Vibo Valentia (on July 3, 2006), in Province of Massa Carrara (on September 23, 2003), in Piedmont, Valle d'Aosta and Lombardy because of the Po flooding (on October 13–16, 2000) and in Soverato (Calabria) 441 mm of rain had fallen and 150 km of coast between Catanzaro and Reggio di Calabria were devastated by bad weather. The worst years were 2014, 2011 and 2000. Perhaps, this is an evidence of climate variability.

The resident population exposed to flood risks in Italy is equal to 2,062,475 inhabitants (3.5%) in the scenario of high hydraulic danger (return time between 20 and 50 years), 6,183,364 inhabitants (10.4%) in the scenario of average danger (time return between 100 and 200 years) and 9,341,533 inhabitants (15.7%) in the scenario of low probability of floods or scenarios of extreme events. The regions with the highest values of allotropic risk population in the medium hydraulic hazard scenario are Emilia-Romagna, Tuscany, Veneto, Lombardy and Liguria.

The families at risk of floods in Italy are 2,648,499 (10.8%) in the medium hydraulic hazard scenario. Emilia-Romagna, Tuscany, Veneto, Lombardy and Liguria present the highest number of families and flood risk in the scenario of average hydraulic hazard. The regions of Emilia-Romagna, Tuscany, Veneto, Lombardy and Piedmont present the highest number of buildings at risk of flooding in the scenario of average hydraulic danger.

The buildings at risk of floods in Italy are 1,351,578 (9.3%) in the scenario with medium hydraulic hazard. The regions of Emilia-Romagna, Tuscany, Veneto, Lombardy and Piedmont present the highest number of buildings at risk of flooding in the scenario of average hydraulic danger.

Table 1 Floods occurred in Italy between 2000 and 2017

Date	Geographical area	Dead
September 9–10, 2017	Livorno and Pisa	8
January 24, 2017	South Italy and Islands	1
September 14, 2015	Emilia	3
November 15, 2014	Genoa, Liguria and Milan	1
November 10, 2014	Liguria	2
October 14, 2014	Tuscany Maremma	2
October 9–10, 2014	Genoa	1
August 2, 2014	Prov. of Treviso	4
May 3, 2014	Prov. of Ancona	1
January 19, 2014	Prov. of Modena	1
November 18, 2013	Sardinia	18
November 12, 2012	Prov. of Grosseto	6
November 11, 2012	Prov. of Massa e Carrara	1
November 22, 2011	Prov. of Messina	3
November 4, 2011	Genoa and province	6
October 25, 2011	La Spezia and Lunigiana	13
June 11, 2011	Prov. of Parma	1
March 3, 2011	Marche and Romagna	5
November 1–2, 2010	Veneto	3
October 5, 2010	Prato and province	3
October 4, 2010	Genoa and Savona	1
September 9, 2010	Costiera Amalfitana	1
October 1, 2009	Province of Messina	36
July 18, 2009	Province of Belluno	2
October 22, 2008	Province of Cagliari	5
May 29, 2008	Villar Pellice (Turin)	4
April 30, 2006	Naples	4
July 3, 2006	Vibo Valentia	4
September 23, 2003	Prov. of Massa Carrara	2
September 8, 2003	Prov. of Taranto	2
August 29, 2003	Prov. of Udine	2
November 6 and 23, 200	Prov. of Imperia and Savona	7
October 13–16, 2000	Piedmont, Valle d'Aosta, Lombardy	23 + 11 missing
September 9, 2000	Soverato (Calabria)	13 + 1 missing

There are 31,137 (15.3%) flooding cultural assets in Italy in the medium hydraulic hazard scenario and reach 39,426 (19.4%) assets in the low-priced scenario. The highest number of cultural assets at risk in the medium hydraulic hazard scenario is recorded in Emilia-Romagna, Veneto, Liguria and Tuscany. Among the municipalities with the highest number of cultural assets at risk of flooding in the scenario of average danger are the cities of art of Venice, Ferrara, Florence, Genoa, Ravenna and Pisa. For the protection of cultural heritage, even the scenarios of low probability of occurrence assume particular importance, considering that in the case of event the damages to the cultural heritage would be priceless and irreversible.

In Fig. 2, the population at risk landslides is highlighted. The populations most affected are those of the Po Valley, as these lands are affected by major Italian waterways such as the Po and its tributaries. In ancient times and even in very recent times, the Po is overflowing, given its nature as a hanging river, i.e. the watercourse flows above the lands crossed. Another population at risk is the one that resides in the areas crossed by the Arno. The flood of the Arno in Florence in 1966 is very well known, when not only houses were damaged, but also a part of the artistic and book heritage. Another area damaged by the floods is that of northeastern Sardinia, where the mouths of the rivers are often invaded by marine waters. The Rome region is also subject to floods due to the presence of surface and underground rivers. The capital itself is often flooded in the streets, because the sewage system does not receive enough rainwater. The same applies to Naples and its region. For decades, Italy has been the victim of floods, even with deaths, but successive governments have done almost nothing to alleviate the health conditions of the population.

3. **Landslides in Italy**

Italy is one of the most affected by landslides countries with 629,808 landslides on an area of 23,700 km^2, equal to 7.9% of the national territory. About a third of the total landslide is due to rapid kinetic phenomena (collapses, rapid mud and debris flows, characterized by high speeds, up to a few meters per second, and by high destructiveness, often with consequences in terms of loss of human lives. One of the first Italian geographers who dealt with landslides was Roberto Almagià in the early twentieth century (Almagià 1907, 1910). The most recent study is that of Walter Palmieri in January 2004, with a good bibliography (https://www.researchgate.net/publication/259358789_La_storia_delle_frane_in_Italia_e_gli_studi_di_Roberto_Almagia 2004 (*author Palmieri W*)). The population at risk of landslides in Italy, residing in high and very dangerous areas, amounts to 1281,970 inhabitants, equal to 2.2 of the total. Campania, Tuscany, Emilia-Romagna and Liguria have the highest values of population at risk landslides. The families at risk in areas of high and very high landslide danger are 538,034, equal to 2.2% of the total.

The cultural heritage at risk of landslides in Italy totaled 37,847, equal to 18.6%. The highest number of cultural heritage sites with high and very dangerous landslides can be seen in Tuscany, Marche, Emilia Romagna, Campania and Liguria. Numbers are the historical *borghi* affected by landslides triggered or reactivated even in recent years, such as the cliff of San Leo (Rimini) Volterra (Pisa) and Civita di Bagnoregio

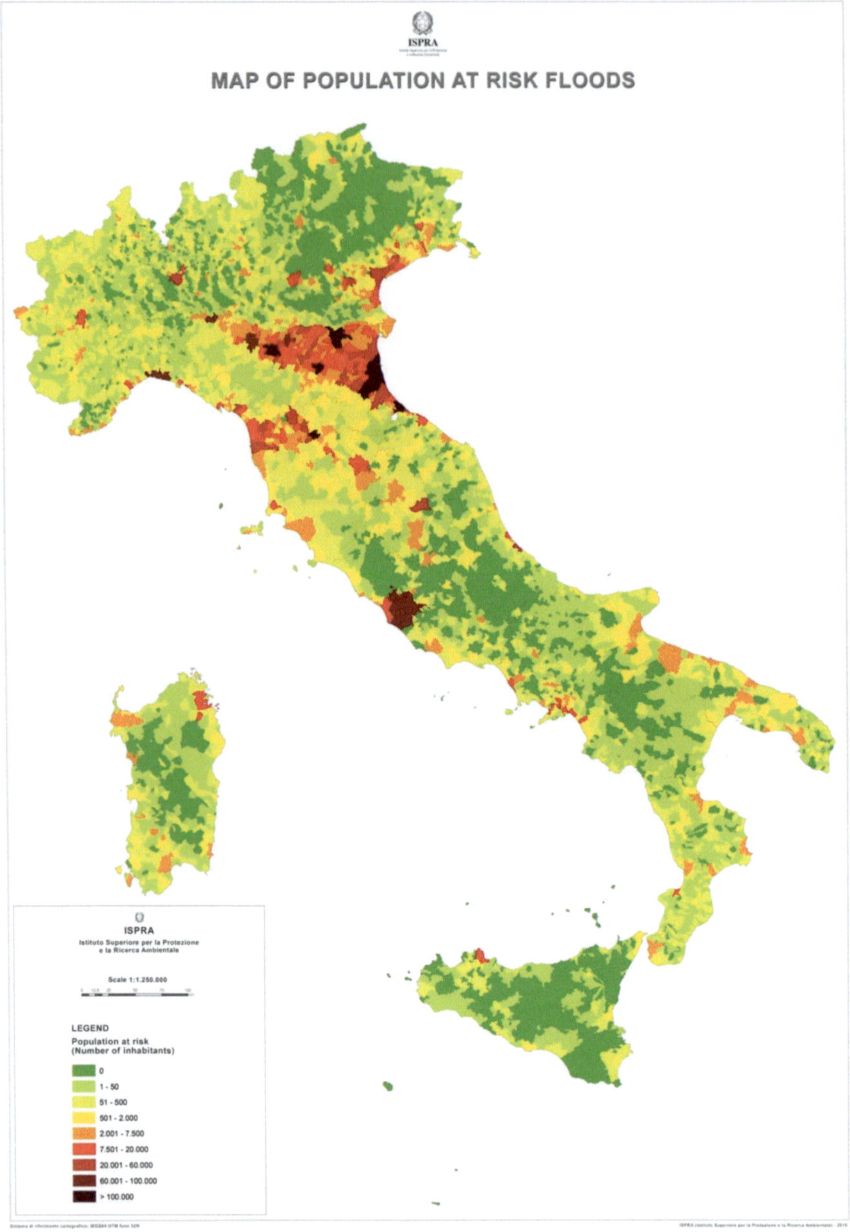

Fig. 2 Map of population at risk floods in Italy. *Source* ISPRA

(Viterbo). In recent decades, several historic centers have been the object of consolidation and reduction of hydrogeological risk, such as in Certaldo (Florence), Todi (Perugia) and Orvieto (Terni).

Buildings at risk in areas of high and very high landslide hazards are 550,723, equal to 3.8% of the total. The regions with the highest number of buildings at risk of landslides in areas of high or very high danger are Campania, Tuscany, Emilia-Romagna and Calabria: On a provincial basis, the provinces of Salerno and Genoa have the highest number of buildings at risk of landslides (ISPRA 2018).

It should be noted that the fragility of the Italian territory for geological, hydrographic and morphological reasons is accentuated by the work of a man who has built houses often abusive in areas at risk, such as in areas near the waterways or even building roads burying the waterways. Even the buildings built at the foot of the ridges are at risk (Table 2).

In Fig. 3, areas with the greatest danger of landslides are highlighted. Compared to the most frequent floods in the plains, landslides are prevalent in mountainous and hilly areas and therefore very intensely distributed throughout the national territory. They constitute the greatest danger especially in winter when they are caused by heavy rains and copious snowfalls. The major problem, besides the instability of the climatic events, is also the improvidence of the man who built in unsuitable land, with unsuitable materials and deforestation, which is no longer able to retain the geologically recent lands.

2 Conclusion

Italy's fragility is due to two essential factors: (i) the frequency of earthquakes and (ii) the irregularity of the rains. Moreover, in the consequence of climate change, temperatures and rains with sudden changes accelerate the frequency and danger of landslides and floods. Added to this is that not only earthquakes but also landslides and ravines are unpredictable and this causes damage to people and property. In recent years, the damages are more considerable.

The latest governments have paid more attention to emergencies than to forecasts, especially since many administrators do not believe in climate change and that often governments lack the ability to make long-term projects.

Not only that but often the controls are not constant or accurate as happened at the recent collapse of the Genoa Bridge. The privately owned motorway company that manages the network does not effectively control the infrastructure nor it is controlled in turn by the responsible ministry.

Table 2 Landslides in Italy 1964–2013 (50 years)

Region	Dead	Missing	Wounded	Total victims
Abruzzo	8	–	3	11
Basilicata	14	–	27	41
Calabria	38	–	150	188
Campania	289	–	410	699
Emilia-Romagna	49	–	79	128
Fiuli-Venezia Giulia	12	–	6	18
Lazio	23	–	81	104
Liguria	37	–	45	82
Lombardy	117	–	121	238
Marche	7	–	4	11
Molise	–	–	4	4
Piedmont	125	8	73	206
Puglia	12	–	4	16
Sardinia	12	–	25	37
Sicily	62	6	298	366
Tuscany	62	1	89	152
Trentino-Alto Adige	349	–	234	583
Umbria	12	–	31	43
Valle d'Aosta	25	–	27	52
Veneto	44	–	20	64
Total	1289	15	1728	3032

Source http://polaris.irpi.cnr.it/cinquantaanni-di-frane-ed-inondazioni-in-Italia1964-2013

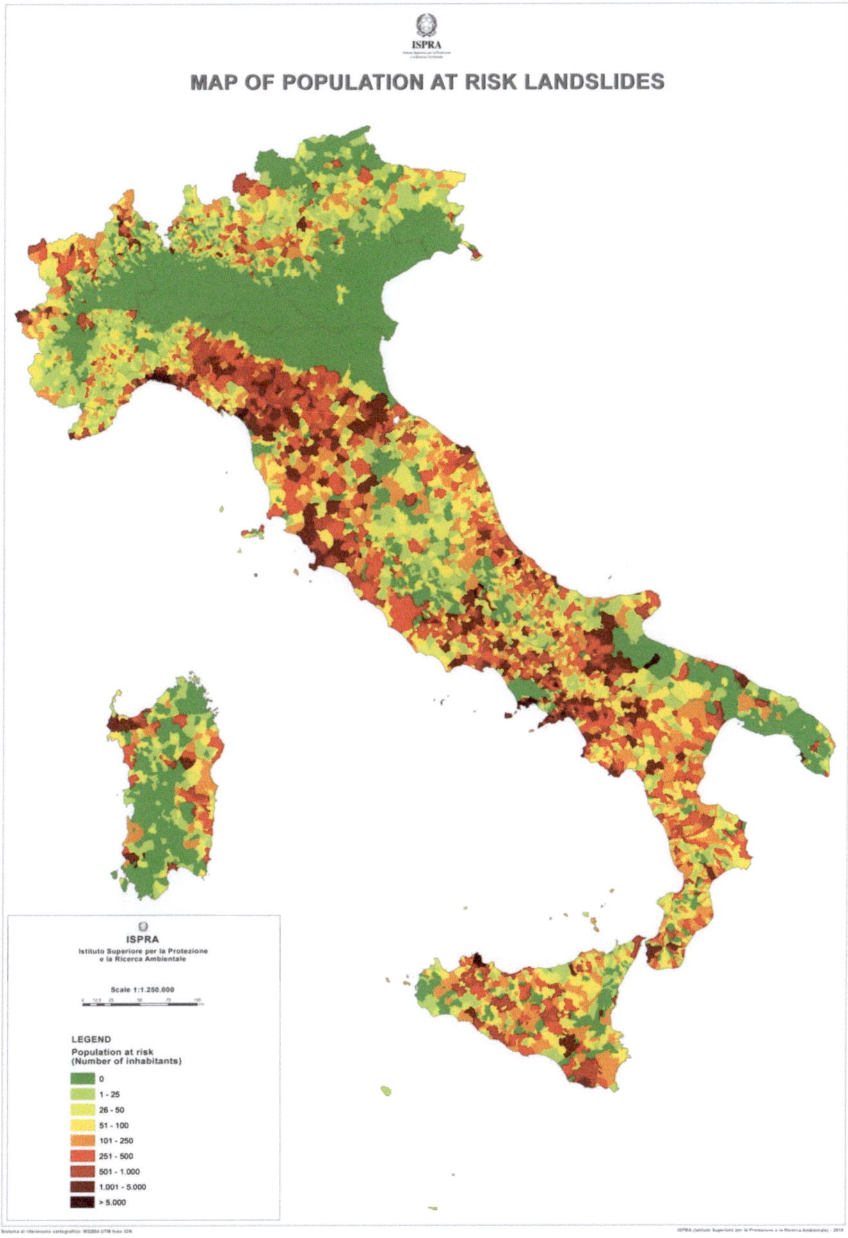

Fig. 3 Map of population at risk landslides in Italy. *Source* ISPRA

References

Almagià, R. (1907, 1910). Studi geografici sulle frane in Italia. In *Memorie della Società Geografica Italiana*, 2 vols. Roma: Società Geografica Italiana.

Batini, G. (1967). *4 November 1966: The River Arno in the Museums of Florence: Galleries, monuments, churches, libraries, archives and masterpieces damaged by the flood*. Florence: Bonechi.

http://polaris.irpi.cnr.it/cinquantaanni-di-frane-ed-inondazioni-in-Italia1964–2013.

http://www.salute.gov.it/caldo.

ISPRA (Istituto Superiore per la Protezione e la Ricerca Ambientale). (2018). *Dissesto idrogeologico in Italia:pericolosità e indicatori di rischio*.

ISTAT (Istituto Nazionale di Statistica). (2010). *Andamento meteo-climatico in Italia*. Anni 2000–2010.

Martin, G. J. (1967, June). The Florence floods: 4 November 1966. *The Geographical Journal, 133*(2), 277–279.

Palmieri, W. (2004). https://www.researchgate.net/publication/259358789_La_storia_delle_frane_in_Italia_e_gli_studi_di_Roberto_Almagia.

Impact of Hurricanes on Mental Health

Olaniyi Olayinka and Muge Akpinar-Elci

Abstract Hurricanes are among the leading natural disasters in the USA and are associated with significant psychological distress and mental disorders. Hence, understanding the mechanism underlying the development of mental health problems during a hurricane is critical. This chapter highlights the main factors that increase the risk for hurricane-related mental health problems. Common mental disorders and popular screening instruments that are available for rapid mental health assessment of a population affected by a hurricane are described. The chapter also emphasizes the importance of disaster mental health management and surveillance during the entire disaster management cycle.

Keywords Mental health · Disasters · Posttraumatic stress disorder · Depression · Psychological stress · Disaster surveillance

1 Introduction

Hurricanes are large, circulating winds with sustained wind speed of 74 miles per hour or higher. In the USA, hurricanes occur in early June to late November, often causing significant morbidity and mortality, as well as destruction of property (Taylor et al. 2012). Regarding cost, a recent analysis by the National Oceanic and Atmospheric Administration (NOAA) found Hurricanes Katrina, Harvey, Maria, Sandy and Irma—the costliest hurricanes in America since 1980—accounted for over US$500 billion in financial loss and damages to the USA (NOAA 2018). The mental health impact of hurricanes on Americans is particularly huge, disproportionately affecting socially vulnerable populations. A pre-post assessment of mental health outcomes related to Hurricane Katrina, for example, revealed a significantly

O. Olayinka (✉)
Department of Psychiatry and Behavioral Sciences, Interfaith Medical Center, Brooklyn, NY 11213, USA
e-mail: olayinkaniyi@gmail.com

M. Akpinar-Elci
School of Community & Environmental Health, College of Health Sciences, Old Dominion University, Norfolk, VA 23529, USA

R. Akhtar (ed.), *Extreme Weather Events and Human Health*,
https://doi.org/10.1007/978-3-030-23773-8_7

higher prevalence of posttraumatic stress disorder (PTSD) and other mental illnesses among participants after the disaster (Rhodes et al. 2010). Recently, a group of mental health clinicians that provided care during Hurricane Harvey, a major hurricane classified as Category 3, reported that over 200 evacuees sought mental health care; 50% reportedly had a secondary psychiatric diagnosis (Shah et al. 2017). Access to psychotropic medications and managing acute suicidality were a few of the challenges first responders faced during Hurricane Harvey. The storm surge and extreme flooding associated with most major hurricanes add to their negative impact on the mental health of affected populations (Akpinar-Elci et al. 2018). As the number and intensity of hurricanes are predicted to increase in the future, partly due to a rising mean global temperature, a corresponding increase in hurricane-related mental health problems is likely. Since Hurricane Galveston made landfall in Southern USA in 1900, Americans have experienced approximately 40 major Hurricanes. In 2017 alone, 10 hurricanes occurred in the USA, two of which were reported to be the most devastating hurricanes to hit the country in the past decade (NOAA 2017). In addition to short-term consequences, the long-term mental health impact of these hurricanes is predicted to be substantial (Shultz and Galea 2017). This chapter provides a general overview of the relationship between hurricane and mental problems, aiming to simplify the complex mechanism underlying the development of disaster-attributable psychological distress.

2 Mental Health Effects of Hurricanes

Several models for explaining the mental health impact of hurricanes exist. As it is for many chronic diseases (e.g., Alzheimer's disease), interaction between multiple risk factors are likely involved in the development of disaster-related mental health problems. In an attempt to simplify this interaction, one might adopt the Risk-Hazard-Exposure-Vulnerability equation, i.e., *Risk* (R) = *Exposure* $(E) \times$ *Vulnerability* $(V) \times$ *Hazard* (H). It is also important to consider this equation in the context of a population's capacity to cope with a hazard. For example, the *risk* for a supposed hurricane-related mental health problem (e.g., PTSD) would depend on the extent of *exposure* of a *vulnerable* population to the *hazard*, hurricane. Indeed, individual-level factors (e.g., gene, resilience) and community-level factors (e.g., level of disaster preparedness and response, availability of social support networks) moderate the risk for developing hurricane-attributable psychological distress. A brief discussion of the components of the risk-exposure equation, in the context of disaster-related mental health, is discussed below.

3 The Hazard—Hurricanes

Hurricanes are a type of natural hazard; hazard is defined as a "process, phenomenon or human activity that may cause loss of life, injury or other health impacts, property damage, social and economic disruption or environmental degradation" (UNISDR 2017). What would make a hurricane more "hazardous" than another? For example, Hurricanes Dolly and Ike, Category 1 and Category 2 hurricanes, respectively, occurred during the same hurricane season. However, Hurricane Ike garnered more media attention because it had more negative impact on human lives and property. Hence, the category of a hurricane, which correlates with its propensity to cause severe damage and disrupt lives, defines how hazardous a hurricane is. According to NOAA, hurricane hazards include high winds, storm surge and flooding (NOAA n.d.).

4 Vulnerability Factors

Despite similar level of exposure, some individuals or populations are more likely than others to develop mental illness following a hurricane. The difference between these two groups may lie in what is considered a "vulnerability factor." World Health Organization (WHO) defines *vulnerability* as "the degree to which a population, individual or organization is unable to anticipate, cope with, resist and recover from the impacts of disasters" (WHO 2002). In general, vulnerable groups reported in the scientific literature include children, pregnant women, the elderly and those suffering from preexisting mental health problems or other chronic diseases. Certainly, these groups are likely to report significant psychological problems during the initial period following a disaster. Other vulnerability factors are also reported in the scientific literature. Thirteen to 16 months after Hurricane Sandy in 2012, a telephone survey of residents living in communities hardest hit by Sandy found that neighborhoods closest to the ocean, parental status, as well as being Asian or Black, were significant vulnerability factors among residents living in the boroughs of Brooklyn and Queens (Gruebner et al. 2015).

5 Exposure

Americans are more likely to be exposed to hurricanes during the Atlantic hurricane season, which starts on June 1 and ends on November 30 of each year. The probability of exposure also increases for people living in areas prone to hurricanes. In its analysis of hurricane data from 1944 to 1999, NOAA found the odds of exposure to a hurricane to be highest among residents of Miami, Florida; Cape Hatteras; North Carolina; San Juan, Puerto Rico; and New Orleans, Louisiana (NOAA 2014). While ***direct*** exposure might be the focus of first responders and public health decision-makers during the immediate aftermath of a hurricane, those ***indirectly*** exposed may suffer negative mental or psychological consequences as well. For example, according to the Diagnostic and Statistical Manual of Mental Disorders, 5th Edition (DSM-5), PTSD may occur in a person who learned that a close family member or friend died or suffered life-threatening injuries during a hurricane. Additionally, single versus repeated exposure to a harmful hurricane, or its effect on others, may cause significant short-lived or lifelong psychological distress. The severity/intensity of exposure is equally crucial when determining the risk for mental illness post-disaster. This is defined by Sir Bradford Hill as *biologic gradient* or dose–response relationship. For example, exposure to a Category 2 hurricane for a protracted period of time is likely to induce psychological distress compared with brief exposure (Hill 2015). Of course, in reality, the relationship between a risk factor/cause and effect could be nonlinear because of the presence of confounding factors. For example, brief exposure to a Category 1 hurricane may cause a mental disorder in a genetically predisposed individual who has a history of childhood trauma (Romens et al. 2015; Ballard et al. 2015; Jansen et al. 2016) (Fig. 1).

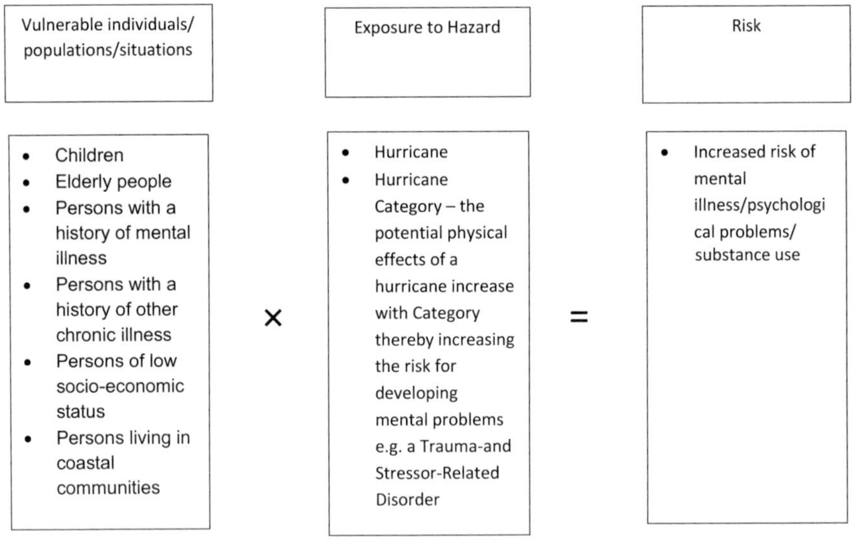

Fig. 1 The disaster risk equation

6 Disaster-Related Mental Health Issues

6.1 Posttraumatic Stress Disorder

PTSD is one of the most reported and widely researched mental disorder among hurricane-impacted populations (Foa et al. 2006; Pietrzak et al. 2012a; North and Pfefferbaum 2013; Fergusson et al. 2014). While very few, well-conducted incidence studies exist, the prevalence of hurricane-related PTSD has been found to range from 2 to 30%, depending on population characteristics, pre-hurricane factors and the period post-hurricane that a population is assessed (CDC 2006; Caramanica et al. 2015; Lowe et al. 2015; Boscarino et al. 2017; Heid et al. 2017). For example, a study found the prevalence of PTSD among police officers and firefighters, 2 months after Hurricane Katrina, was 19 and 22%, respectively (CDC 2006). Some studies have shown that a prior history of disaster-related PTSD and low socioeconomic status increase the risks for PTSD (Caramanica et al. 2015). It is worth mentioning that the prevalence of PTSD appears to decrease months after a hurricane.

Regarding PTSD assessment, most studies use the Primary Care PTSD Screen, the Impact of Events Scale-Revised, the PTSD CheckList or the Trauma Screening Questionnaire survey instruments in screening individuals. Overall, individuals diagnosed with PTSD following a hurricane must have been either directly or indirectly exposed to its traumatic effect and exhibit the following features for at least four weeks: re-experience the event (e.g., through flashbacks), actively avoid thoughts or situations associated with the event, display negative alterations in cognition and mood and suffer from hyperarousal (e.g., hypervigilance or irritable behavior) (American Psychiatric Association 2013).

6.2 Major Depressive Disorder

Clinical features of major depressive disorder (MDD) (e.g., sleep disturbance, depressed mood, suicidal ideation) are prevalent among hurricane-affected populations (Lowe et al. 2015; Akpinar-Elci et al. 2018). Individuals with MDD have five or more depressive symptoms that often interfere significantly with their ability to function socially or at work. The Patient Health Questionnaire (PHQ-9) is the most common standardized screening tool used to identify depression in epidemiologic studies. A survey of residents living in communities impacted by Hurricane Sandy in NYC found that the 13- to 17-month prevalence of probable MDD was 8.9% (Lowe et al. 2015). This is similar to the prevalence of MDD of 8.6% that was reported among 193 Galveston Bay area adult residents who were surveyed 8–20 weeks after Hurricane Ike (Pietrzak et al. 2012b). When the Galveston Bay residents were screened at 12 months after Hurricane Ike, the prevalence of MDD appeared stable compared with a decrease in PTSD rate.

6.3 Other Mental Health Issues

The intense psychological stress experienced during the immediate periods following a hurricane has been associated with various mental health issues (e.g., anxiety disorders) and substance use. However, rapid identification of disaster-related mental disorders can be time-consuming and expensive. Screening instruments that can identify probable DSM disorders are especially important as they inform clinical treatment. The K10/K6 nonspecific distress scale is a validated, easy to administer, screening instrument used in differentiating individuals with DSM-IV disorders from non-cases (Kessler et al. 2003). Of note, screening instruments based on DSM-5 disorders (e.g., the Distress Questionnaire-5) and that are culturally sensitive are being advocated (Batterham et al. 2016; Stolk et al. 2014). Based on the above, anxiety and affective disorders are some of the most commonly reported mental disorders among hurricane survivors. A telephone survey of a representative sample of residents of Hurricane Katrina-affected areas, 5–7 months after the incident, found that the 30-day prevalence of anxiety and affective disorders was approximately 30%. Of these, approximately 10 and 20% met the K6 criteria for a probable severe and mild/moderate mental illness, respectively (Galea et al. 2007).

7 Disaster Mental Health Management and Surveillance

In order to effectively address disaster-related mental health problems, one must adopt a comprehensive approach. This includes activities targeted to the different phases of a disaster which, according to the WHO Disaster Management Cycle, are the predisaster, disaster and post-disaster periods. The goal is to prepare for, mitigate, respond to and quickly recover from the negative health impact of a disaster. For instance, increasing mental health awareness in communities and educating primary health care professionals on the importance of early disease detection and management before disasters strike may increase resilience and reduce risk of disaster-related mental illness in a population. To detect hurricane-related mental disorders, non-traditional sources of data may be sought to augment existing, traditional sources (Olayinka et al. 2017).

Malilay and colleagues identified four, cross-cutting, disaster-related epidemiologic activities that are adaptable in disaster mental health management. These are "rapid needs assessments, health surveillance, tracking and registries and epidemiological investigations" (Malilay et al. 2014). For example, early detection of individuals with hurricane-related psychiatric disorders is critical for rapid allocation of mental health services. Such an approach would invariably reduce the psychological burden of hurricanes, accelerate psychological healing and help affected populations to return to pre-hurricane level of functioning. This is particularly important given that approximately 1 in 4–5 American youths and adults has a mental illness (Bagalman and Napili 2018; Merikangas et al. 2010). Population-based assessment of disaster-related psychiatric disorders, however, is fraught with challenges. This ranges from how cases are identified in affected populations (e.g., screening instruments using differing thresholds/criteria to determine whether a subject is a probable case) to what is considered a true disaster-related mental disorder. For example, psychiatric epidemiologists and public health professionals are not clear whether mental disorders that develop several months following a disaster can be considered disaster-related. Furthermore, unlike the ease with which the physical effects of a disaster can be identified because affected populations are typically within the area where a hurricane makes landfall, the mental health of individuals in distant location may be impacted (e.g., in the case of PTSD). Additionally, areas affected by a hurricane may be inaccessible for rapid mental health surveillance, particularly during the immediate aftermath of such event. Despite the aforementioned, tremendous progress is being made in the field of disaster mental health epidemiology.

8 Conclusion

With the high probability of a hurricane making landfall in the USA annually, the risk for hurricane-related mental disorders remains a major public health concern. The risk for mental illness following a hurricane depends on a complex interaction of factors including the intensity of exposure to hazards associated with the hurricane (e.g., wind speed, flooding) as well as vulnerability factors (e.g., pre-hurricane history of mental illness). There is abundance of evidence showing an increased risk for PTSD, MDD and anxiety disorders among populations exposed to hurricane. Despite the progress made in disaster mental health surveillance, more still needs to be done to stem the negative mental health impact of hurricanes and other natural disasters in general. For example, harmonizing existing screening instruments and developing a robust case definition for hurricane-related mental disorders are lacking . More importantly, making disaster-related mental health a priority through awareness-raising and integration in disaster management policies at the local, national and global levels is crucial.

References

Akpinar-Elci, M., Rose, S., & Kekeh, M. (2018). Well-being and mental health impact of household flooding in Guyana, the Caribbean. *Marine Technology Society Journal, 52*(2).

American Psychiatric Association. (2013). *Diagnostic and statistical manual of mental disorders* (5th ed.). Washington, DC: American Psychiatric Association.

Bagalman, E., & Napili, A. (2018). *Prevalence of mental illness in the United States: Data sources and estimates*. Washington, DC: Congressional Research Service, Library of Congress. Retrieved from https://fas.org/sgp/crs/misc/R43047.pdf.

Ballard, E. D., Van Eck, K., Musci, R. J., Hart, S. R., Storr, C. L., Breslau, N., et al. (2015). Latent classes of childhood trauma exposure predict the development of behavioral health outcomes in adolescence and young adulthood. *Psychological Medicine, 45*(15), 3305–3316.

Batterham, P. J., Sunderland, M., Carragher, N., Calear, A. L., Mackinnon, A. J., & Slade, T. (2016). The Distress Questionnaire-5: Population screener for psychological distress was more accurate than the K6/K10. *Journal of Clinical Epidemiology, 71,* 35–42.

Boscarino, J. A., Hoffman, S. N., Adams, R. E., Figley, C. R., & Solhkhah, R. (2017). Mental health outcomes among vulnerable residents after Hurricane Sandy: Implications for disaster research and planning. *American Journal of Disaster Medicine, 9*(2), 97–106.

Caramanica, K., Brackbill, R. M., Stellman, S. D., & Farfel, M. R. (2015). Posttraumatic stress disorder after Hurricane Sandy among persons exposed to the 9/11 disaster. *International Journal of Emergency Mental Health, 17*(1), 356.

Centers for Disease Control and Prevention. (2006). Health hazard evaluation of police officers and firefighters after Hurricane Katrina–New Orleans, Louisiana, October 17–28 and November 30–December 5, 2005. *MMWR: Morbidity and Mortality Weekly Report, 55*(16), 456-458.

Fergusson, D. M., Horwood, L. J., Boden, J. M., & Mulder, R. T. (2014). Impact of a major disaster on the mental health of a well-studied cohort. *JAMA Psychiatry, 71*(9), 1025–1031.

Foa, E. B., Stein, D. J., & McFarlane, A. C. (2006). Symptomatology and psychopathology of mental health problems after disaster. *Journal of Clinical Psychiatry, 67*(Suppl 2), 15–25.

Galea, S., Brewin, C. R., Gruber, M., Jones, R. T., King, D. W., King, L. A., et al. (2007). Exposure to hurricane-related stressors and mental illness after Hurricane Katrina. *Archives of General Psychiatry, 64*(12), 1427–1434.

Gruebner, O., Lowe, S. R., Sampson, L., & Galea, S. (2015). The geography of post-disaster mental health: Spatial patterning of psychological vulnerability and resilience factors in New York City after Hurricane Sandy. *International Journal of Health Geographics, 14*(1), 16.

Heid, A. R., Pruchno, R., Cartwright, F. P., & Wilson-Genderson, M. (2017). Exposure to Hurricane Sandy, neighborhood collective efficacy, and post-traumatic stress symptoms in older adults. *Aging & Mental Health, 21*(7), 742–750.

Hill, A. B. (2015). The environment and disease: Association or causation? *Journal of the Royal Society of Medicine, 108*(1), 32–37.

Jansen, K., Cardoso, T. A., Fries, G. R., Branco, J. C., Silva, R. A., Kauer-Sant'Anna, M., et al. (2016). Childhood trauma, family history, and their association with mood disorders in early adulthood. *Acta Psychiatrica Scandinavica, 134*(4), 281–286.

Kessler, R. C., Barker, P. R., Colpe, L. J., Epstein, J. F., Gfroerer, J. C., Hiripi, E. et al. (2003). Screening for serious mental illness in the general population. *Archives of General Psychiatry, 60*(2), 184–189.

Lowe, S. R., Sampson, L., Gruebner, O., & Galea, S. (2015). Psychological resilience after Hurricane Sandy: The influence of individual-and community-level factors on mental health after a large-scale natural disaster. *PLoS ONE, 10*(5), e0125761.

Malilay, J., Heumann, M., Perrotta, D., Wolkin, A. F., Schnall, A. H., Podgornik, M. N., et al. (2014). The role of applied epidemiology methods in the disaster management cycle. *American Journal of Public Health, 104*(11), 2092–2102.

Merikangas, K. R., He, J. P., Burstein, M., Swanson, S. A., Avenevoli, S., Cui, L., et al. (2010). Lifetime prevalence of mental disorders in US adolescents: Results from the National Comorbidity Survey Replication–Adolescent Supplement (NCS-A). *Journal of the American Academy of Child & Adolescent Psychiatry, 49*(10), 980–989.

National Oceanic and Atmospheric Administration. (2017). *Extremely active 2017 Atlantic hurricane season finally ends*. Retrieved from http://www.noaa.gov/media-release/extremely-active-2017-atlantic-hurricane-season-finally-ends.

National Oceanic and Atmospheric Administration. (2018). *U.S. Billion-Dollar Weather & Climate Disasters 1980–2018*. Retrieved from https://www.ncdc.noaa.gov/billions/events.pdf.

National Oceanic and Atmospheric Administration. (n.d.). *Hurricane Preparedness—Hazards*. Retrieved from https://www.nhc.noaa.gov/prepare/hazards.php.

NOAA. (2014). *What is my chance of being struck by a tropical storm or hurricane?* Retrieved from http://www.aoml.noaa.gov/hrd/tcfaq/G11.html.
North, C. S., & Pfefferbaum, B. (2013). Mental health response to community disasters: A systematic review. *JAMA, 310*(5), 507–518.
Olayinka, O. O., Bayleyegn, T. M., Noe, R. S., Lewis, L. S., Arrisi, V., & Wolkin, A. F. (2017). Evaluation of real-time mortality surveillance based on media reports. *Disaster Medicine and Public Health Preparedness, 11*(4), 460–466.
Pietrzak, R. H., Southwick, S. M., Tracy, M., Galea, S., & Norris, F. H. (2012a). Posttraumatic stress disorder, depression, and perceived needs for psychological care in older persons affected by Hurricane Ike. *Journal of Affective Disorders, 138*(1), 96–103.
Pietrzak, R. H., Tracy, M., Galea, S., Kilpatrick, D. G., Ruggiero, K. J., Hamblen, J. L., et al. (2012b). Resilience in the face of disaster: Prevalence and longitudinal course of mental disorders following hurricane Ike. *PLoS One, 7*(6), e38964.
Rhodes, J., Chan, C., Paxson, C., Rouse, C. E., Waters, M., & Fussell, E. (2010). The impact of Hurricane Katrina on the mental and physical health of low-income parents in New Orleans. *American Journal of Orthopsychiatry, 80*(2), 237–247.
Romens, S. E., McDonald, J., Svaren, J., & Pollak, S. D. (2015). Associations between early life stress and gene methylation in children. *Child Development, 86*(1), 303–309.
Shah, A. A., Valles, N., Banu, S., Storch, E. A., & Goodman, W. (2017). Meeting the mental health needs of Hurricane Harvey evacuees. *American Journal of Psychiatry, 175*(1), 13–14.
Shultz, J. M., & Galea, S. (2017). Mitigating the mental and physical health consequences of Hurricane Harvey. *JAMA, 318*(15), 1437–1438.
Stolk, Y., Kaplan, I., & Szwarc, J. (2014). Clinical use of the Kessler psychological distress scales with culturally diverse groups. *International Journal of Methods in Psychiatric Research, 23*(2), 161–183.
Taylor, H. T., Ward, B., Willis, M., & Zaleski, W. (2012). *The Saffir-Simpson Hurricane Wind Scale*. Available from http://origin.www.nhc.noaa.gov/pdf/sshws.pdf.
United Nations Office for Disaster Risk Reduction. (2017). *Terminology*. Retrieved from https://www.unisdr.org/we/inform/terminology#letter-h.
World Health Organization. (2002). *Environmental health in emergencies and disasters: a practical guide*. Retrieved from http://www.who.int/water_sanitation_health/hygiene/emergencies/em2002intro.pdf.

Climate Change, Wildfires, Heatwaves and Health Impacts in Australia

Nicolas Borchers Arriagada, David M. J. S. Bowman, Andrew J. Palmer and Fay H. Johnston

Abstract Heat-related extreme events, such as wildfires and heatwaves, have historically imposed a burden on Australian society, and according to rigorous and robust scientific literature, it is expected that there will be an increase in frequency, intensity and duration of these types of natural hazards. Within Australia, wildfires and heatwaves are currently responsible for more than 60% of all direct fatalities related to natural hazards, and it is highly likely that this is an underestimation of all deaths as some health impacts are not routinely quantified (e.g. premature death related to air pollution from wildfire smoke exposure). Deaths attributable to heatwaves and fire smoke pollution are more commonly due to exacerbations of pre-existing health conditions, than to specific direct impacts such as heat stroke. Some groups, such as the elderly, infants and those with pre-existing conditions, tend to be more vulnerable to these impacts. Furthermore, evidence suggests that there are synergistic additional impacts when exposed to high temperature and air pollution and that probably health impacts are considerably underestimated in the case of some specific groups such as those with occupational chronic exposure to fire smoke. To avoid increases in public health effects, society at all levels needs to increase its adaptive capacity. Measures need to be taken from a planning and management perspective through to community response at a local level, adequately focusing resources to include vulnerable sectors and population groups.

Keywords Climate change · Bushfires · Wildfires · Heatwaves · Health · Australia

N. Borchers Arriagada (✉) · A. J. Palmer · F. H. Johnston
Menzies Institute for Medical Research, University of Tasmania, Private Bag 23, Hobart, TAS 7000, Australia
e-mail: nicolas.borchers@utas.edu.au

N. Borchers Arriagada
New South Wales Bushfire Risk Management Research Hub, University of Tasmania, Hobart, TAS, Australia

D. M. J. S. Bowman
School of Natural Sciences, University of Tasmania, Private Bag 55, Hobart, TAS 7001, Australia

R. Akhtar (ed.), *Extreme Weather Events and Human Health*,
https://doi.org/10.1007/978-3-030-23773-8_8

1 Wildfires and Heatwaves and Climate Change: Global Context

Wildfires and heatwaves are important natural hazards that will be exacerbated by climate change. Fire is a process that is complex by nature and is highly variable in how it behaves in space and time; it needs to be ignited, have available fuel to burn and be set in particular weather and topographic conditions that will allow it to successfully spread (Hughes and Alexander 2017). Wildfires, in contrast to prescribed burns, are less predictable processes, with high uncertainty in their behaviour and impacts due to variations in how they start, the fuel being burnt and the existing meteorological conditions that may or may not be beneficial to their spreading.

For heatwaves (or heat events), there is no quantitative universal definition, and different metrics to identify these have been used. According to the World Meteorological Organization and World Health Organization, heatwaves are defined as: "periods of unusually hot and dry or hot and humid weather that have a subtle onset and cessation, a duration of at least two to three days and a discernible impact on human activities. During such periods of hot weather, not only do daytime temperatures reach high values but nocturnal temperatures and humidity levels may also rise well beyond their long-term mean" (McGregor et al. 2015, p. 1). A key feature is a marked departure from typical background weather conditions for the affected geographical region (Nairn and Fawcett 2013). This means, for example, that a heatwave in Tasmania may be considered as moderate weather conditions in another place such as Adelaide, and therefore, the definition of a heatwave is relative to its location.

A wide body of scientific literature supports the fact that global warming will produce more frequent and intense extreme weather events worldwide, and there is very high confidence that these shifting patterns, including those caused by changing temperature and occurrence of heatwaves and fires, have an impact on human health (Bowman et al. 2017; Mora et al. 2017b; Smith et al. 2014). In the Intergovernmental Panel on Climate Change (IPCC) Fifth Assessment Report (AR5), Smith et al. (2014, p. 713) state that there is a "greater risk of injury, disease, and death due to more intensive heatwaves and fires" with some level of climate change.

During a forest fire, the process of combustion produces great amounts of smoke, consequently being an important contributor to air pollution in urban and rural settings. Air pollution is one of the largest threats to humans, one of the main causes of death in the world, and there is no evidence of a safe threshold level of exposure below which no negative effects occur (WHO 2013). It is estimated that outdoor and indoor air pollution (urban and rural sources) is responsible for around 7 million premature annual deaths (1 out of 8 total deaths), mostly from cardiovascular diseases, and from the development of respiratory diseases such as acute respiratory infections (ARIs) and chronic obstructive pulmonary diseases (COPDs) (WHO 2014a). In fact, a study published in 2012 in the Lancet found that for the year 2010, household air pollution from solid fuels was the third leading risk factor for global disease burden (Lim et al. 2012). Air pollution produced by landscape fires is not very different in the way it impacts health, and one of the main differences would be related to exposure. With urban air pollution people usually suffer from chronic exposure, while with landscape fire smoke people might be exposed for short periods of time to very high levels of air pollution. It is estimated that landscape fire smoke is responsible for 340,000 annual deaths globally (Johnston et al. 2012), and for a high number of morbidity effects such as increased respiratory hospital admissions and chronic diseases, asthma attacks, emergency department visits and reduced productivity, among others (Adetona et al. 2016; Black et al. 2017; Duran 2014; Kochi et al. 2010; Liu et al. 2015; Reid et al. 2016; Youssouf et al. 2014). Mortality estimates may have a great variation related to climatic conditions, ranging from 262,000 deaths globally with strong La Nina conditions to 532,000 with strong El Nino conditions (Johnston et al. 2012), and may potentially be underestimated as long-term effects have not been quantified. In fact, landscape fires could well be one of the most relevant emission sources, because of the uncertainty in its nature and its capacity to negatively influence air pollution reduction efforts, producing average increases of annual $PM_{2.5}$ concentration (Larsen et al. 2018) and producing greater impacts in public health (Ford et al. 2018).

According to the World Health Organization (2014b) and assuming no adaptation to climate change, by the year 2030, heat will be responsible for 92,207 (95% confidence intervals: 64,458–121,464) additional deaths globally for people aged 65 and over. This value will rise to 255,486 (191,816–364,002) for the year 2050. Worldwide, a significant amount of people (30%) are currently exposed to extreme heat conditions for at least 20 days per year, and it is expected that by the year 2100, exposure will increase to about 48% in a conservative climate change scenario (Mora et al. 2017b), while severe warming could produce increasingly intolerable conditions (Sherwood and Huber 2010).

These two types of extreme events and their impacts are not necessarily disconnected, and there are often strong associations between them. A heatwave can occur in isolation, but it will increase the risk and likelihood of a severe wildfire (Cardil et al. 2014; Johnston et al. 2011a; Shaposhnikov et al. 2014).

Impacts of extreme weather events related to increased temperatures translate into multiple impacts on human well-being. The objective of this chapter is to integrate research and findings from both government and non-government reports and assessments and peer-reviewed scientific articles to present an overview of the growing health and well-being implications the Australian community will face with increasing risk of wildfires and heatwaves driven by change in climatic conditions. The following sections will present in more detail the health impacts produced by exposure to wildfires and heatwaves, will discuss the conjoined effects of high temperatures and wildfires and will highlight the need to increase adaptation capacities throughout Australian society.

2 Wildfires and Health Impacts in Australia

Wildfires (known as bushfires in Australia) and their consequences have always been a part of the Australian landscape (Hughes and Alexander 2017), and their projected increase in frequency and intensity is of great concern when the effects on human well-being (and health specifically) are considered.

Impacts of wildfires on human well-being are diverse by nature and include:

- Loss of houses and other buildings (Ashe et al. 2009; Blanchi et al. 2014)
- Property damage and destruction (Ashe et al. 2009; Blanchi et al. 2014)
- Threats to water security (Johnston 2009)
- Loss of agricultural crops (Stephenson et al. 2013)
- Fatalities (Blanchi et al. 2014; Haynes et al. 2010)
- Injuries (Cameron et al. 2009)
- Loss of livestock (Whittaker et al. 2012)
- Loss of biodiversity (Stephenson 2010)
- Air pollution emissions and its consequences in human health (Adetona et al. 2016; Cascio 2018; Hyde et al. 2017; Liu et al. 2015; Navarro et al. 2018; Reid et al. 2016)
- Mental health problems (e.g. Post-Traumatic Stress Disorder) (Gibbs et al. 2013; Hughes and Alexander 2017; Reid et al. 2016; Stephenson 2010).

The occurrence of wildfires has a very high impact on Australian society, with a recorded yearly average of 100 fatalities, 3,000 direct injuries and 83 lost homes per year, with an estimated cost of AUD $12,000 million or 1.3% of GDP for 2005 (Ashe et al. 2009). Blanchi et al. (2014) developed and analysed a data set regarding wildfire with loss of life for the period between 1901 and 2011. Complementing this analysis with official fatality figures reported by the Australian Disaster Resilience Knowledge Hub (AIDR 2018), it is estimated that only 7 fire events were responsible for 71% of all civilian fatalities (Table 1) for the period studied.

Although these numbers seem relevant by themselves, impacts of wildfires go further, negatively impacting the quality of the air that we breathe and imposing substantial health burdens on society. But in general, the study of the effects that fire smoke has on human health tends to be challenging (compared to general urban air pollution) for a myriad of reasons that include (1) the unpredictability of episodes, (2) disperse and heterogeneous spatial and temporal distribution of smoke impacts compared to those of other sources, (3) exposure affecting both urban and rural areas with differing levels of population density, (4) higher peak concentrations of $PM_{2.5}$, (5) lack of availability of monitoring data in rural areas affected by fires, (6) exposure of small populations not representative of broader community and (7) short duration of episodes (even on large populations) that don't allow for detection of significant changes (Johnston et al. 2011b; Salimi et al. 2017). Also, the study of health impacts due to exposure to fire smoke has had less attention than those related to urban air pollution, but there is still a considerable amount of literature available, and it has been found that health outcomes are significantly diverse in the nature and size of populations affected as shown in Fig. 1.

Table 1 Major fire events in Australia according to the number of direct civilian fatalities

#	Date of fire	Bushfire	Location	Number of civilian fatalities	Number of homes destroyed
1	Feb/1926–Mar/1926	1926 bushfires	Gippsland, Victoria	60	550
2	Jan/1939	Black Friday bushfires	Victoria	71	1,000
3	14/Jan–14/Feb 1944	1944 bushfires	Victoria	15–20	>500
4	7/Feb/1967	Tasmanian "Black Tuesday" bushfires	Hobart, Tasmania	62	1,400
5	Jan/1969	1969 bushfires	Lara, Victoria	23	230
6	Feb/1983	Ash Wednesday bushfires	South Australia	75	2,000
7	Feb/2009	Black Saturday bushfires	Victoria	173	2,029

Adapted from Blanchi et al. (2014) and AIDR (2018)

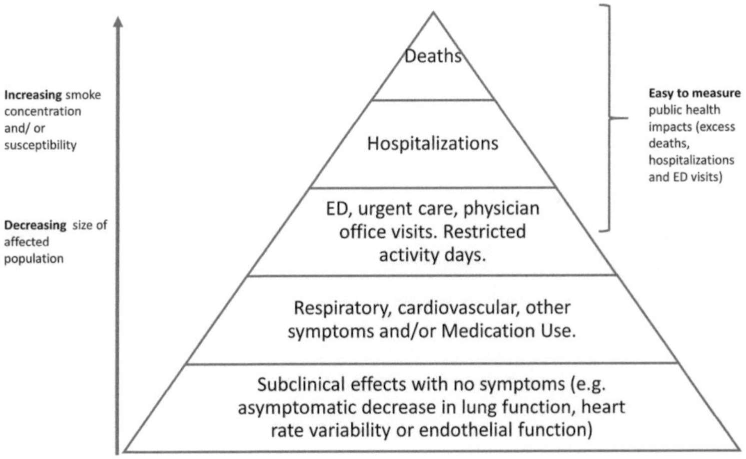

Fig. 1 Total public health impacts of wildfire smoke or $PM_{2.5}$. Adapted from Cascio (2018)

There is robust evidence for an increase in respiratory morbidity and all-cause mortality, and mixed findings in the case of both cardiovascular morbidity and cause-specific mortality due to short-term exposure to fire smoke (Black et al. 2017; Cascio 2018; Reisen et al. 2015; Youssouf et al. 2014). Exacerbation of respiratory health conditions is likely to be the most common immediate impact (Black et al. 2017). Long-term health effects due to exposure to fire smoke should also be taken into account, as some studies have shown that a considerable amount of population may be exposed to an average annual increase of more than 0.75 $\mu g/m^3$ of $PM_{2.5}$ (Cascio 2018; Rappold et al. 2017), in which case public health impacts in terms of morbidity and mortality will likely be considerable.

Epidemiological studies following Australian bushfire smoke events have found positive associations for multiple effects, especially respiratory (Table 2), while results for cardiovascular morbidity and mortality in general have been more variable. In 2010, Morgan et al. (2010) studied fires occurring in Sydney between 1994 and 2002 and found no consistency in associations between wildfire particulate matter (PM_{10}) with cardiovascular admissions and mortality. Later in 2011, Johnston et al. (2011a) analysed fires between 1997 and 2004 and found that smoke events could produce a 5% increase in non-accidental and cardiovascular mortality. Evidence regarding cardiovascular impacts has been strengthened, as Dennekamp et al. (2015) and Haikerwal et al. (2015) have found positive associations between exposure to smoke from the 2006 to 2007 wildfires in Victoria and out-of-hospital cardiac arrest (OHCA) and ischaemic heart disease (IHD). Results seen in Australian studies are consistent with those done elsewhere, where evidence is robust around positive associations between fire smoke and respiratory morbidity, evidence is increasing regarding positive associations between smoke and respiratory infections and all-cause mortality, but there are still important gaps in research with respect to effects on cardiovascular morbidity, specific causes of mortality and other endpoints such as birth outcomes and mental health (Reid et al. 2016).

Similar to the effects of air pollution in general, certain population groups tend to be more vulnerable to these outcomes, such as pregnant women, the elderly, infants, smokers, firefighters and people with pre-existing respiratory or cardiovascular conditions (Hughes and Alexander 2017; Rappold et al. 2017; Youssouf et al. 2014). Australian indigenous people might also be affected more strongly than the rest of the population (Hanigan et al. 2008). Johnston et al. (2007) found effects to be higher in Indigenous people compared to the rest of the population: associations for COPD and asthma were more than double, and they appeared to have a higher risk of being admitted for cardio-respiratory effects. This likely is a reflection of how Indigenous people are at higher risk from environmental hazards such as air pollution (Johnston et al. 2007). Some areas of Australia tend to be at higher risk of suffering bushfires and the largest effects will be experienced in high-density urban areas, where concentrated populations are potentially exposed to fire smoke.

Table 2 Examples of positive associations between fire smoke PM and health effects—Australian studies

Health outcome	Place	Year/period/episode	Reference
Increased respiratory and cardiovascular emergency ambulance dispatches (EAD)	Sydney Greater Metropolitan Region	2004–2015	Salimi et al. (2017)
Increased emergency department (ED) attendances for asthma	Victoria	1 Dec 2006–31 Jan 2007	Haikerwal et al. (2016)
Increased out-of-hospital cardiac arrest (OHCA)	Melbourne	July 2006–June 2007	Dennekamp et al. (2015)
Increased out-of-hospital cardiac arrest (OHCA) and ischaemic heart disease (IHD)	Victoria	1 Dec 2006–31 Jan 2007	Haikerwal et al. (2015)
Increased non-accidental and cardiovascular mortality	Sydney	January 1994–June 2007	Johnston et al. (2011a)
Increased ED attendances for all non-trauma conditions, respiratory conditions, asthma, and chronic obstructive pulmonary disease (COPD)	Sydney	1996–2007	Johnston et al. (2014)
Increased admissions for respiratory infections for Indigenous people	Darwin	Fire seasons between 1996 and 2005 (1 April–30 November)	Hanigan et al. (2008)

3 Heatwaves and Health Impacts in the Australian Context

Australians have been heavily impacted from extreme heat events throughout history. It is a country that is highly vulnerable to changing climatic conditions, and it is expected that the number of hot days, warm nights and heatwaves will increase by the end of the twenty-first century (Cowan et al. 2014). Australian society will experience various types of impacts, such as:

- increased human morbidity and mortality (Cheng et al. 2018; Lindstrom et al. 2013; McGregor et al. 2015; Mora et al. 2017b; Tong et al. 2014; Turner et al. 2013; Wilson et al. 2013)
- increased risk for outdoor activities (Nairn and Fawcett 2013)
- stress for people, animals and plants (Sherwood and Huber 2010)
- increased bushfire risk (Cardil et al. 2014)
- damage to crop and vegetation (Wreford and Neil Adger 2010)
- increased demand for resources (e.g. water, energy, etc.) (Zuo et al. 2015)
- stress on infrastructure such as roads and buildings (McEvoy et al. 2012)
- negative impacts in tourism (Nairn and Fawcett 2013).

Heatwaves will be responsible, in the case of Australasia (Australia and New Zealand), for an estimated additional 217 (132–345) yearly deaths by the year 2030 and for 605 (434–980) by 2050 (WHO 2014b). However, these numbers may increase considerably when in the presence of an extreme heat event. For example, a major heatwave struck Europe during the summer of 2003 and was responsible for over 70,000 deaths (Robine et al. 2008). In Australia, the highest number of recorded deaths from natural hazards are attributable to heatwaves, and even with their relative importance in terms of fatalities, these have not been studied with similar intensity as other natural hazards such as floods and storms (Coates et al. 2014; Lee 2014). Only 11 events (out of 350) are responsible for more than 30% of all recorded heatwave-related deaths for the period 1844–2011 (Table 3). Victoria, New South Wales and South Australia have the highest number of fatalities, with most of these occurring during summer months.

It is probable that many earlier studies underestimated the real impact of heatwaves on Australians, as only uncommon but specific heat-related diseases such as heat stroke were attributed to the heat event (Coates et al. 2014). However, epidemiological studies increasingly use statistical methods to attribute excess hospital admissions, regardless of the specific medical diagnosis, to particular events such as heatwaves, and the pathophysiological processes that lead from elevated body temperature to death through direct toxicity and exacerbations of cardiovascular, renal and metabolic conditions are well characterised (Mora et al. 2017a).

Australian regions are prone to high temperatures, hot summer and extreme heat, and the association of these conditions with adverse health impacts has been well studied (Bi et al. 2011), with a more intense focus on mortality rather than morbidity effects (Dalip et al. 2015; Loughnan et al. 2010) and with less available studies for the southern hemisphere including Australia (Loughnan et al. 2010). Health

Table 3 Significant heat events in Australia, 1844–2011 (Coates et al. 2014)

#	Date of event	Area affected	Total heat-associated deaths
1	January–February 1879	NSW, Vic	22
2	October 1895–January 1896	WA, SA, Vic, Qld, NSW	435
3	January 1906	NSW, SA	28
4	January 1908	Vic, SA, NSW	213
5	January 1939	Vic, SA, NSW	420
6	January 1940	Qld, NSW	65
7	February 1955	Perth (WA)	30
8	January–February 1959	Melbourne (Vic)	145
9	January 1960	Greater Sydney (NSW)	25
10	January 2000	Southeast Qld	22
11	January–February 2009	Vic, SA	432
Total deaths			1837

impacts include dehydration, heat strokes, cardiovascular morbidity effects, renal diseases and death. Recent studies (Table 4) have addressed the associations between heat events and multiple health effects in various cities around Australia and have found increased adverse impacts that range from increased use of health services to increased total mortality.

The very young, the elderly (especially those 75+) and people with pre-existing medical conditions are especially vulnerable, while more socially advantaged groups (in terms of socioeconomic status) tend to have lower impacts (Cheng et al. 2018; Coates et al. 2014; Dalip et al. 2015; Xiao et al. 2017). Other population characteristics such as living in urban or rural areas (Jegasothy et al. 2017), gender (Xiao et al. 2017) and geographical location (Tong et al. 2014) may define the level of vulnerability and adaptability that exists towards heat-related health impacts. Cowan et al. (2014) estimate that changes in frequency and duration of heat events will affect mainly northern tropical regions of Australia, while higher maximum temperatures will be felt in southern Australia.

Table 4 Examples of positive associations between heatwaves and health effects—Australian studies

Health outcome	Place	Year/period/episode	References
Increased health service utilisation	New South Wales	2005–2015	Jegasothy et al. (2017)
Increased mortality in the elderly	Brisbane, Melbourne, Sydney	1988–2009	Tong et al. (2015)
Increased emergency department visits	Brisbane	Summer seasons (December–February) from 2000 to 2012	Toloo et al. (2014)
Increased emergency hospital admissions for renal diseases in children (0–14 years)	Brisbane	1 Jan 1996–31 Dec 2005	Wang et al. (2014)
Increased all-cause, cardiovascular and respiratory mortality and hospital admissions for heat-related injuries, dehydration, and other fluid disorders	Sydney	Mortality (1997–2007) Hospital admissions (1997–2010)	Wilson et al. (2013)
Significantly increased total hospital admissions, emergency department presentations, general medical admissions, and rise in number of deaths	Melbourne	2009 heatwave (Jan/Feb 2009)	Lindstrom et al. (2013)
Increased daily mortality, total emergency department presentations, renal-related emergency department presentations	Perth	1994–2008	Williams et al. (2012)
Increased ambulance call-outs, renal morbidity for elders, heat-related admissions, ischaemic heart disease in the 15–64 age group, total mortality particularly in the 15–64 age group	Adelaide	Summers of 2008 and 2009	Nitschke et al. (2011)

4 Wildfire and Heatwaves: Relative and Combined Effects

Severe fire and heat events have been recorded and will likely increase in the future, producing important environmental, social and economic consequences (Bowman et al. 2017). For Australia, 2017 was its third hottest year on record, including highest historical maximum temperatures during winter and very low rainfall, shaping a perfect scenario for heat-related extreme weather events such as wildfires and heatwaves to develop (Steffen et al. 2018).

Heatwaves and wildfires were responsible for more than 65% of the 8,256 direct fatalities attributed to natural hazards in Australia for the period between 1900 and 2011 (Fig. 2). Heatwaves alone accounted for more fatalities than all other natural hazards (>50% of deaths), while wildfires were responsible for more than 10% of these deaths (Coates et al. 2014), with most wildfire fatalities (>65%) concentrated in the State of Victoria (Blanchi et al. 2014). As we discussed above in the case of wildfires though, it is highly likely that the magnitude of these impacts would increase considerably if indirect deaths caused by exposure to fire smoke were included. Increases in air pollution can have a large public health impact if extensive populations are affected. For example, Horsley et al. (2018) estimated that 183 days of exposure to landscape fire smoke were responsible for 197 premature deaths, 436 cardiovascular hospitalisations and 787 respiratory hospitalisations in a period of 13 years (2001–2013) in Sydney alone. This represents almost 23% of all direct fatalities recorded for Australia in a 111-year period.

As mentioned earlier, days of high temperature (heat events) have a strong influence on increasing the likelihood of large wildfires, especially when combined with high wind. Conditions for fires become highly favourable, and as moisture content decreases the probability for ignition increases, and fires can burn more intensely and rapidly (Cardil et al. 2014). Additionally, there may even be interactions in mortality effects produced by exposure to heatwaves and fire smoke (Katsouyanni et al.

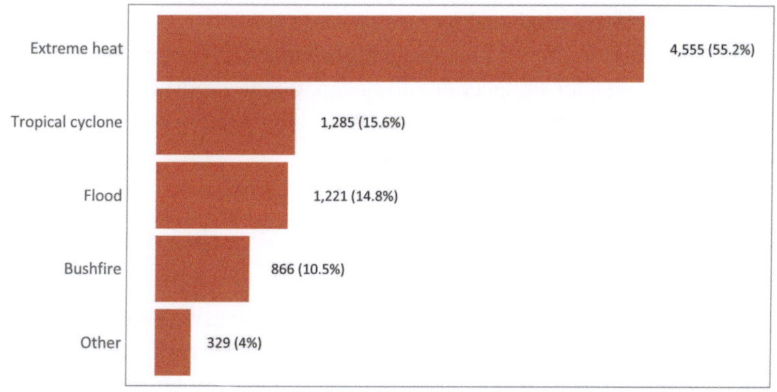

Fig. 2 Total fatalities from bushfires and extreme heat compared with other Australian natural hazards between 1900 and 2011. Adapted from Coates et al. (2014)

1993; Shaposhnikov et al. 2014). In 1987, a major heatwave impacted Greece, and high levels of smoke and ozone were present during the same period. Katsouyanni et al. (1993) found statistically significant interactions between high temperatures (>30 °C) and sulphur dioxide (SO_2), and statistically suggestive interactions for ozone and smoke with high temperatures. In 2010, Moscow experienced a disastrous heatwave with high levels of air pollution simultaneously, with 44 days of continuous heat during major heat event and PM_{10} exceeding 300 $\mu g/m^3$ on multiple days. During this period, there was an estimated near 11,000 excess deaths, almost doubling the expected deaths when considering both phenomena as if they were independent (Shaposhnikov et al. 2014). According to the authors, the interactions between pollution from fire smoke and high temperatures were responsible for more than 2000 of these deaths. Different studies have been undertaken in Australian cities, and there is suggestion for a synergistic effect between high temperature and air pollution (PM_{10}, SO_2 and ozone) and that this impact (excess mortality) is additional to that of the sum of effects from each exposure variable (Dean and Green 2018).

Furthermore, there are fewer studies that relate the effects of occupational exposure (e.g. firefighters) to fire smoke (Adetona et al. 2016). These types of professionals are part of the population segments that are particularly more vulnerable to health effects associated with fire smoke exposure (Youssouf et al. 2014). The exposure to smoke that firefighters experience is very different to that of the rest of the population, and they work very close to the actual sources of emission and are exposed to high concentration levels of particulate matter and other pollutants (Adetona et al. 2016). Some of the effects experienced by firefighters include the increase of pulmonary and systemic inflammation and a decrease in lung function (Youssouf et al. 2014). A study carried on in Australia assessed exposure of firefighters to smoke during prescribed burns and wildfires and found that a considerable proportion of the firefighters had high to very high levels of exposure, and this would result in an increased likelihood of having acute and chronic health effects (Reisen et al. 2011).

It is very likely that the total public health impacts of fire smoke exposure have consistently been underestimated. There appears to an opportunity and need to adequately assess the effects produced by a conjoined exposure to fire smoke and heatwaves, as the synergistic impacts might be higher than the sum of its parts. Also, it seems relevant to consider impacts due to occupational exposure and find ways in which this could be reduced.

5 Adaptation to Heat-Related Extreme Weather Events

As climate change shapes the future of Australia with increasing heat-related extreme weather events, there will be a higher demand for resources, capacities and planning that should be available to approach these issues. There is an urgent need to act regarding future potential natural hazards produced by climate change (Steffen et al. 2018). Wildfires and heatwaves will become a norm for Australian society, and it is necessary that authorities and decision-makers focus adequate amount and type of resources on reducing risks (Clayton et al. 2014; Cleland et al. 2011; Nairn and Fawcett 2013). Actions such as increasing effectiveness of communication channels and levels with the community (Fish et al. 2017), education programs aimed at changing behaviour (Richardson et al. 2012; Walker and Salt 2006), improving emergency response and incorporating wildfire and heatwave events into the planning and management of our built environment and social processes are necessary to improving the adaptive capacity that systems will have in an ever-changing environment (Tong et al. 2014). Resources must be focused on increasing adaptation to climate change, improving and increasing preparedness for bushfire disasters (Johnston 2009) and extreme heat events, including improvement of public health services, identification of vulnerable sectors and population groups and development of early warning systems (Lowe et al. 2016; Smith et al. 2014).

6 Conclusions

As global warming imposes important changes in climate and weather, with increasing temperatures and levels of drought, Australians will experience a higher risk of suffering from extreme wildfires and intense heatwaves. The many impacts will probably be distributed unequally throughout society. Infants, the elderly, labourers and people with pre-existing conditions, lower socioeconomic status or belonging to indigenous groups may be at higher risk of suffering from smoke and heat-related illnesses, while those living at the urban–rural interface are at higher risk of being directly affected by wildfires. Geographically, population exposure and density differ around Australia, and studies suggest that the Southeast of Australia has higher risk of mortality and disease from wildfire and smoke-related impacts while South Australia presents the highest heat-related mortality rates, although most fatalities have concentrated in Victoria, NSW and South Australia. There are some important aspects that have had relatively limited consideration, such as the possible synergistic adverse health impacts of simultaneous exposure to fire smoke and heatwaves, and the more extreme personal exposures of professionals who work outside during heatwaves and bushfire air pollution events. Necessary steps should be taken at federal, state and local levels to ensure safety of communities and to minimise the loss of resources, lives and well-being, especially focusing on the sectors of society that are more vulnerable due to their physical locations, social and economic disadvantage or higher-risk occupational groups.

References

Adetona, O., Reinhardt, T. E., Domitrovich, J., Broyles, G., Adetona, A. M., Kleinman, M. T. … Naeher, L. P. (2016). Review of the health effects of wildland fire smoke on wildland firefighters and the public. *Inhalation Toxicology*. https://doi.org/10.3109/08958378.2016.1145771.

AIDR. (2018). *Australian Disaster Resilience Knowledge Hub: Disasters, Fire—Bushfire. Melbourne: Australian Institute for Disaster Resilience*. Retrieved from https://knowledge.aidr.org.au/search?keywords=&collection=1316&categories=Fire-Bushfire&from=&to=&region=&type=&page=1&results=100.
Ashe, B., McAneney, K. J., & Pitman, A. J. (2009). Total cost of fire in Australia. *Journal of Risk Research*. https://doi.org/10.1080/13669870802648528.
Bi, P., Williams, S., Loughnan, M., Lloyd, G., Hansen, A., Kjellstrom, T. … Saniotis, A. (2011). The effects of extreme heat on human mortality and morbidity in Australia: Implications for public health. *Asia-Pacific Journal of Public Health/Asia-Pacific Academic Consortium for Public Health*. https://doi.org/10.1177/1010539510391644.
Black, C., Tesfaigzi, Y., Bassein, J. A., & Miller, L. A. (2017). Wildfire smoke exposure and human health: Significant gaps in research for a growing public health issue. *Environmental Toxicology and Pharmacology, 55,* 186–195. https://doi.org/10.1016/J.ETAP.2017.08.022.
Blanchi, R., Leonard, J., Haynes, K., Opie, K., James, M., & de Oliveira, F. D. (2014). Environmental circumstances surrounding bushfire fatalities in Australia 1901–2011. *Environmental Science and Policy*. https://doi.org/10.1016/j.envsci.2013.09.013.
Bowman, D. M. J. S., Williamson, G. J., Abatzoglou, J. T., Kolden, C. A., Cochrane, M. A., & Smith, A. M. S. (2017). Human exposure and sensitivity to globally extreme wildfire events. *Nature Ecology & Evolution, 1*(58), 1–6. https://doi.org/10.1038/s41559-016-0058.
Cameron, P. A., Mitra, B., Fitzgerald, M., Scheinkestel, C. D., Stripp, A., Batey, C. … Cleland, H. (2009). Black Saturday: The immediate impact of the February 2009 bushfires in Victoria, Australia. *Medical Journal of Australia*. https://doi.org/10.5694/j.1326-5377.2009.tb02666.x.
Cardil, A., Salis, M., Spano, D., Delogu, G., & Terrén, D. M. (2014). Large wildland fires and extreme temperatures in Sardinia (Italy). *IForest*. https://doi.org/10.3832/ifor1090-007.
Cascio, W. E. (2018). Wildland fire smoke and human health. *Science of the Total Environment*. https://doi.org/10.1016/j.scitotenv.2017.12.086.
Cheng, J., Xu, Z., Bambrick, H., Su, H., Tong, S., & Hu, W. (2018). Heatwave and elderly mortality: An evaluation of death burden and health costs considering short-term mortality displacement. *Environment International*. https://doi.org/10.1016/j.envint.2018.03.041.
Clayton, H., Mylek, M. R., Schirmer, J., Cary, G. J., & Dovers, S. R. (2014). Exploring the use of economic evaluation in Australian wildland fire management decision-making. *International Journal of Wildland Fire*. https://doi.org/10.1071/WF13140.
Cleland, H. J., Proud, D., Spinks, A., & Wasiak, J. (2011). Multidisciplinary team response to a mass burn casualty event: Outcomes and implications. *Medical Journal of Australia*. https://doi.org/10.5694/j.1326-5377.2011.tb03110.x.
Coates, L., Haynes, K., O'Brien, J., McAneney, J., & De Oliveira, F. D. (2014). Exploring 167 years of vulnerability: An examination of extreme heat events in Australia 1844–2010. *Environmental Science and Policy*. https://doi.org/10.1016/j.envsci.2014.05.003.
Cowan, T., Purich, A., Perkins, S., Pezza, A., Boschat, G., & Sadler, K. (2014). More frequent, longer, and hotter heat waves for Australia in the twenty-first century. *Journal of Climate*. https://doi.org/10.1175/JCLI-D-14-00092.1.
Dalip, J., Phillips, G. A., Jelinek, G. A., & Weiland, T. J. (2015). Can the elderly handle the heat? A retrospective case-control study of the impact of heat waves on older patients attending an inner city Australian emergency department. *Asia-Pacific Journal of Public Health*. https://doi.org/10.1177/1010539512466428.
Dean, A., & Green, D. (2018). Climate change, air pollution and human health in Sydney, Australia: A review of the literature. *Environmental Research Letters, 13*(5), 53003. Retrieved from http://stacks.iop.org/1748-9326/13/i=5/a=053003.
Dennekamp, M., Straney, L. D., Erbas, B., Abramson, M. J., Keywood, M., Smith, K. … Tonkin, A. M. (2015). Forest fire smoke exposures and out-of-hospital cardiac arrests in Melbourne, Australia: A case-crossover study. *Environmental Health Perspectives*. https://doi.org/10.1289/ehp.1408436.

Duran, S. (2014). Evidence review: Wildfire smoke and public health risk. Vancouver, British Columbia: BC Centre for Disease Control. Retrieved from http://www.bccdc.ca/resource-gallery/Documents/Guidelines%and%Forms/Guidelines%and%Manuals/Health-Environment/WFSG_EvidenceReview_WildfireSmoke_FINAL_v3_edstrs.pdf.

Fish, J. A., Peters, M. D. J., Ramsey, I., Sharplin, G., Corsini, N., & Eckert, M. (2017). Effectiveness of public health messaging and communication channels during smoke events: A rapid systematic review. *Journal of Environmental Management*. https://doi.org/10.1016/j.jenvman.2017.02.012.

Ford, B., Val Martin, M., Zelasky, S. E., Fischer, E. V, Anenberg, S. C., Heald, C. L., & Pierce, J. R. (2018). Future fire impacts on smoke concentrations, visibility, and health in the contiguous United States. *GeoHealth*. https://doi.org/10.1029/2018GH000144.

Gibbs, L., Waters, E., Bryant, R. A., Pattison, P., Lusher, D., Harms, L. … Forbes, D. (2013). Beyond bushfires: Community, resilience and recovery—A longitudinal mixed method study of the medium to long term impacts of bushfires on mental health and social connectedness. *BMC Public Health*. https://doi.org/10.1186/1471-2458-13-1036.

Haikerwal, A., Akram, M., Del Monaco, A., Smith, K., Sim, M. R., Meyer, M. … Dennekamp, M. (2015). Impact of fine particulate matter ($PM_{2.5}$) exposure during wildfires on cardiovascular health outcomes. *Journal of the American Heart Association*. https://doi.org/10.1161/JAHA.114.001653.

Haikerwal, A., Akram, M., Sim, M. R., Meyer, M., Abramson, M. J., & Dennekamp, M. (2016). Fine particulate matter ($PM_{2.5}$) exposure during a prolonged wildfire period and emergency department visits for asthma. *Respirology*. https://doi.org/10.1111/resp.12613.

Hanigan, I. C., Johnston, F. H., & Morgan, G. G. (2008). Vegetation fire smoke, indigenous status and cardio-respiratory hospital admissions in Darwin, Australia, 1996–2005: A time-series study. *Environmental Health: A Global Access Science Source*. https://doi.org/10.1186/1476-069X-7-42.

Haynes, K., Handmer, J., McAneney, J., Tibbits, A., & Coates, L. (2010). Australian bushfire fatalities 1900–2008: Exploring trends in relation to the "Prepare, stay and defend or leave early" policy. *Environmental Science and Policy*. https://doi.org/10.1016/j.envsci.2010.03.002.

Horsley, J. A., Broome, R. A., Johnston, F. H., Cope, M., & Morgan, G. G. (2018). Health burden associated with fire smoke in Sydney, 2001–2013. *The Medical Journal of Australia, 208*(7), 309–310.

Hughes, L., & Alexander, D. (2017). *Climate change and the Victorian bushfire threat: Update 2017*. Sydney: Climate Council of Australia.

Hyde, J. C., Yedinak, K. M., Talhelm, A. F., Smith, A. M. S., Bowman, D. M. J. S., Johnston, F. H. … Tinkham, W. T. (2017). Air quality policy and fire management responses addressing smoke from wildland fires in the United States and Australia. *International Journal of Wildland Fire, 26*(5), 347–363. Retrieved from https://doi.org/10.1071/WF16154.

Jegasothy, E., McGuire, R., Nairn, J., Fawcett, R., & Scalley, B. (2017). Extreme climatic conditions and health service utilisation across rural and metropolitan New South Wales. *International Journal of Biometeorology*. https://doi.org/10.1007/s00484-017-1313-5.

Johnston, F. H. (2009). Bushfires and human health in a changing environment. *Australian Family Physician*.

Johnston, F. H., Bailie, R. S., Pilotto, L. S., & Hanigan, I. C. (2007). Ambient biomass smoke and cardio-respiratory hospital admissions in Darwin, Australia. *BMC Public Health*. https://doi.org/10.1186/1471-2458-7-240.

Johnston, F., Hanigan, I., Henderson, S., Morgan, G., & Bowman, D. (2011a). Extreme air pollution events from bushfires and dust storms and their association with mortality in Sydney, Australia 1994–2007. *Environmental Research*. https://doi.org/10.1016/j.envres.2011.05.007.

Johnston, F. H., Hanigan, I. C., Henderson, S. B., Morgan, G. G., Portner, T., Williamson, G. J., & Bowman, D. M. J. S. (2011b). Creating an integrated historical record of extreme particulate air pollution events in Australian cities from 1994 to 2007. *Journal of the Air and Waste Management Association*. https://doi.org/10.3155/1047-3289.61.4.390.

Johnston, F. H., Henderson, S. B., Chen, Y., Randerson, J. T., Marlier, M., DeFries, R. S. … Brauer, M. (2012). Estimated global mortality attributable to smoke from landscape fires. *Environmental Health Perspectives, 120*(5), 695–701. https://doi.org/10.1289/ehp.1104422.

Johnston, F. H., Purdie, S., Jalaludin, B., Martin, K. L., Henderson, S. B., & Morgan, G. G. (2014). Air pollution events from forest fires and emergency department attendances in Sydney, Australia 1996–2007: A case-crossover analysis. *Environmental Health: A Global Access Science Source*. https://doi.org/10.1186/1476-069X-13-105.

Katsouyanni, K., Pantazopoulou, A., Touloumi, G., Tselepidaki, L., Moustris, K., Asimakopoulos, D. … Trichopoulos, D. (1993). Evidence for interaction between air pollution and high temperature in the causation of excess mortality. *Archives of Environmental Health*. https://doi.org/10.1080/00039896.1993.9940365.

Kochi, I., Donovan, G. H., Champ, P. A., & Loomis, J. B. (2010). The economic cost of adverse health effects from wildfire-smoke exposure : A review. *International Journal of Wildland Fire, 19,* 803–817. Retrieved from http://dx.doi.org/10.1071/WF09077.

Larsen, A. E., Reich, B. J., Ruminski, M., & Rappold, A. G. (2018). Impacts of fire smoke plumes on regional air quality, 2006–2013. *Journal of Exposure Science and Environmental Epidemiology*. https://doi.org/10.1038/s41370-017-0013-x.

Lee, W. V. (2014). Historical global analysis of occurrences and human casualty of extreme temperature events (ETEs). *Natural Hazards*. https://doi.org/10.1007/s11069-013-0884-7.

Lim, S. S., Vos, T., Flaxman, A. D., Danaei, G., Shibuya, K., Adair-Rohani, H. … Ezzati, M. (2012). A comparative risk assessment of burden of disease and injury attributable to 67 risk factors and risk factor clusters in 21 regions, 1990–2010: A systematic analysis for the Global Burden of Disease Study 2010. *The Lancet*. https://doi.org/10.1016/S0140-6736(12)61766-8.

Lindstrom, S. J., Nagalingam, V., & Newnham, H. H. (2013). Impact of the 2009 Melbourne heatwave on a major public hospital. *Internal Medicine Journal*. https://doi.org/10.1111/imj.12275.

Liu, J. C., Pereira, G., Uhl, S. A., Bravo, M. A., & Bell, M. L. (2015). A systematic review of the physical health impacts from non-occupational exposure to wildfire smoke. *Environmental Research*. https://doi.org/10.1016/j.envres.2014.10.015.

Loughnan, M. E., Nicholls, N., & Tapper, N. J. (2010). The effects of summer temperature, age and socioeconomic circumstance on Acute Myocardial Infarction admissions in Melbourne, Australia. *International Journal of Health Geographics*. https://doi.org/10.1186/1476-072X-9-41.

Lowe, R., García-Díez, M., Ballester, J., Creswick, J., Robine, J. M., Herrmann, F. R., & Rodó, X. (2016). Evaluation of an early-warning system for heat wave-related mortality in Europe: Implications for sub-seasonal to seasonal forecasting and climate services. *International Journal of Environmental Research and Public Health*. https://doi.org/10.3390/ijerph13020206.

McEvoy, D., Ahmed, I., & Mullett, J. (2012). The impact of the 2009 heat wave on Melbourne's critical infrastructure. *Local Environment*. https://doi.org/10.1080/13549839.2012.678320.

McGregor, G. R., Bessemoulin, P., Ebi, K. L., & Menne, B. (2015). *Heatwaves and health: Guidance on warning-system development*. Geneva, Switzerland.

Mora, C., Counsell, C. W. W., Bielecki, C. R., & Louis, L. V. (2017a). Twenty-seven ways a heat wave can kill you: Deadly heat in the era of climate change. *Circulation: Cardiovascular Quality and Outcomes*. https://doi.org/10.1161/CIRCOUTCOMES.117.004233.

Mora, C., Dousset, B., Caldwell, I. R., Powell, F. E., Geronimo, R. C., Bielecki, C. R. … Trauernicht, C. (2017b). Global risk of deadly heat. *Nature Climate Change*. https://doi.org/10.1038/nclimate3322.

Morgan, G., Sheppeard, V., Khalaj, B., Ayyar, A., Lincoln, D., Jalaludin, B. … Lumley, T. (2010). Effects of bushfire smoke on daily mortality and hospital admissions in Sydney, Australia. *Epidemiology*. https://doi.org/10.1097/EDE.0b013e3181c15d5a.

Nairn, J., & Fawcett, R. (2013). *Defining heatwaves: Heatwave defined as a heat-impact event servicing all community and business sectors in Australia*. CAWCR Technical Report 551.5250994.

Navarro, K., Schweizer, D., Balmes, J., & Cisneros, R. (2018). A review of community smoke exposure from wildfire compared to prescribed fire in the United States. *Atmosphere*. https://doi.org/10.3390/atmos9050185.

Nitschke, M., Tucker, G. R., Hansen, A. L., Williams, S., Zhang, Y., & Bi, P. (2011). Impact of two recent extreme heat episodes on morbidity and mortality in Adelaide, South Australia: A case-series analysis. *Environmental Health: A Global Access Science Source*. https://doi.org/10.1186/1476-069X-10-42.

Rappold, A. G., Reyes, J., Pouliot, G., Cascio, W. E., & Diaz-Sanchez, D. (2017). Community Vulnerability to Health Impacts of Wildland Fire Smoke Exposure. *Environmental Science and Technology*. https://doi.org/10.1021/acs.est.6b06200.

Reid, C. E., Brauer, M., Johnston, F. H., Jerrett, M., Balmes, J. R., & Elliott, C. T. (2016). Critical review of health impacts of wildfire smoke exposure. *Environmental Health Perspectives, 124*(9), 1334–1343. https://doi.org/10.1289/ehp.1409277.

Reisen, F., Duran, S. M., Flannigan, M., Elliott, C., & Rideout, K. (2015). Wildfire smoke and public health risk. *International Journal of Wildland Fire*. https://doi.org/10.1071/WF15034.

Reisen, F., Hansen, D., & Meyer, (Mick) P. (2011). Exposure to bushfire smoke during prescribed burns and wildfires: Firefighters' exposure risks and options. *Environment International*. https://doi.org/10.1016/j.envint.2010.09.005.

Richardson, L. A., Champ, P. A., & Loomis, J. B. (2012). The hidden cost of wildfires: Economic valuation of health effects of wildfire smoke exposure in Southern California. *Journal of Forest Economics*. https://doi.org/10.1016/j.jfe.2011.05.002.

Robine, J. M., Cheung, S. L. K., Le Roy, S., Van Oyen, H., Griffiths, C., Michel, J. P., & Herrmann, F. R. (2008). Death toll exceeded 70,000 in Europe during the summer of 2003. *Comptes Rendus—Biologies*. https://doi.org/10.1016/j.crvi.2007.12.001.

Salimi, F., Henderson, S. B., Morgan, G. G., Jalaludin, B., & Johnston, F. H. (2017). Ambient particulate matter, landscape fire smoke, and emergency ambulance dispatches in Sydney, Australia. *Environment International*. https://doi.org/10.1016/j.envint.2016.11.018.

Shaposhnikov, D., Revich, B., Bellander, T., Bedada, G. B., Bottai, M., Kharkova, T. … Pershagen, G. (2014). Mortality related to air pollution with the Moscow heat wave and wildfire of 2010. *Epidemiology*. https://doi.org/10.1097/EDE.0000000000000090.

Sherwood, S., & Huber, M. (2010). An adaptability limit to climate change due to heat stress. *Pnas*. https://doi.org/10.1073/pnas.0913352107/-/DCSupplemental.www.pnas.org/cgi/doi/10.1073/pnas.0913352107.

Smith, K. R., Woodward, A., Campbell-Lendrum, D., Chadee, D., Honda, Y., Liu, Q. … Genova, R. (2014). *Human health: Impacts, adaptation, and co-benefits*. Climate Change 2014: Impacts, Adaptation, and Vulnerability. Part A: Global and Sectoral Aspects. Contribution of Working Group II to the Fifth Assessment Report of the Intergovernmental Panel on Climate Change [Field CB, Barros VR, Dokken DJ, Mach KJ, Ma. https://doi.org/10.1017/CBO9781107415379.016.

Steffen, W., Rice, M., & Alexander, D. (2018). *2017: Another record-breaking year for heat and extreme weather*. Sydney: Climate Council of Australia.

Stephenson, C. (2010). *A literature review on the economic, social and environmental impacts of severe bushfires in south-eastern Australia*. (V. D. of S. and Environment & E. M. Australia, Eds.). Melbourne: Victorian Government Department of Sustainability and Environment.

Stephenson, C., Handmer, J., & Betts, R. (2013). Estimating the economic, social and environmental impacts of wildfires in Australia. *Environmental Hazards*. https://doi.org/10.1080/17477891.2012.703490.

Toloo, G. S., Yu, W., Aitken, P., FitzGerald, G., & Tong, S. (2014). The impact of heatwaves on emergency department visits in Brisbane, Australia: A time series study. *Critical Care (London, England)*. https://doi.org/10.1186/cc13826.

Tong, S., FitzGerald, G., Wang, X. Y., Aitken, P., Tippett, V., Chen, D. … Guo, Y. (2015). Exploration of the health risk-based definition for heatwave: A multi-city study. *Environmental Research*. https://doi.org/10.1016/j.envres.2015.09.009.

Tong, S., Wang, X. Y., Yu, W., Chen, D., & Wang, X. (2014). The impact of heatwaves on mortality in Australia: A multicity study. *BMJ Open*. https://doi.org/10.1136/bmjopen-2013-003579.

Turner, L. R., Connell, D., & Tong, S. (2013). The effect of heat waves on ambulance attendances in Brisbane, Australia. *Prehospital and Disaster Medicine*. https://doi.org/10.1017/S1049023X13008789.

Walker, B., & Salt, D. (2006). Resilience thinking: Sustaining ecosystems and people in a changing world. Washington, DC: Island Press.

Wang, X.-Y., Barnett, A., Guo, Y.-M., Yu, W.-W., Shen, X.-M., & Tong, S.-L. (2014). Increased risk of emergency hospital admissions for children with renal diseases during heatwaves in Brisbane, Australia. *World Journal of Pediatrics: WJP*. https://doi.org/10.1007/s12519-014-0469-x.

Whittaker, J., Handmer, J., & Mercer, D. (2012). Vulnerability to bushfires in rural Australia: A case study from East Gippsland, Victoria. *Journal of Rural Studies*. https://doi.org/10.1016/j.jrurstud.2011.11.002.

WHO. (2013). *Review of evidence on health aspects of air pollution—REVIHAAP Project: Technical Report*. Copenhagen, Denmark: WHO.

WHO. (2014a). *Burden of disease from Ambient Air pollution for 2012*. Retrieved from http://www.who.int/phe/health_topics/outdoorair/databases/FINAL_HAP_AAP_BoD_24March2014.pdf?ua=1.

WHO. (2014b). *Quantitative risk assessment of the effects of climate change on selected causes of death, 2030s and 2050s*. WHO. ISBN: 978-92-4-150769-1.

Williams, S., Nitschke, M., Weinstein, P., Pisaniello, D. L., Parton, K. A., & Bi, P. (2012). The impact of summer temperatures and heatwaves on mortality and morbidity in Perth, Australia 1994–2008. *Environment International*. https://doi.org/10.1016/j.envint.2011.11.011.

Wilson, L. A., Gerard Morgan, G., Hanigan, I. C., Johnston, F. H., Abu-Rayya, H., Broome, R. … Jalaludin, B. (2013). The impact of heat on mortality and morbidity in the Greater Metropolitan Sydney Region: A case crossover analysis. *Environmental Health: A Global Access Science Source*. https://doi.org/10.1186/1476-069X-12-98.

Wreford, A., & Neil Adger, W. (2010). Adaptation in agriculture: Historic effects of heat waves and droughts on UK agriculture. *International Journal of Agricultural Sustainability*. https://doi.org/10.3763/ijas.2010.0482.

Xiao, J., Spicer, T., Jian, L., Yun, G. Y., Shao, C., Nairn, J. … Weeramanthri, T. S. (2017). Variation in population vulnerability to heat wave in Western Australia. *Frontiers in Public Health*. https://doi.org/10.3389/fpubh.2017.00064.

Youssouf, H., Liousse, C., Roblou, L., Assamoi, E. M., Salonen, R. O., Maesano, C., … Annesi-Maesano, I. (2014). Non-accidental health impacts of wildfire smoke. *International Journal of Environmental Research and Public Health*. https://doi.org/10.3390/ijerph111111772.

Zuo, J., Pullen, S., Palmer, J., Bennetts, H., Chileshe, N., & Ma, T. (2015). Impacts of heat waves and corresponding measures: A review. *Journal of Cleaner Production*. https://doi.org/10.1016/j.jclepro.2014.12.078.

Climate Adaptation of Sea-Level Rise in Hong Kong

Yun Fat Lam and Shimul Roy

Abstract Hong Kong is a coastal city with 733 km long of coastline. With the high-rise buildings residing millions of people, it is highly susceptible to the impacts of sea-level rise (SLR) and storm surge. It is observed that the average sea level has been steadily increased at a rate of ~3.1 cm per decade. As the sea level rise is expected to be exacerbated in the end of this century, due to climate change, it would be important for the local government to implement adaptation measures for combating this issue. In 2013, the Hong Kong government initiated a comprehensive review on SLR caused by climate change and its implications on design of coastal structure, attempting to update the existing Port Works Division Manual (PWDM). In the study, the IPCC AR5 was used as the projection scenario for estimating future SLR in Hong Kong. Pattern scaling was applied to normalize the relationship between the local SLR and global SLR from different AR5 scenarios (i.e., RCP scenarios), producing a 2D SLR pattern for the South China Sea. It is recommended that the height of the coastal structures should be increased by 0.46, 0.56, 0.58 and 0.78 m for accounting the rise in the mean sea level under RCP2.6, 4.5, 6.0 and 8.5 scenarios, respectively. For the worst-case scenario (i.e., RCP8.5), the construction cost associated with the changes in the coastal structures (e.g., public pier structure and vertical block work seawall) would be increased by 1.3–1.9%.

Keywords Climate change · SLR · Pattern scale · Sea-level projection · Thermal expansion

Y. F. Lam (✉)
Department of Geography, University of Hong Kong, Pokfulam, Hong Kong
e-mail: yunlam@hku.hk

Y. F. Lam · S. Roy
School of Energy and Environment, City University of Hong Kong, Kowloon Tong, Hong Kong

R. Akhtar (ed.), *Extreme Weather Events and Human Health*,
https://doi.org/10.1007/978-3-030-23773-8_9

1 Introduction

Climate change is a burning issue for the present world. It is evident that the increase in greenhouse gases (GHGs) has contributed to a large part of the increase in the earth's average temperature, resulting in an escalation of the polar ice-caps and glaciers melting, and sea-level rise. It modifies the distribution of fresh water by influencing regional precipitations and affecting the ecosystem (IPCC 2007). Currently, sea level is rising at an unprecedented rate, potentially threatens billions of people who are living in the coastal regions (Hardy 2003). In Hong Kong, the government was aware of the potential influence of climate change. An inter-departmental working group had been formed to evaluate its impacts, as sea-level rise and extreme weathers (e.g., typhoon) were identified as the major impacts. In 2010, the Civil Engineering and Development Department had conducted the first review on the influence of sea-level rise and recommended suitable adaptation strategies under the Intergovernmental Panel on Climate Change (IPCC) Fourth Assessment Report (AR4) (Arup 2012). A follow-up study based on the latest IPCC Fifth Assessment Report (AR5) projections was conducted in 2013 to review the existing Port Works Design Manual (PWDM) and implement necessary adaptation measures for the coastal structures (Arup 2016). In the study, the rise of local mean sea level and increase in average wind speed were identified to cause the increase in maximum sea level height during the storm surge condition (Chow et al. 2017). In this chapter, we summarize the factors influencing global sea-level rise and sea-level changes in the twenty-first century; different projections of the global sea-level rise; factors influencing local sea level; overview of pattern scaling and projection of sea level rise in Hong Kong; and changes in the design of coastal infrastructures.

2 Climate Change and Sea-Level Rise

Climate change is one of the most serious issues for the human environment to exist on the planet earth. It is evident that the global climate is changing dramatically since the earth formed more than 4.5 billion years ago. Scientific studies identified the increase in greenhouse gases (GHGs) is the main cause behind the global climate change in this century, where heat is being trapped and affects the earth environment and its components (biotic and abiotic). Billions of people living in the coastal cities around the globe are susceptible to the impacts of climate change. These impacts include sea-level rise (SLR), increase in coastal erosion and intensification of tropical cyclones and storm surges. It has been reported in the IPCC AR4, many countries in Asia regions and marine-time continents are highly vulnerable to these impacts. The effects of SLR not only influence agriculture, forest ecosystem and freshwater resources, but also produce immense socioeconomic damages to human settlements and coastal infrastructures. It increases the costs for coastal embankments/protection from flooding. In the IPCC AR5, Representative Concentration Pathways (RCPs) are

Table 1 Projection of global mean SLR from AR5 (IPCC 2013)

Scenario	Mean SLR (m)	
	2046–2065[a]	2081–2100[a]
RCP2.6	0.17–0.32	0.26–0.55
RCP4.5	0.19–0.33	0.32–0.63
RCP6.0	0.18–0.32	0.33–0.63
RCP8.5	0.22–0.38	0.45–0.82

[a]Projections at 5–95% model range

used to represent different greenhouse gas concentration (not emissions) trajectories, and the numerical values of RCP (e.g., RCP8.5) represent a continuous rise in radiative forcing to the designated levels (W m^{-2}) at the end of the twenty-first century (e.g., 8.5 W m^{-2}). A bigger value means a higher level of GHGs in the atmosphere by 2100. In the report, it is projected that the global SLR will be increased noticeably at the end of this century, regardless of which RCP is being occurred. Table 1 shows the projected global mean SLR in the mid- and late-twenty-first century for different RCP scenarios obtained from the ensemble mean of General Circulation Models (GCM). In RCP4.5 and RCP6.0, the projected ranges of mean SLR are almost at the same levels for the mid- and late-twenty-first century, respectively, while much higher range of global mean SLR (i.e., 0.45–0.82 m) is observed in RCP8.5 (high GHGs emission). The result in RCP8.5 mimics the impacts of SLR if no or minimal GHGs mitigation is being in place.

When the IPCC AR5 projected the global mean sea level would rise in between 0.2 and 0.8 m in the twenty-first century (Bilbao et al. 2015), some studies have suggested that the projected SLR could be much higher (0.8 m or even 2 m) due to the accelerated loss of ice-sheets (Pfeffer et al. 2008; Lowe et al. 2009). The recent studies based on semi-empirical approach are summarized in Fig. 1. Rahmstorf (2007) stated that by 2100, the projected SLR could be 1.4 m above the 1990 level. The reason behind the lower rate of SLR in the IPCC AR4 is probably due to the limitations in GCM. For instance, the contribution from ice-sheets melting in Greenland and Antarctic assumes to be nearly zero or melted at a lower rate than the actual observed values (Rahmstorf 2010). For AR5, it has been reported that the projection of SLR is also larger than the one in AR4 with similar pattern and regional distribution. This discrepancy is mainly attributed by the improvement of model representation on land-ice components in GCM. Although the projected SLR in AR5 may not be perfect and its projections are still subjected to some levels of uncertainties, it is still the most widely accepted projections for future adaptation and mitigation planning.

2.1 *Factors Influencing Global Sea-Level Rise*

SLR is one of the major consequences of climate change directly contributed from the global warming. Compared to the previous century, the global mean sea level

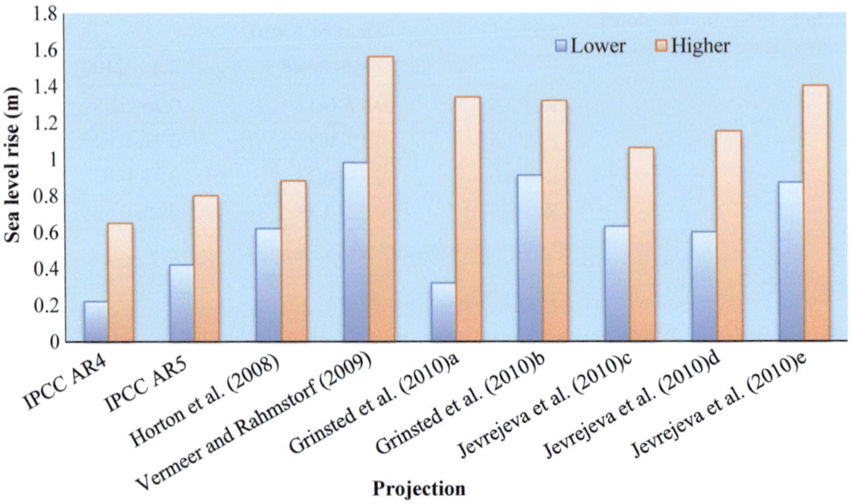

Fig. 1 Projected global MSL (5–95% range) for the twenty-first century

(MSL) has changed significantly. Church and White (2006) estimated a global SLR at a rate of 1.7 ± 0.3 mm year^{-1} for the twentieth century with 0.6–0.9 °C increase. However, during the period from 1993 to 2010, the global mean SLR was intensified to 2.3–3.4 mm year^{-1}, corresponding to the change of 1.5–1.8 °C per century (Blunden et al. 2018). According to the AR5 projections, by 2100, global MSL will rise by ~0.8 m mainly contributed from the warming of the global ocean and increased rate of glaciers melting. Figure 2 summarizes the factors that influence the global SLR. These factors include thermal expansion of the global ocean, melting of glaciers, polar ice-caps and ice-sheets, input from land water reservoirs, and vertical land movement. It is clear that thermal expansion and melting of glaciers, ice-caps and ice-sheets are the two key factors responsible for the global SLR. It contributes to 30–55 and 15–35% of global SLR, respectively (IPCC 2007; IPCC 2013; Hallegatte et al. 2011; Nicholls and Cazenave 2010). The combination of these effects has found to constitute more than 75% of global mean SLR observed since the 1970s (IPCC 2013). According to IPCC, the projected SLR at the end of the twenty-first

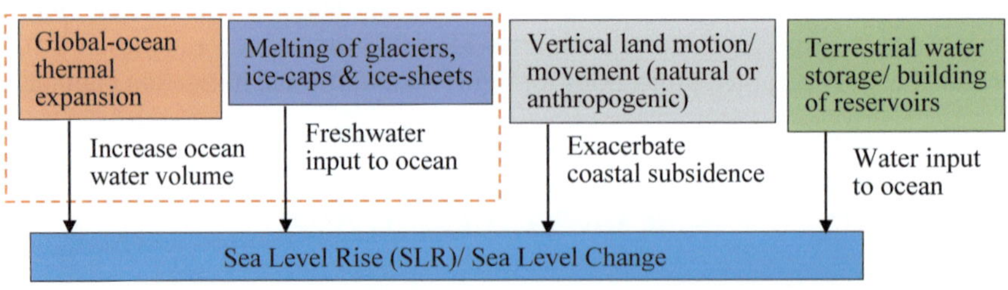

Fig. 2 Factors influencing global SLR (Dash box: factors were considered in AR5)

century due to the thermal expansion is ranged from 0.11 to 0.38 m. It should be noted that regardless of the change of GHG emissions or projected scenarios in the twenty-first century, the global MSL is still expected to rise continuously in response to the slow feedback from the effects of global ocean thermal expansion and melting of glaciers and ice-sheet from the last century.

3 Understanding of Local Sea-Level Rise in Hong Kong

3.1 Climate and Geophysical Characteristics of Hong Kong

Hong Kong is located at the tip of the Pearl River Delta (PRD) in Southern China with a total area of 1,106 km^2. It lies between 22.3964°N and 114.1095°E with the sub-tropical climate and influenced by Asian monsoons. Daily average temperature ranges from 12 to 31 °C. In winter, the temperature may drops to 10 °C while in summer it may exceed 35 °C. In winter, prevailing synoptic winds from the north and northeast brings relatively cold air while in summer it comes from the southwest. Generally, June to August are considered as the wettest months, while January and December are the driest months. The annual average rainfall ranges from 1,400 to 3,000 mm, whereas most of it occurs in between May and September, and it frequently experiences tropical cyclones and results in substantial precipitation and storm surges (Lam 2018). Geographically, Hong Kong has a complex terrain, which is surrounded by mountains and ocean (Wang et al. 2001). There are many faults and joints in Hong Kong's bedrock that play a vital role in rivers and streams formation. Although there are no large rivers present, the river system in Hong Kong still contributes appreciable freshwater to Pearl River.

3.2 Historical Observation of Mean Sea Level

From the records of different tide gauge stations, it is unambiguous that since the 1990s the sea level in Hong Kong is rising noticeably. Table 2 shows the rate of mean SLR in different tide gauge stations including North Point and Quarry Bay (NPQB), Tai Po Kau, Waglan Island, and Tsim Bei Tsui (Li and Mok 2012). Both NPQB and Tai Po Kau are located in the eastern part of Hong Kong and highly exaggerated by storm surges from typhoons (Wong et al. 2003). The results show that a significant rise in sea level has been observed in different parts of Hong Kong (e.g., central, eastern and western) with a range from 2.6 to 4.2 mm $year^{-1}$. These values are much higher than the global mean SLR of 1.9 mm $year^{-1}$ projected by the IPCC during the period 1901–2010 (IPCC 2013). Recent observation data show that the local SLR from NPQB (Victoria Harbour) is rising at a rate of 3.0 cm per decade from 1954 to 2017. However, this rising trend is not constant throughout the period as shown in Fig. 3. From 1954 to 1986, a falling trend was observed followed by a significant rising from 1987 to 2000, afterward a falling trend was observed again, and from 2011 the sea level has started to rise again. Similar trend is also observed in other stations (e.g., Tai Po Kau in the Tolo Harbour) of Hong Kong. Overall, the trend analysis reveals that the estimated rate of local SLR from observation could be easily affected by the time period selection. As a result, the linear projection of local SLR using observation could be subjected to appreciable bias. This is clearly reflected in the data of Waglan Island, where a short period (1982–2010) was used and resulted in much higher rate of mean SLR (55% higher than the other two stations). Hence, linear projection is not recommended to be used for projecting long-term local SLR (e.g., 2050 or 2100).

Table 2 Rate of mean SLR in Hong Kong (CEDD 2013)

Station	Period	Rate of mean SLR (mm year^{-1})
North Point/Quarry Bay	1954–2010	2.6 ± 0.5
Tai Po Kau	1963–2010	2.7 ± 0.5
Waglan Island	1982–2010	4.2 ± 2.0
Tsim Bei Tsui	1974–2010	No noticeable trend
IPCC reference	1901–2010	1.9

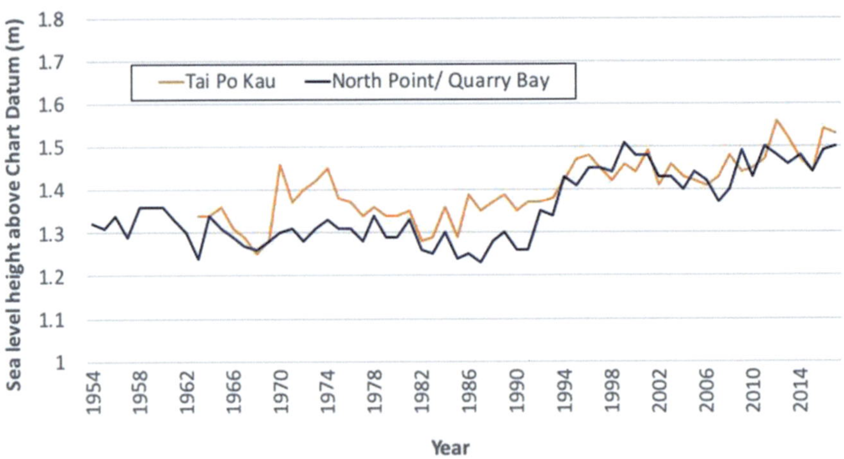

Fig. 3 Observed annual MSL in Hong Kong (adopted from HKO 2018)

3.3 Short-Term Influence on Local Sea Level from Tropical Cyclones

In Hong Kong, storm surge caused by the tropical cyclones or typhoons has a profound influence on the short-term SLR, which is referred to as the temporary rise of sea level resulted from the integrated effects of low pressure and strong winds (HKO 2018). Due to its geographical location, Hong Kong is subjected to the tropical cyclones from the Western North Pacific and the South China Sea, which cause storm surges and therefore have profound impacts on SLR or flooding. Hong Kong experienced around 6–7 tropical cyclones every year (Lee and Wong 2007), which increase sea flooding caused by storm surges and consequently results in coastal inundation, coastal erosion, saltwater intrusion in land and freshwater system. Studies showed that additional 0.5–1.0 m rise in MSL due to the storm surges resulted from past tropical cyclones affect Hong Kong, which assumes to be high enough to cause sea flooding when occurs near the astronomical high tide (Lee et al. 2010). In extreme cases, the sea level could rise 3.0 m that may result in severe sea flooding and coastal inundation. For instance, Typhoon 'Hagupit' in 2008 carried a storm surge of 1.4 m at the Victoria Harbor, which caused a sea-level rise of 3.53 m (Lee et al. 2010) thereby threatened the community and caused huge property damages and economic loss. As the increase in MSL in the future climate, the SLR impacts would be intensified. Therefore, it is necessary to have more stringent regulations on the construction of coastal structure to against the potential devastated impact from the combination of SLR with storm surge.

3.4 Factors Influencing Local Mean Sea-Level Rise

The local SLR can be exacerbated by both global SLR and localized factors such as tectonic movement or sedimentation process (IPCC 2013). In the process of estimating local SLR in the future, it is crucial to take into account the climate change-induced global SLR as part of the impacts. In general, the global SLR is served as the baseline of mean SLR while other local factors (e.g., changes in surface wind pattern) are added on top of it to reflect the local influence. Figure 4 summarizes the factors

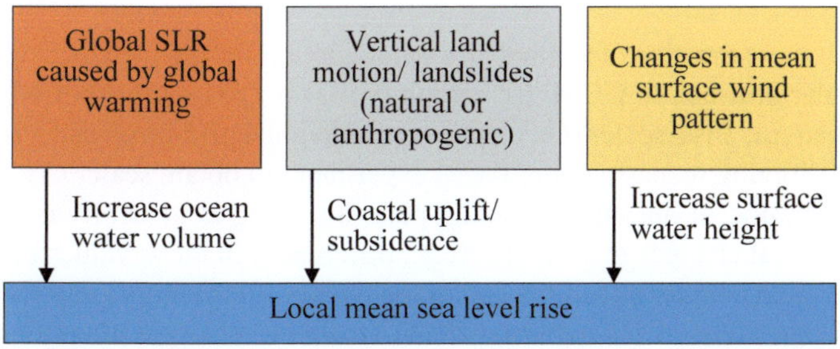

Fig. 4 Factors influencing local mean SLR

that influence the local SLR. These factors include the global SLR resulted from global warming, vertical land movement or earth's tectonic movement, and changes in surface wind pattern. Among all factors, the vertical land motion or movement along the coastal areas is considered as one of the crucial factors. The movement can either caused by natural phenomena (e.g., tectonic activity) or anthropogenic processes (e.g., mining, oil drill and excessive groundwater extraction). According to the IPCC AR5 (IPCC 2013), land subsidence in the coastal areas is a common phenomenon that leads to sea-level changes. In Hong Kong, it has been reported that the land subsidence is at a rate of 4–5 mm year^{-1} (1960–2005) observed in North Point and Quarry Bay (NPQB) station. The land subsidence adds an additional rise of sea-level relative to the land surface, and it would have a huge implication to the future MSL and design of coastal structure (Ding et al. 2004).

4 Projection of Local Sea-Level Rise Using Pattern Scaling in Hong Kong

4.1 *Pattern Scaling Technique, Its Assumption and Applications*

In this study, pattern scaling was applied to normalize the results of mean SLR from different GCM to yield a single estimate of local mean SLR for each of the IPCC-projected scenarios. Below summarizes the basic theory and its limitations of pattern scaling. The pattern scaling is a technique used to construct climate scenarios by connecting the Simple Climate Models (SCMs) to the GCM (Mitchell 2003), which was first used in the IPCC TAR (Third Assessment Report) for developing climate scenarios for 2030 (Mitchell et al. 1990). The pattern-scaling approach is first presented by Santer et al. (1990) and was employed to equilibrium experimentations where the foremost assumption was that there is a linear relationship between the climate system and radiative forcing (Santer et al. 1990). However, the recent method in pattern scaling as suggested by Mitchell et al. (1990) can be defined as an effort to estimate the difference in a variable (V) as shown in the following equation:

$$V^{*}_{xijy} = S_{xy} \cdot V'_{zij} \tag{1}$$

where $V*$ is the estimation of anomaly for a variable; x is the selected forcing scenario; i is particular grid-box in GCM; j is month or season; y is year or period; V' is the response pattern; S is a scaler (i.e. response at individual grid-box); and z is the same or different forcing scenario under GCM experiment to obtain scaler (S).

The estimation of an anomaly using pattern scaling is generally employed by calculating local changes in variables (e.g., differences in mean values between the end of twentieth century and at the end of the twenty-first century) relative to global average change (Tebaldi and Arblaster 2014). However, the key assumption of pattern scaling is that there is a linear relationship between the local changes in a variable (e.g., temperature, precipitation, sea-level height) due to climate change and the global anomaly in climate variables (Mitchell 2003). For sea-level rise projection, it is assumed that the rise in local sea level is linearly related to the changes in the global mean sea level (Bilbao et al. 2015; Chow et al. 2017; Tebaldi and Arblaster 2014; Mitchell 2003). By using the feature of pattern scaling, it allows one to perform easy interpolation for the period even when the GCM output is not available, making the interpolated data highly flexible for use in engineering application.

4.2 Sea-Level Rise Projection in Hong Kong

The GCM outputs are so far considered as the reliable source of information used for the projection of climate scenarios; however, in the case of local-scale climate change simulation, it shows comparatively poor performances compared to the historical observed data and has limitations in the direct use of GCM time series to analyze extreme climate event. Different downscaling methods such as dynamic downscaling, statistical downscaling and pattern scaling are available for climate change projections. According to IPCC (2013), for different climate change impact studies at local scales, these downscaling techniques are employed and there is high confidence that downscaling techniques provide better results for regional or local scales having highly varying topography. Although the dynamic downscaling technique can provide better accuracy at finer scales, the computational demand is high, which makes it highly challenging for analyzing uncertainties for various scenarios and GCM. On the contrary, statistical downscaling and pattern scaling are cost-effective and efficient in computation and can provide more accuracy on local MSL projection from the GCM outputs (CEDD 2013).

In this study, the pattern scaling has been applied to downscale the global SLR information from GCM outputs to derive local mean SLR projection from the AR5 scenarios. More than 13 GCM outputs from various model configurations have passed the basic screening process and were incorporated into pattern scaling. The process of pattern scaling involves two steps. The first step is to calculate the ratio of local SLR in relation with the global mean SLR from all selected GCM outputs using Eq. (2), expressed as relative sea surface height (DZOS). This value reveals whether the local SLR will be higher, equal or lower than the global mean SLR value. Figure 5 shows the DZOS values from 4 of the selected GCM (i.e., CSIRO-MK3.0, GFDL2.1, MIROC-H and HADCM3). In these cases, the DZOS values were separately determined. The normalized value of SLR has the unit of cm/cm and is ranged from 0.64 to 2.3 cm/cm for eastern China. High variability is observed in between the model cases, while only limited variability is found in between grid locations, indicating that ensemble process is an important step to reduce uncertainties for the projected SLR.

$$\mathrm{DZOS}_{ij} = \frac{(\mathrm{ZOS}_{ij,y} - \mathrm{ZOS}_{ij,\mathrm{base\ year}}) + \Delta\mathrm{GMSLR}}{\Delta\mathrm{GMSLR}} \quad (2)$$

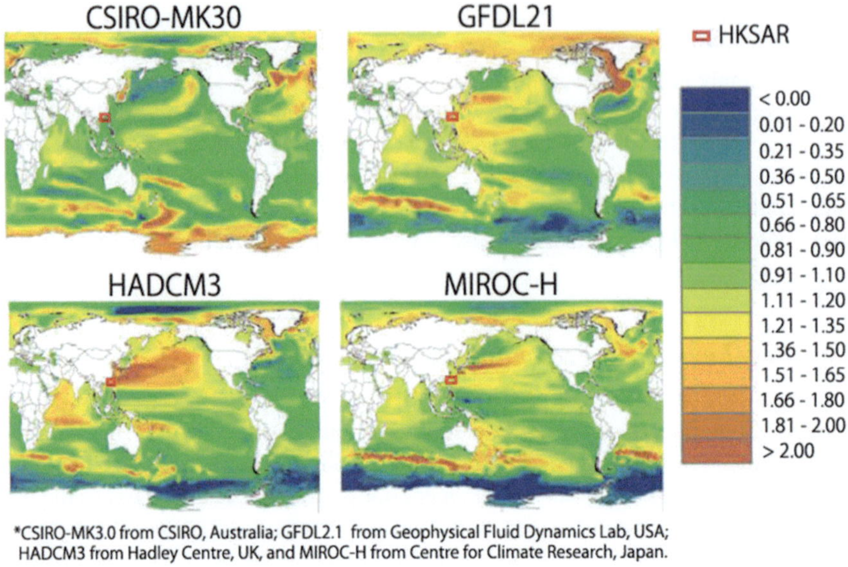

Fig. 5 Relative sea surface height from pattern scaling (cm/cm) (adopted from CEED 2013)

where DZOS is the relative sea surface height; ΔGMSLR is the change in global MSL and is calculated by $\Delta\text{GMSLR} = \text{ZOSTOGA}_y - \text{ZOSTOGA}_{\text{base year}}$; ZOSTOGA is the total change in global mean SLR due to thermostatic change; ZOS is the local sea surface height; *ij* is the longitude and latitude position; *y* is the target year; and "base year" is assumed to be 1996 (average value of 1986–2005).

The second step involves using the obtained DZOS value from each selected GCM and multiplied it with the change in global MSL (ΔGSLR) to obtain the projected SLR (ΔSLR), as shown in Eq. (3). All local SLR results are then averaged to produce the ensemble mean SLR for the projection scenarios. Figure 6 shows the projected change in MSL in Hong Kong for different time horizons of the twenty-first century (e.g., 2010, 2030, 2050 and 2100). It should be noted that the projected mean SLR in Hong Kong may reach 0.78 m while under the RCP2.6, RCP4.5 and RCP6.0, it may reach 0.46, 0.56 and 0.58 m, respectively. These values are on the high side of the reported values shown in Table 1, meaning that Hong Kong would experience a higher SLR than most of the places around the world.

$$\Delta\text{SLR}_{ij} = \text{DZOS}_{ij} \times \Delta\text{GSLR}_{ij} \tag{3}$$

where ΔSLR_{ij} is the calculated SLR; DZOS_{ij} is the relative sea surface height; ΔGSLR is the change in global sea level; and *ij* is the longitude and latitude position.

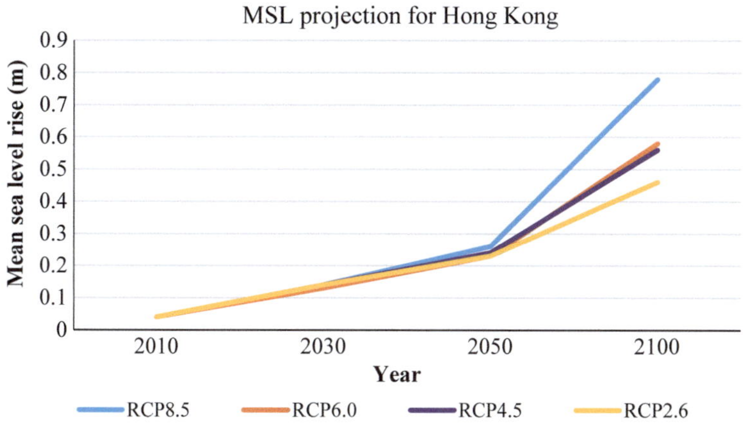

Fig. 6 Mean SLR projections for Hong Kong (CEED 2013)

4.3 Coastal Structures in Hong Kong

The Civil Engineering and Development Department (CEDD) of Hong Kong SAR is responsible for the development of the coastal structures and its design based on the changing circumstances due to the climate impacts (e.g., rising sea-level height). Observation records on SLR from the previous decades indicate a rising trend of mean sea level, which draws attention in recent times to develop or design new coastal structures to cope with the adverse impacts of long-term SLR. The CEDD Port Works Division is the responsible party to oversee different coastal structures including public pier, sloping rock armor seawall, vertical blockwork seawall and rubble mound breakwater. These coastal structures have their distinct construction and maintenance costs, which depends on their design parameters, materials used and construction lifetime. In this section, we summarize the change of design criteria in the Port Works Division Manual (PWDM) and estimated cost associated with such changes. Table 3 shows the suggested changes to the design parameters for the projected SLR (RCP8.5) of 0.33 m for the year 2062. The selection of 2062 as the base year, instead of 2100, is due to the fact that the design lifetime of a typical coastal structure is 50 years, while the 0.33 m is obtained directly from the local MSL projection without adding the contribution of land subsidence projection. It has been observed that the spatial pattern of land subsidence is not clear and not yet been fully understood. Therefore, it was not included in the final SLR estimate. If more evidence is found in the future, another call of review should be conducted in the future.

In Table 3, it is recommended that the proposed deck level in public pier has to raise from 4.30 to 4.63 m to take into account for the local SLR caused by climate change. The change in deck height induces extra wave lift and water forces to the structure; hence, structure reinforcement is required to maintain the stability of the elevated deck. For vertical block work seawall, the crest height is needed to be increased to

Table 3 Design changes in selected coastal structures due to local sea-level rise (Chow et al. 2017)

Parameters	Units	Year 2012	Year 2062
Design water level	mPD	3.40	3.73
Deck level	mPD	4.30	4.63
Pile cut-off level	mPD	2.50	2.83
Pile-steel liner	m	12.5	12.8
Pile-concrete infill	m	12.5	12.8
Pile-concrete reinforcement	%	2.50	3.00
Fender top	mPD	4.00	4.33
Fender length	m	3.65	3.98
Crest height	mPD	5.5	5.86
Filter layer (vol.)	m^3	22.1	23.0
Rock fill (vol.)	m^3	163.5	176.2
Geotextile (vol.)	m^2/m	58.8	60.8
Sand fill (vol.)	m^3	264.5	260.6
Reclamation general fill (vol.)	m^3	133.2	148.1

mPD Meter in principal datum, *vol.* volume

sustain the wave overtopping and the structure of the block work seawall has to be wider as well as the volume of armor layer, rock fill, sand fill and geotextile volume need to be increased for better stabilization of the structure. It is expected that, due to changes, the construction and maintenance costs of these selected coastal structures will increase. The percentage of costs in construction and maintenance has increased by 1.9 and 1.3% for the public pier structure and vertical block work seawall, respectively (Chow et al. 2017).

5 Conclusion

The consequences of climate change such as SLR and extreme weather events have become a major threat for the human and infrastructures for many coastal regions around the world. In Hong Kong, due to its geographical location, it is subjected to frequent tropical cyclones and storm surges. As the storm surge would be strengthened by the effect of SLR results from climate change, it is important to quantify its effects and therefore incorporated into the coastal structure design. The projected SLR using pattern scaling from RCP8.5 has been determined as 0.33 m and 0.78 m for 2062 and 2100, respectively. The selection of 2062 is based on 50-year design lifetime of a typical coastal structure. The required changes in the design standards include (1) raising structure height by 0.33 m, and (2) increasing the structure strength to sustain additional force resulted from building a taller structure. At last, it has been determined that the additional construction and maintenance cost due to these changes for public pier structure and vertical block work seawall would increase the overall cost by 2% or less.

Acknowledgements The work was partially supported by the Guy Carpenter Asia-Pacific Climate Impact Centre, City University of Hong Kong, HKSAR (Project No. 9360126), and the contract research from Ove Arup and Partners (Hong Kong) Limited. The authors also gratefully acknowledge Civil Engineering and Development Department, and Hong Kong Observatory for providing the required data.

References

Arup. (2012). Review of studies on climate change and its implications on the design of coastal structures—Feasibility study. Final Report, CE 46/2011(CE).

Arup. (2016). Further review of studies on climate change and its implications on the design of coastal structures—Feasibility study. Final Report, PW 8/2013.

Bilbao, R. A., Gregory, J. M., & Bouttes, N. (2015). Analysis of the regional pattern of sea level change due to ocean dynamics and density change for 1993–2099 in observations and CMIP5 AOGCMs. *Climate Dynamics, 45*(9–10), 2647–2666.

Blunden, J., Arndt, D. S., & Hartfield, G. (Eds.). (2018). State of the climate in 2017. *Bulletin of the American Meteorological Society, 99*(8), Si–S332. https://doi.org/10.1175/2018bamsstateoftheclimate.1.

CEDD (Civil Engineering and Development Department). (2013). Review of studies on climate change and its Implications on the design of coastal structures. Final Report, REP/005/02.

Chow, A., Leung, T., & Lee, F. (2017). Benefit-cost analysis on coastal structures design for climate change adaptation in Hong Kong. *Coastal Engineering Journal, 59*(02), 1740005.

Church, J. A., & White, N. J. (2006). A 20th century acceleration in global sea-level rise. *Geophysical Research Letters, 33*(1).

Ding, X., Zheng, D., Wong, W. T., Li, K. W., Wu, C., & Zhong, P. (2004). Recent sea level variations in Southern China from tide gauge observations. In *Proceedings of the Asia-Pacific Space Geodynamics Symposium*, Singapore, 6–8 July 2004, pp. 126–136.

Hallegatte, S., Ranger, N., Mestre, O., Dumas, P., Corfee-Morlot, J., Herweijer, C., & Wood, R. M. (2011). Assessing climate change impacts, sea level rise and storm surge risk in port cities: A case study on Copenhagen. *Climatic Change, 104*(1), 113–137.

Hardy, J. T. (2003). *Climate change: Causes, effects, and solutions*. Wiley & Sons.

HKO. (2018). *Storm surge records in Hong Kong during the passage of tropical cyclones*. https://www.hko.gov.hk/wservice/tsheet/pms/stormsurgedb_e.htm. Accessed August 30, 2018.

IPCC. (2007). *Climate change 2007: Synthesis report. Contribution of working groups I, II and III to the fourth assessment report of the intergovernmental panel on climate* (104 pp). Geneva, Switzerland: IPCC.

IPCC. (2013). *Climate change 2013: The physical science basis. Contribution of working group I to the fifth assessment report of the intergovernmental panel on climate* (1535 pp). Cambridge, United Kingdom and New York, NY, USA: Cambridge University Press.

Lam, Y. F. (2018). Climate change and air Quality in South-eastern China: Hong Kong study. In *Climate change and air pollution* (pp. 181–196). Cham: Springer.

Lee, T. C., & Wong, C. F. (2007). Historical storm surges and storm surge forecasting in Hong Kong. In *JCOMM Scientific and Technical Symposium on Storm Surges* (Vol. 158, pp. 477–478).

Lee, B. Y., Wong, W. T., Woo, W. C. (2010, April). Sea-level rise and storm surge–impacts of climate change on Hong Kong. In *HKIE Civil Division Conference* (pp. 12–14).

Li, K. W., & Mok, H. Y. (2012). Long term trends of the regional sea level changes in Hong Kong and the adjacent waters. In *Asian and Pacific Coasts* 2011 (pp. 349–359).
Lowe, J., Howard, T., Pardaens, A., Tinker, J., Holt, J., Wakelin, S., et al. (2009). UK Climate Projections science report: Marine and coastal projections.
Mitchell, T. D. (2003). Pattern scaling: An examination of the accuracy of the technique for describing future climates. *Climatic Change, 60*(3), 217–242.
Mitchell, J. F. B., Manabe, S., Tokioka, T., & Meleshko, V. (1990). Equilibrium climate change. In J. T. Houghton, G. J. Jenkins, & J. J. Ephraums (Eds.), *Climate change: The IPCC scientific assessment*. Cambridge: Cambridge University Press.
Nicholls, R. J., & Cazenave, A. (2010). Sea-level rise and its impact on coastal zones. *Science, 328*(5985), 1517–1520.
Pfeffer, W. T., Harper, J. T., & O'Neel, S. (2008). Kinematic constraints on glacier contributions to 21st-century sea-level rise. *Science, 321*(5894), 1340–1343.
Rahmstorf, S. (2007). A semi-empirical approach to projecting future sea level rise. *Science, 315*(5810), 368. https://doi.org/10.1126/science.1135456.
Rahmstorf, S. (2010). A new view on sea level rise. *Nature Reports Climate Change, 4*(4), 44–45.
Santer, B. D., Wigley, T. M., Schlesinger, M. E., & Mitchell, J. F. (1990). Developing climate scenarios from equilibrium GCM results. Report/Max-Planck-Institut für Meteorologie, 47.
Tebaldi, C., & Arblaster, J. M. (2014). Pattern scaling: Its strengths and limitations, and an update on the latest model simulations. *Climatic Change, 122*(3), 459–471.
Wang, T., Wu, Y. Y., Cheung, T. F., & Lam, K. S. (2001). A study of surface ozone and the relation to complex wind flow in Hong Kong. *Atmospheric Environment, 35*(18), 3203–3215.
Wong, W. T., Li, K. W., & Yeung, K. H. (2003). Long term sea level change in Hong Kong. *Hong Kong Meteorological Society Bulletin, 13*(1–2), 24–40.

Heat-Related Mortality/Morbidity in East Asia

Yasushi Honda and Daisuke Onozuka

Abstract Heat causes (or exacerbates) various illness, including heat stroke, circulatory diseases, respiratory diseases, infectious diseases, accidents and suicides. Because of these diverse impacts on health, we evaluated heat-related excess mortality using all-cause mortality as the outcome and statistical model in its definition; the excess risk beyond the minimum mortality temperature (=MMT) is regarded as the heat-related excess mortality. Based on the whole Japanese data for about 4 decades of observation, we found the MMT can be estimated using 84th percentile of daily maximum temperature. Using this finding, we performed a projection of heat-related excess mortality; with no adaptation, the world's heat-related excess deaths attributable to climate change was more than 90,000 in 2030 and 255,000 in 2050. Autonomous adaptation, i.e., MMT shift along with warming, has been observed in some countries; we also took this phenomenon into account. With the autonomous adaptation, the future impact would be smaller, but the speed of adaptation is still unknown, and further research is needed.

Keywords Excess mortality · Minimum mortality temperature · Distributed lag nonlinear model · Future projection · General circulation model · Adaptation

1 Introduction

Heat is a killer. Heat is a disaster. This is what we have learned from summer 2018 in Japan. In the southern half of Japan, including Tokyo, the prevalence of air-conditioners is close to 100%. Still, the number of deaths due to heatstroke was 210 in Tokyo in 2010, when the number of days with a daily maximum temperature higher than 35 °C was 13—the death toll was 35 in 2017 and 29 in 2016 for which 35+ °C days were 3 and 2, respectively. This year, 2018, Tokyo experienced very hot summer and had 5 of 35+ °C days in July only, resulted in 96 heatstroke deaths, according to

Y. Honda (✉)
University of Tsukuba, Tsukuba, Japan
e-mail: honda.yasushi.fn@u.tsukuba.ac.jp

D. Onozuka
National Cerebral and Cardiovascular Center Research Institute, Suita, Japan

R. Akhtar (ed.), *Extreme Weather Events and Human Health*,
https://doi.org/10.1007/978-3-030-23773-8_10

one of the national newspapers (Sankei Shinbun 2018). Compared with the death toll of 41 due to Hokkaido Earthquake in 2018 that caused the whole Hokkaido Prefecture power blackout and the death toll of 9 due to Typhoon no. 21 in 2018 that caused the shutdown of Kansai Airport; this year's heat wave should be regarded as a huge disaster. More striking fact is that heatstroke is only a small fraction of heat-related mortality. Based on the temperature-mortality relation analyses, the contribution of heat stroke to heat-related mortality is small (Honda et al. 1995); circulatory or respiratory diseases contribute the most (Gasparrini et al. 2012; Honda et al. 1995). In addition, causes that used to be considered non-heat-related can also be heat-related. For example, it had not been obvious that suicide mortality is related to day-to-day heat variability, but now we have several papers that showed this relation (Kim et al. 2016; Likhvar et al. 2011; Page et al. 2007).

In this chapter, we first introduce how diverse the heat impacts are in the morbidity section. Then, in the mortality section, we concentrate on all-cause mortality to avoid underestimation of heat effect.

2 Heat-Related Morbidity

2.1 Communicable Diseases

Several studies have assessed the public health implications of temperature-related infectious disease morbidity. In Japan, previous studies have indicated an association between high temperatures and a significant increase in the risk of infectious diseases, such as *Mycoplasma pneumoniae* pneumonia (Onozuka et al. 2009) infectious gastroenteritis, (Onozuka et al. 2010) mumps, (Onozuka and Hashizume 2011a), hand-foot-mouth disease (HFMD), (Onozuka and Hashizume 2011b) and tuberculosis (Onozuka and Hagihara 2015). A recent study in three East Asian cities of Japan, China and South Korea has also shown that temperature was highly correlated with the diversity of airborne bacteria (Lee et al. 2017). In China, high temperatures are associated with the increased risk in the incidence of dengue (Shen et al. 2015), bacillary dysentery (Cheng et al. 2017; Li et al. 2016a, b; Xu et al. 2017) infectious diarrhea, (Zhou et al. 2013) and HFMD (Huang et al. 2018; Zhang et al. 2016; Zhao et al. 2017a, b). A recent study in China has also suggested that the association between temperature and HFMD varies across China, and the future impact of climate change on HFMD will vary as well (Zhao et al. 2018). In Hong Kong Special Administrative Region (SAR), admissions for respiratory and infectious diseases increased during extreme heat (Chan et al. 2013). In Taiwan, there was an increase in medical services for gastroenteritis when average daily temperature is above 30 °C (Huang et al. 2011). In Vietnam, high temperatures were associated with hospital admissions due to gastroenteritis in young children (Phung et al. 2015). Recent review studies have shown that high temperatures were significantly associated with increased incidence of gastrointestinal infections (Ghazani et al. 2018) and HFMD (Cheng et al. 2018). Additionally, people living in subtropical regions and middle-income areas had a higher risk of HFMD morbidity (Cheng et al. 2018). In Pacific island countries, the highest-priority climate-sensitive health risks included compromised safety and security of water and food, vector-borne diseases, and zoonoses (McIver et al. 2016). These findings indicate that heat events can have a substantial impact on the incidence of various infectious diseases in East Asia and suggest the importance for the development and implementation of public health policies and strategies to control heat-related infectious diseases.

2.2 *Non-communicable Diseases*

Emerging evidence suggests associations between heat and use of emergency ambulance transports, emergency department visits and hospitalization for various diseases. In Japan, a nationwide study has shown that extreme heat was associated with an increased risk in the incidence of emergency transport for all-cause, cardiovascular diseases, and respiratory diseases, and demonstrated differences in spatial and temporal variations (Onozuka and Hagihara 2016). Another nationwide study in Japan has shown that extremely high temperature is associated with an increased risk of "out-of-hospital-cardiac-arrest" (OHCA) (Onozuka and Hagihara 2017a). Another recent study in Japan has also demonstrated spatiotemporal homogeneity in the risk of OHCA during periods of extremely high temperature (Onozuka and Hagihara 2017b). Previous studies in Japan and Korea have also reported that hot temperature was associated with good neurologic outcome of OHCA (Cho et al. 2016; Fukuda et al. 2014). In China, there was significant association between extreme heat and emergency department visits and emergency ambulance dispatches in Beijing (Song et al. 2018), Shanghai, (Sun et al. 2014; Zhang et al. 2014) and Huainan (Cheng et al. 2016). In Beijing, China, there was a U-shaped association between temperature and risk of emergency department visits that is independent of air pollution and humidity (Zhao et al. 2017a, b). Another study in China has shown that high temperatures could increase the risk of OHCA (Niu et al. 2016). A recent study conducted in six Chinese cities showed that extreme hot temperatures increase the risk of out-of-hospital coronary deaths (Chen et al. 2014). Another study in Beijing, China, reported that daily hospital emergency department visits due to dog bites also increased due to extremely hot temperature (Zhang et al. 2017). A study in Beijing, China has also indicated that high temperatures significantly increased the risk of occurrence of accidental casualties (Ma et al. 2016). In Guangzhou, China, hot weather affected work-related injury, and significant associations were observed for male and middle-aged workers, workers in small- and medium-sized enterprises, and those working in the manufacturing sector (Sheng et al. 2018). In Changsha, China, maternal exposure to extreme heat days during pregnancy increased the risk of pneumonia in the offspring (Miao et al. 2017). A previous study in China also reported that there was a clear positive relationship between maximum temperature and heat-related illness (Gu et al. 2016). A multi-city study in China indicated that extreme heat is a risk factor for preterm birth (Guo et al. 2018). In Chongqing, China, there were strong associations between extremely high temperature and heat wave and heatstroke (Li et al. 2017). In a coastal city of China, heat waves had a substantial and delayed effect on heat stroke, heat cramp and heat exhaustion (Bai et al. 2014). In Hong Kong, China, significant nonlinear and delayed associations of hot temperatures were observed with pneumonia in the elderly (Qiu et al. 2016). Another study in Hong Kong SAR, high temperatures are associated with an significant increase in the incidence of mental disorder hospitalization (Chan et al. 2018). In Taiwan, the annual first extreme heat event of 99th percentile temperature was associated with higher emergency room visits for all causes and circulatory diseases (Wang et al. 2012). Another study in Taiwan has shown that there was an increase in medical services for skin and eye diseases when average daily temperature is above 30 °C (Huang et al. 2011). In Korea, a previous study found a relationship between temperature and acute myocardial infarction occurrence during heat (Lee et al. 2014). Another study in Korea has reported that hot temperatures were associated with a higher incidence of ischemic stroke, especially in the older age ($\geq$65) group and in men (Han et al. 2015). Recent studies in Korea have also reported that extremely high temperature is associated with an increase in the risk of hospital admission for acute kidney injury (Kim et al. 2018; Lim et al. 2018). A multi-city study in Korea reported that high temperature was responsible for an attributable risk for mental disease, and the burden was higher in the elderly (Lee et al. 2018). Another study conducted in seven major cities in Korea reported that the risk of OHCA was significantly increased with heat waves, and excess OHCA events primarily occurred during the afternoon when the temperature was high (Kang et al. 2016). In Vietnam, heat waves were significantly associated with an increase in all causes and infectious admissions, and cardiovascular and respiratory admissions were not significantly increased after a heat wave event (Phung et al. 2017). A previous study in Vietnam has also shown that the risk of cardiovascular hospital admissions increased during heat wave events (Phung et al. 2016). A recent multi-city study in Vietnam has shown that there were the heterogeneous magnitudes of temperature-related hospitalization across districts (Phung et al. 2018). Another study in Vietnam has also shown that heat waves have increased the risk of hospital admissions for mental disorders (Trang et al. 2016). In Pacific island countries, the highest-priority climate-sensitive health risks included trauma from extreme weather events, heat-related illnesses, respiratory illnesses, psychosocial ill-health, non-communicable diseases, population pressures, and health system deficiencies (McIver et al. 2016). These results could help public health officials predict temperature-related emergency ambulance transports, emergency department visits, and hospitalization, and better prepare for the impact of climatic change on these emergency events. These findings may be useful for public health officials to predict temperature-related emergency events to prepare for the effects of climatic change through public health intervention strategies, such as timely public health and medical advice, and improvements to housing and urban planning, early warning systems, health education, and healthcare system preparedness.

2.3 Human Behavior

In China, a cross-sectional study suggested that having a positive attitude toward sunstroke prevention and engaging in more preventive practices to avoid heat exposure had a protective interaction effect on reducing the prevalence of heat-related illnesses (Li et al. 2016a, b). Another qualitative study in Jinan, China, suggested that summer heat poses a great threat to city bus drivers' physiological and psychological health and knowledge about heat-related illness may protect them from dangerous heat, and air-conditioning is regarded as a strong protector in minimizing negative health effects during heat exposure (Zhou et al. 2014). In Hong Kong, China, help-seeking behavior such as call frequency among females appeared to be more sensitive to high temperatures, while calls among males were more sensitive to cold temperatures (Chan et al. 2011). Another study in Hong Kong, China, has also indicated that females, elderly, people who did not live alone, relatively high-income people, and those without medical histories of heart disease, hypertension, stroke, and diabetes were more sensitive to extreme weather condition (Wong et al. 2015). However, heat is a complex phenomenon resulting from the synergistic effects of air temperature, humidity and ventilation levels, radiation loads and metabolic activity; therefore, air temperature alone is seldom the reason for heat stress and heat-related health effects (McGregor and Vanos 2017). Further studies are needed to address these issues.

3 Heat-Related Mortality

3.1 Definition

As shown in the previous section, epidemiological studies showed that there are many types of diseases that are related to heat but that cannot be identified if we simply rely on the known physiological mechanisms. Thus, to avoid possible underestimation of the impact, we address all-cause mortality to evaluate the short-term heat effect. Using statistical model, we define the heat-related excess mortality as follows.

When we plot the relation between daily maximum temperature and all-cause mortality, we usually see the relation as shown in Fig. 1. This is an example of Tokyo, 2010–2015, but this V- or U-shaped relation can be observed all over the world with few exceptions (Gasparrini et al. 2015a, b). The temperature at which the mortality is lowest is regarded as minimum mortality temperature (=MMT). The mortality at the MMT can be considered to be the lowest risk of this population in terms of the ambient temperature. Then, if the temperature exceeds this MMT, the mortality becomes higher. Thus, we can define the heat-related excess mortality as the shaded area as shown in Fig. 1.

As for the temperature metric, we used daily maximum temperature, but the correlation coefficient between daily maximum temperature and daily average temperature is very high, and either index can be used. Hereafter, we use daily maximum

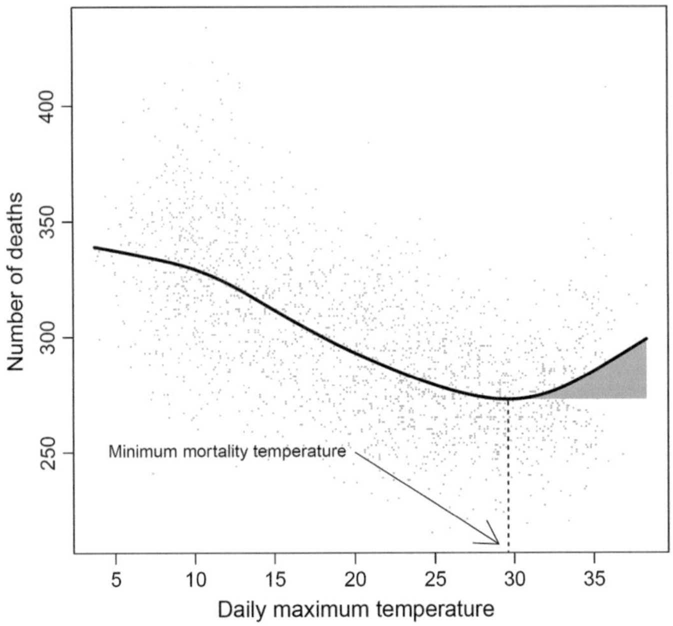

Fig. 1 Relation between daily maximum temperature and mortality (Tokyo, Japan, 2010–2015)

temperature for temperature metric. Other weather factors are of concern. One of the biggest concerns may be humidity effect. We decided not to include humidity in our model, because there is no good model for humidity that can be used for global projection. Wind speed and solar radiation would also be related to heat-related mortality/morbidity. These factors are important to outdoor risk. People working outside maybe heavily affected by these factors, but they are usually under close supervision. Thus, they may suffer from heatstroke, but they rarely die.

3.2 *Future Projection of Heat-Related Mortality*

The basis for the projection model is the temperature–mortality relation model. Figure 1 is one of the examples. The regression line of Fig. 1 seems good, because it allows nonlinear relation. In addition, Armstrong developed the nonlinear temperature–mortality model, i.e., distributed lag nonlinear model (=dlnm), that incorporated the lag effect in evaluating temperature–mortality relation (Armstrong 2006). Using the same model and the multi-country multi-city data, Gasparrini et al. confirmed that the lag effect of heat exists; they also showed that the lag effect of heat is usually short, i.e., less than a week (Gasparrini et al. 2015a, b). Thus, in our calculation of future projection of heat-related mortality, we decided to use dlnm that allows nonlinear relation and that considers the lag effect of heat. To evaluate the heat effect, we accumulated the heat effect across the lag days.

Our assessment of heat-related mortality in the future was done as one of the health impacts of climate change project lead by World Health Organization (Hales et al. 2014). Because most of the victims of the other impacts were children, we restricted our projection to 65+ years old age group to avoid possible double count.

Use of dlnm allows us to develop heat–mortality relation model. However, one of the biggest obstacles in developing a global projection model was that MMT varies with climate; warmer areas have higher MMT and colder areas have lower MMT (Curriero et al. 2002). Although we knew MMT is related to climate, there was no good model to estimate the MMT. After an exhaustive search, we found that 84th percentile of daily maximum temperature can be used for MMT (Honda et al. 2014). Using this finding, we developed the global projection model based on 47 prefectures in Japan. To note, it was not certain at the time whether or not the relation is V-shaped in tropical areas where there is little temperature difference throughout the year. However, we are now certain that the similar relation can be found even in tropical areas such as the Philippines (Dang et al. 2016; Dang et al. 2018; Seposo et al. 2015).

The detailed description of the projection can be found in the WHO report (Hales et al. 2014). In brief, the projection used 5 global circulation models, i.e., BCM2, EGMAM1, EGMAM2, EGMAM3, IPCM4 and assumed A1B emission and socio-economic scenario and calculated the impacts for 2030 and 2050. Table 1 shows the "attributable deaths due to climate change". Even without climate change, heat-related mortality occurs. Here we subtracted the heat-related mortality without climate change from the heat-related mortality with climate change. Thus, for example, in high-income Asia-Pacific region, the expected heat-related mortality in 2030 is, on average, 3383 deaths more than the situation without climate change.

Table 1 shows that the hardest-hit areas are among Asian countries. This is partly because the population size is large. Also it is obvious that the impact is not restricted to low- and middle-income countries. This is noteworthy because other impacts such as undernutrition, diarrheal disease, and vector-borne diseases, are mainly among low- and middle-income countries; high-income countries suffer little. This fact is important to demonstrate that high-income countries need to be serious about mitigation and adaptation actions, not only because they are responsible for the large emission of greenhouse gases but also because their own citizens are susceptible to climate change.

After this 2014 report (=WHO2014), World Health Organization provided the impact report by country. See http://www.who.int/globalchange/resources/country-profiles/en/ for country-specific impacts.

Table 1 Climate change-attributable heat-related excess number of deaths by region, without adaptation[a] (cited from WHO2014 report with permission by WHO with ID: 267350)

Region	2030	2050
Asia Pacific, high income	3383 (2375–4106)	6221 (4339–8280)
Asia, central	1752 (847–2282)	4886 (2850–5656)
Asia, east	19,323 (13,080–23,740)	47,367 (29,689–70,528)
Asia, south	21,648 (15,974–25,653)	62,821 (48,133–83,447)
Asia, south-east	6739 (4269–9089)	22,517 (17,174–32,887)
Australasia	217 (132–345)	605 (434–980)
Caribbean	281 (193–431)	862 (550–1314)
Europe, central	2279 (1563–4244)	4373 (2461–8184)
Europe, eastern	4988 (2899–8185)	8745 (5576–14,469)
Europe, western	6261 (2644–12,412)	14,148 (8942–25,840)
Latin America, Andean	540 (332–753)	2142 (1689–3100)
Latin America, central	2293 (1481–2989)	7704 (6138–11,251)
Latin America, southern	972 (690–1612)	2377 (1769–3386)
Latin America, tropical	1808 (1330–2707)	5912 (3727–10,181)
North America, high income	7288 (4986–8609)	16,076 (12,488–21,152)
North Africa/Middle East	4997 (3184–5837)	18,688 (12,122–22,936)
Oceania	52 (44–71)	217 (177–341)
Sub-Saharan Africa, central	921 (717–1119)	4107 (3399–5277)
Sub-Saharan Africa, eastern	3266 (2828–4448)	13,713 (10,055–19,295)
Sub-Saharan Africa, southern	671 (384–911)	1970 (1469–2700)
Sub-Saharan Africa, western	2529 (1716–3391)	9971 (7890–13,365)
World	92,207 (64,458–121,464)	255,486 (191,816–364,002)

[a]ENSEMBLE means, with low and high estimates in brackets

3.3 *Unsolved Issues*

As we have shown above, MMT is higher for warmer areas. Because susceptibility to heat is less for warmer areas because of this phenomenon, MMT difference can be regarded as adaptation. In this adaptation, no adaptation policy has been involved. For this reason, we call it "autonomous adaptation." Because this area difference has occurred, it is natural to think that the climate change would raise the MMT along with the 84th percentile value rise over time. Our preliminary analysis suggested that this MMT rise had been occurring. For this reason, we incorporated this chronological "autonomous adaptation" in the WHO2014 report. However, we had no good model to estimate the speed of autonomous adaptation. Thus, we considered 100% adaptation when the MMT is at the level of 84th percentile value of daily maximum temperature at the time, and 50% adaptation if the MMT is the mid-point of no adaptation case and the 100% adaptation case.

In the WHO2014 report, we did not take seasonal change into account. As Gasparrini et al. reported, heat impact of the same strength is larger when it occurs early summer than in late summer (Gasparrini et al. 2016). This fact is important for adaptation strategy, but for future projections, average impact for entire summer should be sufficient.

Although heat impact should be alleviated with adaptation actions anyway, policymakers are concerned about net effect of global warming, i.e., whether or not increased heat impact would be offset by the improvement of cold impact. Gasparrini and coworkers reported that, if the risk function stays as it is today, the net impact of climate change on heat and cold related mortality will be mostly adverse (Gasparrini et al. 2017). Other report suggests that the cold impact would not be alleviated by climate change (Kinney et al. 2015).

Another issue that would affect the impact estimation is the change in susceptibility; the risk for a certain temperature is mostly decreasing (Gasparrini et al. 2015a, b). The exception was UK, Australia and Korea, where they did not find risk reduction. Another paper on this issue showed the risk reduction in Korea (Chung et al. 2017). The discrepancy occurred because the former used 2006 to show the recent situation; after 2006, the risk declined in Korea, which was captured by the latter. Still, UK showed no sign of risk reduction, and we need more research to incorporate the risk reduction in the projection.

3.4 Future Direction

As stated above, research on autonomous adaptation and on chronological risk reduction are the issues many researchers are investigating. Other issues include the following:

As Dang et al. showed, the temperature–mortality relation curve for the central area of Ho Chi Minh City was almost identical to that for the surrounding area, and that the effect of urban heat island effect in terms of heat-related mortality was attributed to temperature distribution difference (Dang et al. 2018). Up until now, global projection has not incorporated urban heat island effect, but because the city population will occupy 68% of the global population in 2050 (https://population.un.org/wup/Publications/Files/WUP2018-KeyFacts.pdf) and ignoring urban heat island impact would substantially underestimate the heat-related mortality in the future. For this projection, understanding of heat-island impact in some other cities and development of downscaling method that can provide valid sub-city level surface temperatures.

Co-benefit analysis is very important for policy makers. In this regard, co-benefit by reducing short-lived climate pollutants is also necessary to be studied. For this purpose, development of impact models that incorporate temperature and air pollutants is the first step. Although some researchers controlled for temperature in evaluating air pollution effect, deeper understanding of the inter-relation between the air pollutants and temperature is necessary for co-benefit analyses.

Acknowledgements This chapter is based on some studies supported by the Environment Research and Technology Development Fund (S-14) of the Environmental Restoration and Conservation Agency.

References

Armstrong, B. (2006). Models for the relationship between ambient temperature and daily mortality. *Epidemiology, 17*(6), 624–631. https://doi.org/10.1097/01.ede.0000239732.50999.8f.

Bai, L., Ding, G., Gu, S., Bi, P., Su, B., Qin, D., et al. (2014). The effects of summer temperature and heat waves on heat-related illness in a coastal city of China, 2011–2013. *Environmental Research, 132,* 212–219. https://doi.org/10.1016/j.envres.2014.04.002.

Chan, E. Y., Goggins, W. B., Kim, J. J., Griffiths, S., & Ma, T. K. (2011). Help-seeking behavior during elevated temperature in Chinese population. *Journal of Urban Health, 88*(4), 637–650. https://doi.org/10.1007/s11524-011-9599-9.

Chan, E. Y., Goggins, W. B., Yue, J. S., & Lee, P. (2013). Hospital admissions as a function of temperature, other weather phenomena and pollution levels in an urban setting in China. *Bulletin of the World Health Organization, 91*(8), 576–584. https://doi.org/10.2471/BLT.12.113035.

Chan, E. Y. Y., Lam, H. C. Y., So, S. H. W., Goggins, W. B., Ho, J. Y., Liu, S., & Chung, P. P. W. (2018). Association between ambient temperatures and mental disorder hospitalizations in a subtropical city: A time-series study of Hong Kong special administrative region. *International Journal of Environmental Research and Public Health, 15*(4). https://doi.org/10.3390/ijerph15040754.

Chen, R., Li, T., Cai, J., Yan, M., Zhao, Z., & Kan, H. (2014). Extreme temperatures and out-of-hospital coronary deaths in six large Chinese cities. *Journal of Epidemiology and Community Health, 68*(12), 1119–1124. https://doi.org/10.1136/jech-2014-204012.

Cheng, Q., Bai, L., Zhang, Y., Zhang, H., Wang, S., Xie, M., et al. (2018). Ambient temperature, humidity and hand, foot, and mouth disease: A systematic review and meta-analysis. *Science of the Total Environment, 625,* 828–836. https://doi.org/10.1016/j.scitotenv.2018.01.006.

Cheng, J., Xie, M. Y., Zhao, K. F., Wu, J. J., Xu, Z. W., Song, J., et al. (2017). Impacts of ambient temperature on the burden of bacillary dysentery in urban and rural Hefei, China. *Epidemiology and Infection, 145*(8), 1567–1576. https://doi.org/10.1017/S0950268817000280.

Cheng, J., Xu, Z., Zhao, D., Xie, M., Zhang, H., Wang, S., & Su, H. (2016). The burden of extreme heat and heatwave on emergency ambulance dispatches: A time-series study in Huainan, China. *Science of the Total Environment, 571,* 27–33. https://doi.org/10.1016/j.scitotenv.2016.07.103.

Cho, E. J., Shin, S. D., Jeong, S., Kwak, Y. H., & Suh, G. J. (2016). The effect of atmosphere temperature on out-of-hospital cardiac arrest outcomes. *Resuscitation, 109,* 64–70. https://doi.org/10.1016/j.resuscitation.2016.10.004.

Chung, Y., Noh, H., Honda, Y., Hashizume, M., Bell, M. L., Guo, Y. L., & Kim, H. (2017). Temporal changes in mortality related to extreme temperatures for 15 cities in Northeast Asia: Adaptation to heat and maladaptation to cold. *American Journal of Epidemiology, 185*(10), 907–913. https://doi.org/10.1093/aje/kww199.

Curriero, F. C., Heiner, K. S., Samet, J. M., Zeger, S. L., Strug, L., & Patz, J. A. (2002). Temperature and mortality in 11 cities of the eastern United States. *American Journal of Epidemiology, 155*(1), 80–87.

Dang, T. N., Seposo, X. T., Duc, N. H., Thang, T. B., An, D. D., Hang, L. T., et al. (2016). Characterizing the relationship between temperature and mortality in tropical and subtropical cities: A distributed lag non-linear model analysis in Hue, Viet Nam, 2009–2013. *Global Health Action, 9,* 28738. https://doi.org/10.3402/gha.v9.28738.

Dang, T. N., Van, D. Q., Kusaka, H., Seposo, X. T., & Honda, Y. (2018). Green space and deaths attributable to the urban heat island effect in Ho Chi Minh City. *American Journal of Public Health, 108*(S2), S137–S143. https://doi.org/10.2105/AJPH.2017.304123.

Fukuda, T., Ohashi, N., Doi, K., Matsubara, T., Kitsuta, Y., Nakajima, S., & Yahagi, N. (2014). Impact of seasonal temperature environment on the neurologic prognosis of out-of-hospital cardiac arrest: A nationwide, population-based cohort study. *Journal of Critical Care, 29*(5), 840–847. https://doi.org/10.1016/j.jcrc.2014.03.034.

Gasparrini, A., Armstrong, B., Kovats, S., & Wilkinson, P. (2012). The effect of high temperatures on cause-specific mortality in England and Wales. *Occupational and Environmental Medicine, 69*(1), 56–61. https://doi.org/10.1136/oem.2010.059782.

Gasparrini, A., Guo, Y., Hashizume, M., Kinney, P. L., Petkova, E. P., Lavigne, E., et al. (2015a). Temporal variation in heat-mortality associations: A multicountry study. *Environmental Health Perspectives, 123*(11), 1200–1207. https://doi.org/10.1289/ehp.1409070.

Gasparrini, A., Guo, Y., Hashizume, M., Lavigne, E., Tobias, A., Zanobetti, A., et al. (2016). Changes in susceptibility to heat during the summer: A multicountry analysis. *American Journal of Epidemiology, 183*(11), 1027–1036. https://doi.org/10.1093/aje/kwv260.

Gasparrini, A., Guo, Y., Hashizume, M., Lavigne, E., Zanobetti, A., Schwartz, J., et al. (2015b). Mortality risk attributable to high and low ambient temperature: A multicountry observational study. *The Lancet, 386*(9991), 369–375. https://doi.org/10.1016/s0140-6736(14)62114-0.

Gasparrini, A., Guo, Y., Sera, F., Vicedo-Cabrera, A. M., Huber, V., Tong, S., et al. (2017). Projections of temperature-related excess mortality under climate change scenarios. *Lancet Planet Health, 1*(9), e360–e367. https://doi.org/10.1016/s2542-5196(17)30156-0.

Ghazani, M., FitzGerald, G., Hu, W., Toloo, G. S., & Xu, Z. (2018). Temperature variability and gastrointestinal infections: A review of impacts and future perspectives. *International Journal of Environmental Research and Public Health, 15*(4). https://doi.org/10.3390/ijerph15040766.

Gu, S., Huang, C., Bai, L., Chu, C., & Liu, Q. (2016). Heat-related illness in China, summer of 2013. *International Journal of Biometeorology, 60*(1), 131–137. https://doi.org/10.1007/s00484-015-1011-0.

Guo, T., Wang, Y., Zhang, H., Zhang, Y., Zhao, J., Wang, Y., et al. (2018). The association between ambient temperature and the risk of preterm birth in China. *Science of the Total Environment, 613–614,* 439–446. https://doi.org/10.1016/j.scitotenv.2017.09.104.

Hales, S., Kovats, S., Lloyd, S., & Campbell-Lendrum, D. (Eds.). (2014). *Quantitative risk assessment of the effects of climate change on selected causes of death, 2030s and 2050s.* Geneva, Switzerland: World Health Organization.

Han, M. H., Yi, H. J., Kim, Y. S., & Kim, Y. S. (2015). Effect of seasonal and monthly variation in weather and air pollution factors on stroke incidence in Seoul, Korea. *Stroke, 46*(4), 927–935. https://doi.org/10.1161/STROKEAHA.114.007950.

Honda, Y., Kondo, M., McGregor, G., Kim, H., Guo, Y. L., Hijioka, Y., et al. (2014). Heat-related mortality risk model for climate change impact projection. *Environmental Health and Preventive Medicine, 19*(1), 56–63. https://doi.org/10.1007/s12199-013-0354-6.

Honda, Y., Ono, M., Sasaki, A., & Uchiyama, I. (1995). Relationship between daily high temperature and mortality in Kyushu, Japan. *Nihon Koshu Eisei Zasshi, 42*(4), 260–268.

Huang, J., Chen, S., Wu, Y., Tong, Y., Wang, L., Zhu, M., et al. (2018). Quantifying the influence of temperature on hand, foot and mouth disease incidence in Wuhan, Central China. *Scientific Reports, 8*(1), 1934. https://doi.org/10.1038/s41598-018-20318-z.

Huang, L. Y., Wang, Y. C., Liu, C. M., Wu, T. N., Chou, C. H., Sung, F. C., & Wu, C. C. (2011). Water outage increases the risk of gastroenteritis and eyes and skin diseases. *BMC Public Health, 11*, 726. https://doi.org/10.1186/1471-2458-11-726.

Kang, S. H., Oh, I. Y., Heo, J., Lee, H., Kim, J., Lim, W. H., et al. (2016). Heat, heat waves, and out-of-hospital cardiac arrest. *International Journal of Cardiology, 221*, 232–237. https://doi.org/10.1016/j.ijcard.2016.07.071.

Kim, Y., Kim, H., Honda, Y., Guo, Y. L., Chen, B. Y., Woo, J. M., & Ebi, K. L. (2016). Suicide and ambient temperature in East Asian countries: A time-stratified case-crossover analysis. *Environmental Health Perspectives, 124*(1), 75–80. https://doi.org/10.1289/ehp.1409392.

Kim, E., Kim, H., Kim, Y. C., & Lee, J. P. (2018). Association between extreme temperature and kidney disease in South Korea, 2003–2013: Stratified by sex and age groups. *Science of the Total Environment, 642,* 800–808. https://doi.org/10.1016/j.scitotenv.2018.06.055.

Kinney, P. L., Schwartz, J., Pascal, M., Petkova, E., Tertre, A. L., Medina, S., & Vautard, R. (2015). Winter season mortality: Will climate warming bring benefits? *Environmental Research Letters, 10*(6), 064016. https://doi.org/10.1088/1748-9326/10/6/064016.

Lee, S., Lee, H., Myung, W., Kim, E. J., & Kim, H. (2018). Mental disease-related emergency admissions attributable to hot temperatures. *Science of the Total Environment, 616–617,* 688–694. https://doi.org/10.1016/j.scitotenv.2017.10.260.

Lee, S., Lee, E., Park, M. S., Kwon, B. Y., Kim, H., Jung, D. H., et al. (2014). Short-term effect of temperature on daily emergency visits for acute myocardial infarction with threshold temperatures. *PLoS ONE, 9*(4), e94070. https://doi.org/10.1371/journal.pone.0094070.

Lee, J. Y., Park, E. H., Lee, S., Ko, G., Honda, Y., Hashizume, M., et al. (2017). Airborne bacterial communities in three East Asian cities of China, South Korea, and Japan. *Scientific Reports, 7*(1), 5545. https://doi.org/10.1038/s41598-017-05862-4.

Li, Y., Li, C., Luo, S., He, J., Cheng, Y., & Jin, Y. (2017). Impacts of extremely high temperature and heatwave on heatstroke in Chongqing, China. *Environmental Science and Pollution Research, 24*(9), 8534–8540. https://doi.org/10.1007/s11356-017-8457-z.

Li, J., Xu, X., Ding, G., Zhao, Y., Zhao, R., Xue, F., et al. (2016a). A cross-sectional study of heat wave-related knowledge, attitude, and practice among the public in the Licheng District of Jinan City, China. *International Journal of Environmental Research and Public Health, 13*(7). https://doi.org/10.3390/ijerph13070648.

Li, K., Zhao, K., Shi, L., Wen, L., Yang, H., Cheng, J., et al. (2016b). Daily temperature change in relation to the risk of childhood bacillary dysentery among different age groups and sexes in a temperate city in China. *Public Health, 131,* 20–26. https://doi.org/10.1016/j.puhe.2015.10.011.

Likhvar, V., Honda, Y., & Ono, M. (2011). Relation between temperature and suicide mortality in Japan in the presence of other confounding factors using time-series analysis with a semiparametric approach. *Environmental Health and Preventive Medicine, 16*(1), 36–43. https://doi.org/10.1007/s12199-010-0163-0.

Lim, Y. H., So, R., Lee, C., Hong, Y. C., Park, M., Kim, L., & Yoon, H. J. (2018). Ambient temperature and hospital admissions for acute kidney injury: A time-series analysis. *Science of The Total Environment, 616–617*, 1134–1138. https://doi.org/10.1016/j.scitotenv.2017.10.207.

Ma, P., Wang, S., Fan, X., & Li, T. (2016). The impacts of air temperature on accidental casualties in Beijing, China. *International Journal of Environmental Research and Public Health, 13*(11). https://doi.org/10.3390/ijerph13111073.

McGregor, G. R., & Vanos, J. K. (2017). Heat: A primer for public health researchers. *Public Health.* https://doi.org/10.1016/j.puhe.2017.11.005.

McIver, L., Kim, R., Woodward, A., Hales, S., Spickett, J., Katscherian, D., et al. (2016). Health impacts of climate change in Pacific Island countries: A regional assessment of vulnerabilities and adaptation priorities. *Environmental Health Perspectives, 124*(11), 1707–1714. https://doi.org/10.1289/ehp.1509756.

Miao, Y., Shen, Y. M., Lu, C., Zeng, J., & Deng, Q. (2017). Maternal exposure to ambient air temperature during pregnancy and early childhood pneumonia. *Journal of Thermal Biology, 69,* 288–293. https://doi.org/10.1016/j.jtherbio.2017.09.001.

Niu, Y., Chen, R., Liu, C., Ran, P., Chen, A., Chen, X., & Kan, H. (2016). The association between ambient temperature and out-of-hospital cardiac arrest in Guangzhou, China. *Science of The Total Environment, 572*, 114–118. https://doi.org/10.1016/j.scitotenv.2016.07.205.

Onozuka, D., & Hagihara, A. (2015). The association of extreme temperatures and the incidence of tuberculosis in Japan. *International Journal of Biometeorology, 59*(8), 1107–1114. https://doi.org/10.1007/s00484-014-0924-3.

Onozuka, D., & Hagihara, A. (2016). Spatial and temporal variation in emergency transport during periods of extreme heat in Japan: A nationwide study. *Science of the Total Environment, 544,* 220–229. https://doi.org/10.1016/j.scitotenv.2015.11.098.

Onozuka, D., & Hagihara, A. (2017a). Extreme temperature and out-of-hospital cardiac arrest in Japan: A nationwide, retrospective, observational study. *Science of the Total Environment, 575,* 258–264. https://doi.org/10.1016/j.scitotenv.2016.10.045.

Onozuka, D., & Hagihara, A. (2017b). Spatiotemporal variation in heat-related out-of-hospital cardiac arrest during the summer in Japan. *Science of the Total Environment, 583,* 401–407. https://doi.org/10.1016/j.scitotenv.2017.01.081.

Onozuka, D., & Hashizume, M. (2011a). Effect of weather variability on the incidence of mumps in children: A time-series analysis. *Epidemiology and Infection, 139*(11), 1692–1700. https://doi.org/10.1017/S0950268810002967.

Onozuka, D., & Hashizume, M. (2011b). The influence of temperature and humidity on the incidence of hand, foot, and mouth disease in Japan. *Science of the Total Environment, 410–411,* 119–125. https://doi.org/10.1016/j.scitotenv.2011.09.055.

Onozuka, D., Hashizume, M., & Hagihara, A. (2009). Impact of weather factors on *Mycoplasma pneumoniae* pneumonia. *Thorax, 64*(6), 507–511. https://doi.org/10.1136/thx.2008.111237.

Onozuka, D., Hashizume, M., & Hagihara, A. (2010). Effects of weather variability on infectious gastroenteritis. *Epidemiology and Infection, 138*(2), 236–243. https://doi.org/10.1017/S0950268809990574.

Page, L. A., Hajat, S., & Kovats, R. S. (2007). Relationship between daily suicide counts and temperature in England and Wales. *British Journal of Psychiatry, 191,* 106–112. https://doi.org/10.1192/bjp.bp.106.031948.

Phung, D., Chu, C., Rutherford, S., Nguyen, H. L. T., Do, C. M., & Huang, C. (2017). Heatwave and risk of hospitalization: A multi-province study in Vietnam. *Environmental Pollution, 220*(Pt A), 597–607. https://doi.org/10.1016/j.envpol.2016.10.008.

Phung, D., Chu, C., Tran, D. N., & Huang, C. (2018). Spatial variation of heat-related morbidity: A hierarchical Bayesian analysis in multiple districts of the Mekong Delta Region. *Science of the Total Environment, 637–638,* 1559–1565. https://doi.org/10.1016/j.scitotenv.2018.05.131.

Phung, D., Guo, Y., Thai, P., Rutherford, S., Wang, X., Nguyen, M., et al. (2016). The effects of high temperature on cardiovascular admissions in the most populous tropical city in Vietnam. *Environmental Pollution, 208*(Pt A), 33–39. https://doi.org/10.1016/j.envpol.2015.06.004.

Phung, D., Rutherford, S., Chu, C., Wang, X., Nguyen, M., Nguyen, N. H., et al. (2015). Temperature as a risk factor for hospitalisations among young children in the Mekong Delta area, Vietnam. *Occupational and Environmental Medicine, 72*(7), 529–535. https://doi.org/10.1136/oemed-2014-102629.

Qiu, H., Sun, S., Tang, R., Chan, K. P., & Tian, L. (2016). Pneumonia hospitalization risk in the elderly attributable to cold and hot temperatures in Hong Kong, China. *American Journal of Epidemiology, 184*(8), 555–569. https://doi.org/10.1093/aje/kww041.

Sankei Shinbun. (2018, August 1st). Ninety-six heat stroke deaths in July in Tokyo 23 Wards: Four times as high as the last year. *Sankei Shinbun*. Retrieved from https://www.sankei.com/affairs/news/180801/afr1808010021-n1.html.

Seposo, X. T., Dang, T. N., & Honda, Y. (2015). Evaluating the effects of temperature on mortality in Manila City (Philippines) from 2006-2010 using a distributed lag nonlinear model. *International Journal of Environmental Research and Public Health, 12*(6), 6842–6857. https://doi.org/10.3390/ijerph120606842.

Shen, S. Q., Wei, H. X., Fu, Y. H., Zhang, H., Mo, Q. Y., Wang, X. J., et al. (2015). Multiple sources of infection and potential endemic characteristics of the large outbreak of dengue in Guangdong in 2014. *Scientific Reports, 5,* 16913. https://doi.org/10.1038/srep16913.

Sheng, R., Li, C., Wang, Q., Yang, L., Bao, J., Wang, K., et al. (2018). Does hot weather affect work-related injury? A case-crossover study in Guangzhou, China. *International Journal of Hygiene and Environmental Health, 221*(3), 423–428. https://doi.org/10.1016/j.ijheh.2018.01.005.

Song, X., Wang, S., Li, T., Tian, J., Ding, G., Wang, J., et al. (2018). The impact of heat waves and cold spells on respiratory emergency department visits in Beijing, China. *Science of the Total Environment, 615,* 1499–1505. https://doi.org/10.1016/j.scitotenv.2017.09.108.

Sun, X., Sun, Q., Yang, M., Zhou, X., Li, X., Yu, A., et al. (2014). Effects of temperature and heat waves on emergency department visits and emergency ambulance dispatches in Pudong New Area, China: A time series analysis. *Environmental Health, 13,* 76. https://doi.org/10.1186/1476-069X-13-76.

Trang, P. M., Rocklov, J., Giang, K. B., Kullgren, G., & Nilsson, M. (2016). Heatwaves and hospital admissions for mental disorders in Northern Vietnam. *PLoS ONE, 11*(5), e0155609. https://doi.org/10.1371/journal.pone.0155609.

Wang, Y. C., Lin, Y. K., Chuang, C. Y., Li, M. H., Chou, C. H., Liao, C. H., & Sung, F. C. (2012). Associating emergency room visits with first and prolonged extreme temperature event in Taiwan: A population-based cohort study. *Science of The Total Environment, 416*, 97–104. https://doi.org/10.1016/j.scitotenv.2011.11.073.

Wong, H. T., Chiu, M. Y., Wu, C. S., Lee, T. C., & Senior Citizen Home Safety, A. (2015). The influence of weather on health-related help-seeking behavior of senior citizens in Hong Kong. *International Journal of Biometeorology, 59*(3), 373–376. https://doi.org/10.1007/s00484-014-0831-7.

Xu, C., Li, Y., Wang, J., & Xiao, G. (2017). Spatial-temporal detection of risk factors for bacillary dysentery in Beijing, Tianjin and Hebei, China. *BMC Public Health, 17*(1), 743. https://doi.org/10.1186/s12889-017-4762-1.

Zhang, W., Du, Z., Zhang, D., Yu, S., & Hao, Y. (2016). Quantifying the adverse effect of excessive heat on children: An elevated risk of hand, foot and mouth disease in hot days. *Science of the Total Environment, 541,* 194–199. https://doi.org/10.1016/j.scitotenv.2015.09.089.

Zhang, Y., Yan, C., Kan, H., Cao, J., Peng, L., Xu, J., & Wang, W. (2014). Effect of ambient temperature on emergency department visits in Shanghai, China: A time series study. *Environmental Health, 13*, 100. https://doi.org/10.1186/1476-069x-13-100.

Zhang, Y., Zhao, Q., Zhang, W., Li, S., Chen, G., Han, Z., & Guo, Y. (2017). Are hospital emergency department visits due to dog bites associated with ambient temperature? A time-series study in Beijing, China. *Science of The Total Environment, 598*, 71–76. https://doi.org/10.1016/j.scitotenv.2017.04.112.

Zhao, Q., Li, S., Cao, W., Liu, D. L., Qian, Q., Ren, H., et al. (2018). Modeling the present and future incidence of pediatric hand, foot, and mouth disease associated with ambient temperature in Mainland China. *Environmental Health Perspectives, 126*(4), 047010. https://doi.org/10.1289/EHP3062.

Zhao, D., Wang, L., Cheng, J., Xu, J., Xu, Z., Xie, M., et al. (2017a). Impact of weather factors on hand, foot and mouth disease, and its role in short-term incidence trend forecast in Huainan City, Anhui Province. *International Journal of Biometeorology, 61*(3), 453–461. https://doi.org/10.1007/s00484-016-1225-9.

Zhao, Q., Zhang, Y., Zhang, W., Li, S., Chen, G., Wu, Y., et al. (2017b). Ambient temperature and emergency department visits: Time-series analysis in 12 Chinese cities. *Environmental Pollution, 224,* 310–316. https://doi.org/10.1016/j.envpol.2017.02.010.

Zhou, L., Xin, Z., Bai, L., Wan, F., Wang, Y., Sang, S., et al. (2014). Perceptions of heat risk to health: A qualitative study of professional bus drivers and their managers in Jinan, China. *International Journal of Environmental Research and Public Health, 11*(2), 1520–1535. https://doi.org/10.3390/ijerph110201520.

Zhou, X., Zhou, Y., Chen, R., Ma, W., Deng, H., & Kan, H. (2013). High temperature as a risk factor for infectious diarrhea in Shanghai, China. *Journal of Epidemiology, 23*(6), 418–423.

Thunderstorms During Pollen Season as Risk Factors for Allergic Respiratory Diseases and Severe Asthma

Gennaro D'Amato and Maria D'Amato

Abstract Thunderstorm-asthma outbreaks are characterized, mostly at the beginning of thunderstorms, by a rapid increase of visits for asthma to general practitioners or hospital emergency departments. Pollen grains can be carried by thunderstorms at ground level with release of allergenic biological aerosols of paucimicronic size derived from the cytoplasm of pollens ruptured or not, which can penetrate deep into lower airways. In other words, there is evidence that under wet conditions or during thunderstorms, pollen grains, in part after rupture by osmotic shock, release into the atmosphere their content, including respirable, allergen-carrying cytoplasmic starch granules (0.5–2.5 micron) or other paucimicronic components that can reach lower airways, inducing asthma reactions in pollinosis patients. Subjects without asthma symptoms, but affected by seasonal rhinitis, can experience an asthma attack. No unusual levels of air pollution were noted at the time of the epidemics, but there was a strong association with high atmospheric concentrations of pollen grains such as grasses or other allergenic plant species such as Parietaria in Mediterranean areas and in some cases also Alternaria. However, subjects affected by pollen allergy should be informed about a possible risk of asthma attack at the beginning of a thunderstorm during pollen season. In particular, during the first 20–30 min of a thunderstorm, patients suffering from pollen allergy may inhale a high concentration of the allergenic material released by pollen that is dispersed into the atmosphere, which in turn can induce asthmatic reactions that can be severe.

G. D'Amato (✉)
Division of Respiratory and Allergic Diseases, Department of Respiratory Diseases, High Specialty Hospital "A. Cardarelli", Naples, Italy
e-mail: gdamatomail@gmail.com

School of Specialization in Respiratory Diseases, University of Napoli Federico II, Naples, Italy

Medical School and School of Specialization in Respiratory Diseases of University of Naples "Federico II", Rione Sirignano, 10-80121 Naples, Italy

M. D'Amato
First Division of Pneumology, High Speciality Hospital "V. Monaldi" and University, "Federico II" Medical School, Naples, Italy

R. Akhtar (ed.), *Extreme Weather Events and Human Health*,
https://doi.org/10.1007/978-3-030-23773-8_20

Keywords Bronchial asthma · Severe asthma · Near-fatal asthma · Thunderstorm-related asthma · Meteorological factors and asthma · Prevention of Thunderstorm-related asthma · Asthma-related deaths

There are observations that thunderstorms occurring during pollen season can induce severe asthma attacks in pollinosis patients (Andrew et al. 2017, p. 5636; Bellomo et al. 1992, p. 834; D'Amato et al. 2007a, p. 976; 2007b, p. 11; 2015, p. 1; 2016a, p. 434; 2017, p. 1786; 2018, p. 17; Hajat et al. 1997, p. 639).

According to current climate change scenarios, there will be an increase in intensity and frequency of heavy rainfall episodes, including thunderstorms, over the next few decades, which can be expected to be associated with an increase in the number and severity of asthma attacks, both in adults and in children (Andrew et al. 2017, p. 5636; D'Amato 2002, p. 30; D'Amato et al. 2005, p. 1113; 2007a, p. 976; b, p. 11; 2011, p. 120; 2012, 82; 2015, p. 1; 2016a, p. 434; 2017, p. 1786; 2018, p. 17).

Associations between thunderstorms and asthma attacks have been identified in multiple locations around the world (Andrew et al. 2017, p. 5636; Bellomo et al. 1992, p. 834; D'Amato et al. 2007a, p. 976; 2007b, p. 11; 2015, p. 1; 2016a, p, 434; 2017, p. 1786; 2018, p. 17). The so-called thunderstorm asthma is characterized by asthma outbreaks, possibly caused by the dispersion of more respirable allergenic particles derived from pollen and spores (Pulimood et al. 2007, p. 610).

Thunderstorms have been linked to asthma epidemics, especially during the pollen seasons, and there are descriptions of asthma outbreaks associated with thunderstorms, which occurred in several cities, mostly in Europe (Birmingham and London in the UK and Napoli in Italy) and Australia (Melbourne and Wagga Wagga) (Andrew et al. 2017, p. 5636; Bellomo et al. 1992, p. 834; D'Amato et al. 2016b, p. 390). The thunderstorm-asthma outbreaks are characterized, at the beginning of thunderstorms, by a rapid increase of visits for asthma to general practitioners or hospital emergency departments. Subjects without asthma symptoms but affected by seasonal rhinitis can experience an asthma attack. No unusual levels of air pollution were noted at the time of the epidemics, but there was a strong association with high atmospheric concentrations of pollen grains such as grasses or other allergenic plant species. However, subjects affected by pollen allergy should be informed about a possible risk of asthma attack at the beginning of a thunderstorm during pollen season.

On 21 November 2016 in Melbourne, there was a dramatic event with 10 deaths and 9000 patients who needed medical treatments in emergency departments of Melbourne Hospital for asthma attacks (Andrew et al. 2017, p. 5636; D'Amato et al. 2017, p. 1786; Lee et al. 2017, p. 146; Lindstrom et al. 2017, p. 235).

This has been the worst event of thunderstorm-asthma in Melbourne. It arose from an extraordinary association of environmental factors with a very unusual weather occurrence of wind and torrential rain combined with a high pollen count (grass pollen airborne count of more than 100 pollens for cubic meter), which dispersed a high quantity of pollens and allergenic submicronic particles derived from pollens across the city.

In Melbourne, ambulances and hospitals experienced a high demand during a few hours on 21 November, with grass allergens dispersed over a very large geographical area of Victoria. The rapid onset of the medical emergency and its consequences were unprecedented in intensity, and it tested the capacity of the Melbourne health system to be ready for this type of medical emergency. However, demand management strategies were insufficient to manage such a widespread and rapid onset event, with ambulance resources quickly depleted and police officers had to be used to conduct welfare checks (Andrew et al. 2017, p. 5636; Lee et al. 2017, p. 146; Lindstrom et al. 2017, p. 235).

The event in Melbourne greatly surpassed the previous epidemic in London of June 1994 that was the largest documented outbreak before the Victoria epidemic (Table 1).

One of the first observations regarding thunderstorms and asthma outbreaks was provided by Packe and Ayres (1985, p. 199) at the East Birmingham Hospital (Birmingham, UK) on 6 and 7 July 1983. These authors described a remarkable increase in the number of asthma emergency department admissions during the hours of a thunderstorm. In a 36-h period, 26 asthma cases were treated in the emergency department, compared with a daily average of two or three cases in the days preceding the outbreak.

Another asthma outbreak occurred in London, UK, coinciding with a heavy thunderstorm on 24 June 1994, when a large increase in the number of visits for asthma at the emergency departments of London and the south-west of England was observed (Davidson et al. 1996, p. 601; Venables et al. 1997, p. 725; Levy and Bryden 2007, p. 69). Several patients, who were not known to be asthmatics or were affected only by seasonal rhinitis, experienced an asthma attack. During a 30-h period from 6 pm on 24 June 1994, 640 patients with asthma or other airways disease (283 of whom were not known to be asthmatic and 403 who were affected only by seasonal rhinitis) attended several emergency departments, nearly 10 times the expected number of 66 patients. In total, 104 patients were admitted (including five to an intensive care unit); 574 patient admissions were attributable to the thunderstorm.

Other asthma outbreaks during thunderstorms have been described in Australia. In Melbourne, other than the dramatic outbreak of 21 November 2016, two large asthma outbreaks (rapid increase in hospital or general practitioner visits for asthma) coincided with thunderstorms (Andrew et al. 2017, p. 5636; Lindstrom et al. 2017, p. 235).

In south-eastern Australia, Marks et al. (2001, p. 468; 2007, p. 530) observed that the incidence of excess hospital attendances for asthma during late spring and summer was strongly linked to the occurrence of thunderstorm outflows and demonstrated that the arrival of a thunderstorm outflow was accompanied by a large increase in the concentration of ruptured pollen grains in ambient air.

Thunderstorm-related asthma was observed in Naples, Italy, on 3 June 2004 (D'Amato et al. 2007a, p. 976; 2016a, p. 434; 2017, p. 1786), when five adults and one child received treatment in emergency departments. One patient was admitted to an intensive care unit for a very severe bronchial obstruction and acute respiratory insufficiency following a sudden thunderstorm. All individuals were outdoors when

Table 1 Examples of thunderstorm-associated asthma outbreaks (D'Amato et al. 2007a, p. 976; 2007b, p. 11; 2015, p. 1; 2016a, p. 434; 2017, p. 1786; 2018, p. 17)

Year	Country	Observations
1983	UK	26 sudden cases of asthma attacks in relation to thunderstorms
1992	Australia	Late spring thunderstorms in Melbourne can trigger epidemics of asthma attacks (five- to tenfold rise)
1997	UK	Asthma or other airways disease-related hospital visits. 640 cases attended during a 30-h period in June 1994, nearly 10 times the expected number
1992–2000	Canada	18,970 hospital ED asthma visits among children 2–15 years of age. Summer thunderstorm activity was associated with an OR of 1.35 (95% CI 1.02–1.77) relative to summer periods with no activity
1993–2004	USA	215,832 asthma ED visits; 24350 of these visits occurred on days following thunderstorms. There was a significant association between daily counts of asthma ED visits and thunderstorm occurrence Asthma visits were 3% higher on days following thunderstorms
2000	Australia	Asthma visits during thunderstorms. History of hay fever and allergy to ryegrass are strong predictors for asthma exacerbation during thunderstorms in spring
2001	Australia	The incidence of excess hospital attendances for asthma during late spring and summer was strongly linked to the occurrence of thunderstorm outflows
2002	UK	A case–control study of 26 patients presenting to Cambridge University Hospital with asthma after a thunderstorm. Alternaria alternata sensitivity is a compelling predictor of epidemic asthma in patients with seasonal asthma and grass pollen allergy and is likely to be the important factor in thunderstorm-related asthma
2004	Italy	Six cases of thunderstorm-related asthma because of pollen (Paretaria)
2010	Italy	20 cases of thunderstorm-related asthma because of pollen (olive tree)
2010	Australia	Epidemics of "thunderstorm asthma" that occurred in Melbourne during spring 2010. The approach of spring, together with high winter rainfall in and around Melbourne that heralds another severe pollen season, raises the risk of allergic rhinitis and asthma in pollen-sensitive individuals
2016	Australia	Epidemics of thunderstorm asthma in Melbourne with 10 deaths and 9000 emergency department visits

the thunderstorm struck. In one severe case, a female sensitized only to Parietaria pollen allergen soon began to show symptoms of intense dyspnea, which gradually worsened. She was taken to hospital where she was intubated and given high intravenous doses of corticosteroids. She was discharged a few days later. This patient had previously suffered from seasonal asthma but had been asthma-free for the past few years and did not need continuous therapy. None of the other five subjects took anti-allergic and/or anti-asthma drugs regularly. All six patients were sensitized with allergic respiratory symptoms upon exposure to Parietaria pollen but were not sensitized to grasses. Parietaria is an Urticacea that is widespread in the Naples area of Italy with a spring and summer pollen season that is, in part, coexistent with that of grasses. During the thunderstorm, the concentration of airborne Parietaria pollen grains was particularly high, with a peak of 144 grains/m^3 being recorded on 3 June 2004. Air pollution levels for both gaseous and particulate components based on the hourly concentrations of nitric dioxide, ozone and respirable particulate matter were not particularly high in Naples on 3 and 4 June 2004. Subjects with sensitization to Parietaria who were indoors in Naples with the windows closed during the night between 3 and 4 June 2004 did not experience asthma attacks. No moulds or viruses were involved in the Naples epidemics.

Other outbreaks and/or case reports have been described in Barletta (Italy), Cartagena (Spain), Atlanta (USA) and Canada (Toronto).

A similar phenomenon has been suggested for moulds after the observation of a possible key role of sensitization to *Alternaria* species in thunderstorm-related asthma (Pulimood et al. 2007, p. 610).

Although much remains to be discovered about the relationship between an increase in the number of asthma attacks and thunderstorms, reasonable evidence exists in favour of a causal link between them in patients suffering from pollen allergy. The most prominent hypotheses for thunderstorm-related asthma are linked with bio-aerosols and involve the role of rainwater in promoting the release of respirable particulate matter (D'Amato et al. 2016a, p. 434). Pollen grains can be carried by thunderstorms at ground level, where pollen rupture would be increased with the release of allergenic biological aerosols of paucimicronic size, derived from the cytoplasm and which can penetrate deep into lower airways. In other words, there is evidence that under wet conditions or during thunderstorms, pollen grains may, after rupture by osmotic shock, release into the atmosphere part of their content, including respirable, allergen-carrying cytoplasmic starch granules (0.5–2.5 lm) or other paucimicronic components that can reach lower airways, inducing asthma reactions in pollinosis patients.

These allergens can likely penetrate deeper into the lung, provoking more severe symptoms. It has been suggested that grass pollen starch granules are the most likely cause of associations between thunderstorms and asthma (D'Amato et al. 2016a, p. 434). Suphioglu et al. (1992, p. 6569) showed that ryegrass pollen grains contain a large amount of starch granules coated with allergens. After being ruptured in rainwater by osmotic shock, each grain can release 700 starch granules, which are small enough to penetrate the airways and trigger asthma attacks in previously sensitized subjects. Later Taylor et al. (2002, p. 569) hypothesized that the turbulent front of the advancing outflow releases more pollen from flowering grasses and grass pollen may release large amounts of paucimicronic allergenic particles, namely cytoplasmatic starch granules containing grass allergens (allergen bearing starch granules), after rupture by osmotic shock during thunderstorms.

A recent manuscript in Nature Climate Change (Finney et al. 2018, p. 210) suggests that lightning may decrease in the tropics under climate change. However, it seems that science on effects of climate change is quite dynamic and uncertain in relation to lightning/thunderstorm intensity and frequency.

Table 2 Aspects of Epidemics of Thunderstorm-Associated Allergic Asthma in the World (D'Amato et al. 2007a, p. 976; 2007b, p. 11; 2015, p. 1; 2016a, p. 434; 2017, p. 1786; 2018, p. 17)

• There is a link between storms and asthma epidemics in patients with pollen allergy during pollen seasons with the appearance of symptoms during the first 20–30 min of a storm
• Thunderstorm-related epidemics are limited to late spring and summer (in Europe, USA and Canada from April to end of June and in Australia from October to December), when pollen and/or mould counts are high
• There are no descriptions of allergic symptoms in individuals with allergy to pollens and moulds who are indoors with the windows closed during a storm
• It is possible that sudden cold and/or electric charges trigger asthma attacks in allergic subjects
• Individuals with allergic rhinitis only and no previous asthma can experience broncho-constriction, sometimes severe, during thunderstorms
• Subjects with pollen allergy need be informed about a possible risk of asthma attack at the beginning of a thunderstorm during pollen season
• Individuals who experience rhinitis and asthma during a storm are not usually taking suitable anti-inflammatory treatment. It is important to have correct anti-asthma treatment by using bronchodilators and corticosteroids by inhalation at increasing dosage if there is a need
• The world's worst recorded thunderstorm asthma attack was on 21 November 2016 in Melbourne, where nine subjects died and more than 8500 were hospitalized in Victoria. It caused many people, including those who had no history of asthma or respiratory issues, to experience mild to severe breathing difficulties and near-fatal asthma
• Any serious asthma attack during a thunderstorm can be life-threatening and can induce near-fatal asthma and death
The health consequences of thunderstorm asthma may be prevented by meteorological forecast, by the correct use by patients of adequate anti-allergic and anti-asthma therapy and by avoiding the outdoors at the start of a storm during pollen season

In summary (Table 2) the occurrence of these epidemics is closely linked to thunderstorms and they are limited to late spring and summer when there are high levels of airborne pollen grains. There is a close temporal association between the arrival of the thunderstorm, a major rise in the concentration of pollen grains and the onset of epidemics. As a consequence, subjects affected by pollen allergy should be alert to the danger of being outdoors during a thunderstorm in the pollen season (D'Amato et al. 2017, p. 1786).

References

Andrew, E., Nehme, Z., Bernard, S., Abramson, M. J., Newbigin, E., Piper, B., et al. (2017, December 13) Stormy weather: A retrospective analysis of demand for emergency medical services during epidemic thunderstorm asthma. *BMJ, 359*, j5636. https://doi.org/10.1136/bmj.j5636.

Bellomo, R., Gigliotti, P., Treloar, A., Holmes, P., Suphioglu, C., Singh, M. B., et al. (1992). Two consecutive thunderstorm associated epidemic of asthma in Melbourne. *Medical Journal of Australia, 156*, 834–837.

Celenza, A., Fothergill, J., Kupek, E., & Shaw, R. J. (1996). Thunderstorm associated asthma: A detailed analysis of environmental factors. *BMJ, 312,* 604–607.

Dales, R. E., Cakmak, S., Judek, S., Dann, T., Coates, F., Brook, J. R., et al. (2003). The role of fungal spores in thunderstorm asthma. *Chest, 123,* 745–750.

D'Amato, G. (2001). Airborne paucimicronic allergen-carrying particles and seasonal respiratory allergy. *Allergy, 56,* 1109–1111.

D'Amato, G. (2002). Environmental urban factors (air pollution and allergens) and the rising trends in allergic respiratory diseases. *Allergy, 57,* 30–33.

D'Amato, G., Annesi Maesano, I., Molino, A., Vitale, C., & D'Amato, M. (2017). Thunderstorm-related asthma attacks. *The Journal of Allergy and Clinical Immunology, 139*(6), 1786–1787. https://doi.org/10.1016/j.jaci.2017.03.003. Epub March 23, 2017.

D'Amato, G., Cecchi, L., & Annesi-Maesano, I. (2012). A trans-disciplinary overview of case reports of thunderstorm-related asthma outbreak and relapse. *European Respiratory Review, 21,* 82–87.

D'Amato, G., Cecchi, L., Bonini, S., Nunes, C., Annesi-Maesano, I., Behrendt, H., et al. (2007a, September). Allergenic pollen and pollen allergy in Europe. *Allergy, 62*(9), 976–990.

D'Amato, G., Cecchi, L., & Liccardi, G. (2008). Thunderstorm-related asthma: Not only grass pollen and spores. *Journal of Allergy and Clinical Immunology, 120,* 530–536.

D'Amato, G., Liccardi, G., D'Amato, M., & Holgate, S. T. (2005). Environmental risk factors and allergic bronchial asthma. *Clinical and Experimental Allergy, 35,* 1113–1124.

D'Amato, G., Liccardi, G., & Frenguelli, G. (2007b). Thunderstorm-asthma and pollen allergy. *AllergyJan, 62*(1), 11–16.

D'Amato, G., Holgate, S. T., Pawankar, R., Ledford, D. K., Cecchi, L., Al-Ahmad, M., et al. (2015,). Meteorological conditions, climate change, new emerging factors and asthma and related allergic disorders. A statement of the World Allergy Organitation. *World Allergy Organization Journal, 8*(1), 25. https://doi.org/10.1186/s40413-015-0073-0.

D'Amato, G., Rottem, M., Dahl, R., Blaiss, M., Ridolo, E., & Cecchi, L. (2011). Annesi-Maesano and the WAO special committee on climate change and allergy. *Climate Change, Migration and Allergic Respiratory Diseases: An Update for the Allergist. World Allergy Organ Journal, 4,* 120–125.

D'Amato, G., Vitale, C., Lanza, M., Molino, A., & D'Amato, M. (2016a, October). Climate change, air pollution and allergic respiratory diseases: an update. *Current Opinion in Allergy and Clinical Immunology, 16*(5), 434–40.

D'Amato, G., Vitale, C., D'Amato, M., Cecchi, L., Liccardi, G., Molino, A., et al. (2016b, March). Thunderstorm-related asthma: What happens and why. *Clinical & Experimental Allergy, 46*(3), 390–396. https://doi.org/10.1111/cea.12709.

D'Amato, M., Cecchi, L., Annesi-Maesano, I., & D'Amato, G. (2018, January 17). News on climate change, air pollution and allergic trigger factors of asthma. *Journal of Investigational Allergology & Clinical Immunology.* https://doi.org/10.18176/jiaci.0228.

Davidson, A. C., Emberlin, J., Cook, A. D., & Venables, K. M. (1996). Thames region accident and emergency trainees association. A major outbreak of asthma associated with a thunderstorm. *BMJ, 312,* 601–604.

Finney, D. L., Doherty, R. M., Wild, O., Stevenson, D. S., MacKenzie, I. A., & Blyth, A. M. (2018). A projected decrease in lightning under climate change. *Nature Climate Change, 8,* 210–213.

Girgis, S. T., Marks, G. B., Downs, S. H., Kolbe, A., Car, G. N., & Paton, R. (2000). Thunderstorm associated asthma in an inland town in southeastern Australia. Who is at risk? *European Respiratory Journal, 16,* 3–8.

Grinn-Gofroń, A., & Strzelczak, A. (2013). Changes in concentration of *Alternaria and Cladosporium* spores during summer storms. *International Journal of Biometeorology, 57,* 759–68. (Select item 292376042).

Hajat, S., Goubel, S. A., & Haines, A. (1997). Thunderstorm-associated asthma: The effect on GP consultations. *British Journal of General Practice, 147*(423), 639–641.

Lee, J., Kronborg, C., O'Hehir, R. E., & Hew, M. (2017, November). Who's at risk of thunderstorm asthma? The ryegrass pollen trifecta and lessons learnt from the Melbourne thunderstorm epidemic. *Respiratory Medicine, 132*, 146–148. https://doi.org/10.1016/j.rmed.2017.10.012. Epub October 18, 2017.

Levy, M. L., & Bryden, C. (2007). Thunderstorm asthma. *British Journal of Primary Care Nursing, 1,* 69–71.

Lindstrom, S. J., Silver, J. D., & Sutherland, M. F. (2017). Thunderstorm asthma outbreak of November 2016: A natural disaster requiring planning. *The Medical Journal of Australia, 207,* 235–237. https://doi.org/10.5694/mja17.00285.

Marks, G. B., & Bush, R. K. (2007). It's blowing in the wind: New insights into thunderstorm-related asthma. *Journal of Allergy and Clinical Immunology, 120*(3), 530–532.

Marks, G. B., Colquhoun, J. R., Girgis, S. T., Hjelmroos Koski, M., Treoloar, A. B. A., Hansen, P., et al. (2001). Thunderstorm outflows preceding epidemics of asthma during spring and summer. *Thorax, 56,* 468–471.

Newson, R., Strachan, D., Archibald, E., Emberlin, J., Hardaker, P., & Collier, C. (1998). Acute asthma epidemics, weather and pollen in England, 1987–1994. *European Respiratory Journal, 11,* 694–701.

Packe, G. E., & Ayres, J. G. (1985). Asthma outbreak during a thunderstorm. *Lancet, 2,* 199–204.

Pulimood, T. B., Corden, J. M., Bryden, C., Sharples, L., & Nasser, S. M. (2007). Epidemic asthma and the role of the fungal mould *Alternaria alternata*. *Journal of Allergy and Clinical Immunology, 120,* 610–617.

Suphioglu, C., Singh, M. B., Taylor, P., Knox, R. B., Bellomo, R., Holmes, P., et al. (1992). Mechanism of grass-pollen-induced asthma. *Lancet, 339,* 569–572.

Taylor, P. E., Hagan, R., Valenta, R., & Glovsky, M. M. (2002). Release of allergens in respirable aerosols: a link between grass pollen and asthma. *Journal of Allergy and Clinical Immunology, 109,* 51–56.

Venables, K. M., Allitt, U., Collier, C. G., Emberlin, J., Greig, J. B., Hardaker, P. J., et al. (1997). Thunderstorm-related asthma—Epidemic of 24/25. *Clinical and Experimental Allergy, 27,* 725–736.

Wardman, A. E., Stefani, D., & MacDonald, J. C. (2002). Thunderstorm-associated asthma or shortness of breath epidemic: A Canadian case report. *Canadian Respiratory Journal, 9,* 267–270.

Developing Countries

Sea Level Rise and Its Socio-economic Impacts: A Case Study in Mumbai, India

Gandharva Pednekar and S. Siva Raju

Abstract Coastal zone supports flora and fauna biodiversities and is endowed with a wide variety of habitats. Both marine and terrestrial processes have their influence on coastal zones. As coastal zones are often under pressure from both anthropogenic activities and natural processes, coastal zones are one of the most fragile, dynamic and productive ecosystems. Sea Level Rise (SLR) is a by-product of climate change on a global scale. Though climate change has happened in the past, the rate of current levels of change is unparallel and cannot be explained, without highlighting human contribution as a major reason. Global commerce and trade relies heavily on historic marine transportation routes and port cities. In the era of globalization, sea-level rise becomes a significant concern at individual, local government as well as to global economy levels. India has a coastline measuring upto 7515 km. Around 35 percent of Indians live within 100 km of the Indian coast. Given the varying levels of sea rise predictions, it is important to understand the concentration of population in the coastal regions, and they will become the victims of adverse effects of sea-level rise. The paper focuses on sea-level rise (SLR), its effects on coastal communities and socio-economic impacts in India, with a special focus on the island city of Mumbai.

Keywords Sea level rise · Global warming · Displacement · Urban infrastructure · Biodiversity · Mitigation

1 Introduction

In a phenomenon unprecedented in 65 million years at least, the ocean has absorbed about 30 per cent of the anthropogenic carbon dioxide, resulting in ocean acidification and changes to carbonate chemistry. Survival, calcification, growth, development and abundance of a broad range of taxonomic groups (i.e., from algae to fish) are at risk with substantial evidence of predictable trait-based sensitivities. There will

G. Pednekar (✉)
School of Development Studies, Tata Institute of Social Sciences, Mumbai, India
e-mail: gandharva.pednekar@tiss.edu

S. Siva Raju
Tata Institute of Social Sciences, Mumbai, India

R. Akhtar (ed.), *Extreme Weather Events and Human Health*,
https://doi.org/10.1007/978-3-030-23773-8_11

be a wide range impacts on marine organisms, ecosystems, as well as sectors such as aquaculture and fisheries due to ocean warming and acidification (corresponding to global warming of 1.5 °C of global warming) (IPCC 2018). The coast is defined as a spatial zone where land, sea and atmosphere interact and influence each other continuously. Both marine and terrestrial processes have their influence on coastal zones. As they are often under pressure from both anthropogenic activities and natural processes, coastal zones are one of the most fragile, dynamic and productive ecosystem. Coastal zone supports floral and faunal biodiversity and is endowed with a wide variety of habitats such as coral reefs, mangroves, sea grasses, sand dunes, mudflats, salt marshes, estuaries, lagoons etc.

In Indian scenario, as per Rao et al. (2008), frequent tropical cyclones and associated floods and tidal surges occur regularly in coastal Andhra Pradesh leading to loss of life and property in the region. An area of about 565 km^2 would be submerged under the new low-tide level along the coastal Andhra Pradesh, if the sea level rises by 0.59 m as predicted by IPCC (2007). The 2004-tsunami that devastated the Indian coasts underlined the role of geomorphology and coastal slope in sea level rise. At Nagapattinam in the southern state of Tamil Nadu, the low swales behind shore-parallel dune ridges claimed several lives due to lateral flows from tidal inlets or breaches in dune ridges. Longer penetration of tsunami inland due to gentle slope of the coastland was also observed (ISRO 2012). ISRO also calculated the Coastal Vulnerability Index for the state of Gujarat, which in on the west coast of India, and found that geomorphologically only 29.24 per cent of the coast represented by rocky/cliffy and indented sections comes under very-low to low categories, while 34.53 per cent of the Gujarat coast is very high to high vulnerable and 36.04 per cent of Gujarat coast is moderate vulnerable.

Studies (IPCC 2013; Parris et al. 2012; Sweet et al. 2017) suggest that the average rate sea-level rise of 3.4 mm per year is observed in globally, which will increase in the coming decades. According to the U.N. Atlas of the Oceans (2016), eight of the ten most populous cities—Tokyo, Mexico, Mumbai, Sao Paulo, New York, Shanghai, Lagos, Los Angeles, Calcutta and Buenos Aires—are in coastal regions. The paper focuses on sea-level rise (SLR), its effects on coastal communities and socio-economic impacts, with a special focus on the island city of Mumbai.

2 Sea-Level Rise (SLR)

According to the National Oceanic and Atmospheric Administration (NOAA 2017a, b, c, d), there is variation in sea-level rise, as some coastal communities experience a greater increase in sea level. As per Mimura (2013), there are two significant causes to global sea-level rise: thermal expansion and the melting of land-based ice. Complicated geologic processes such as isostatic rebound and groundwater pumping are also to be attributed for the same (Eggleston and Pope 2013). Since the beginning of the twentieth century, global sea surface temperatures have increased at an average rate of 0.13 °F per decade (1901–2015; NOAA 2016). Though climate change has happened in the past, the rate of current levels of change is unparalleled and cannot be explained without bearing in mind human contribution as a major reason (NCA 2014). NOAA (2015) estimates that 90 per cent energy of heat-trapping greenhouse gases has been stored in the ocean and as water warms, it expands. The other main cause of global sea-level rise is the melting of land-based ice sheets, mainly found on Greenland and Antarctica. Their melting can lead to sea-level rise in metres, as together they contain more than 99 per cent of the freshwater ice on Earth (NSIDC 2018). Tropical storms, prolonged rain, fluctuations in ocean currents and moon phases also play an important role in the spatial-temporal variability in the height of the ocean other than thermal expansion and the decay of glaciers.

Sea level can be divided into three broad categories according to the variations in the range of temporal and spatial scales. These are occurrences of sea-level extremes associated with storm surges and tides, Global Mean Sea Level (GMSL) and the regional variation about this mean. Since the late nineteenth century, GMSL has been rising at low rates of change that characterized the previous two millennia (Church et al. 2013). Instrumental drift in the observing satellite system (Watson et al. 2015) and volcanoes (Fasullo et al. 2016) may be the reasons for slowing in the reported rate over the last two decades (Cazenave et al. 2014), accounting for the former results in rates (1993 to mid-2014) of between 2.6 and 2.9 mm yr-1 (Watson et al. 2015). Thermal expansion, glacier and ice-sheet mass loss, as well as fresh water storage on land, are the main contributors towards the same (Church et al. 2013; Watson et al. 2015), and their attribution is dominated by anthropogenic forcing since 1970 ($15 \pm 55\%$ before 1950, $69 \pm 31\%$ after 1970) (Slangen et al. 2016).

As per IPCC (2018), by the end of the century, GMSL rise will be around 0.1 m less in a 1.5 °C world as compared to a 2 °C warmer world. Up to 10.4 million fewer people are exposed to the impacts of sea level globally in 2100 at 1.5 °C as compared to 2 °C. Reduced sea-level rise will enable greater opportunities for adaptation though the sea-level rise will continue beyond 2100, as instabilities exist for both the Greenland and Antarctic ice sheets that could result in multi-metre rises in sea level on centennial to millennial timescales.

From Fig. 1, we can observe the estimated, observed and predicted global sea-level rise during the period 1800–2100. Different colour bands reflecting different data are red band—1800–1890—estimates from proxy data; pink band—uncertainty; blue band—1880–2009—shows tide gauge data; and green band—1993–2012—satellite observations. The figure predicts future scenarios of SLR ranging from 0.66 to 6.6 ft in 2100.

3 Concentration of Population in the Coastal Regions

Given the varying levels of sea rise predictions, it is further more important to understand the concentration of population in the coastal regions, as they will become the victims of adverse effects of sea-level rise.

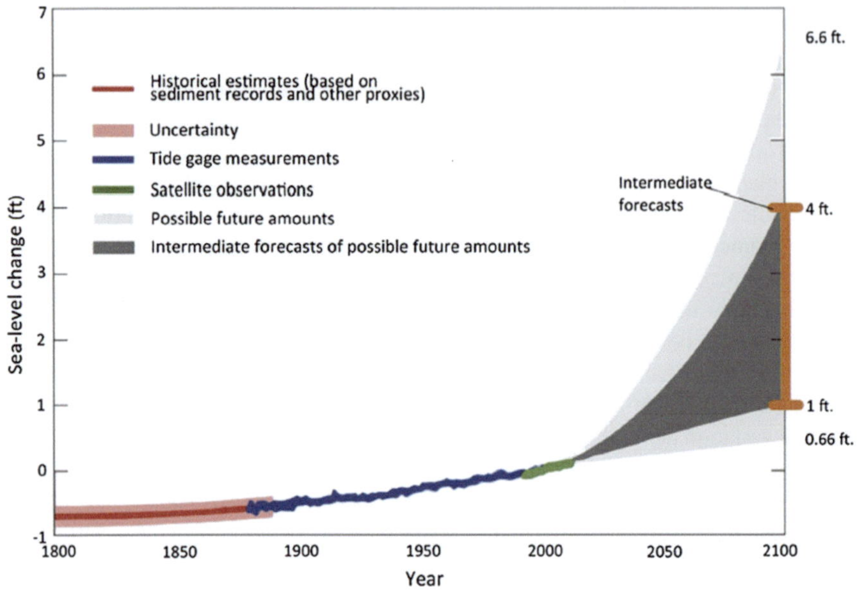

Fig. 1 Sea-level rise predictions. *Source* Khan (2018)

(1) Favourable climatic conditions—Coastal zones have moderating influence of land and sea breezes and the temperature rage also remains low.
(2) Contact point settlement—coastal habitats are also termed as contact point settlements, as they are located at such a place, where the residents were involved in more than one primary economic activities such as fishing as well as agricultural.
(3) Favourable locations during colonial period—coastal locations were developed as ports for import-export of materials. The colonial rulers found it safer to reside, as the local people were unknown to them. The colonial rulers built forts and isolated themselves from the local people. The coastal climate was moderate, and it was easier for colonial rulers to assimilate with the new environment.
(4) Legacy of the colonial period—underdeveloped/developing countries depended on the developed countries for the technological support in colonial era. When the developing countries became independent, they used the infrastructure developed by the Europeans. The port cities became million cities due to internal migration. Impact of these cities spread along with the coast e.g. the sphere of influence of Mumbai extends from Dahanu in the north to Alibag in the south along the coast of Arabian Sea.
(5) Development of tourism—tourism is considered as invisible export and contributes substantially to the developing countries like Singapore and Mauritius. Many new tourist destinations developed along the coast, thus increasing the concentration of population in the coastal areas.

(6) Migration—as the port cities developed at a rapid pace, the influx of migrants to cater to industries also grew. As port cities have limited space to grow horizontally, they started to grow vertically by construction of skyscrapers. But the cost of living became exorbitant, leading to the creation of slums, where the majority of the migrant population settled.

Indian port towns like Kolkata, Chennai, Mumbai, Kochi, Vishakhapatnam were commercial hubs during mercantile period. During the colonial period, their growth became more rapid, as they were the transport linkages with rest of the country to source out raw material. With the growth in quaternary economic activities such as service sector post-liberalization, the yesteryear port cities like Chennai, Kolkata and Delhi developed into metropolitan cities, while Mumbai achieved the status of a megalopolis. All the states on the west coast of India have more than 35% of its population living in urban area, while on east coast there is a gradual progression in urban population among states (Census 2011).

4 SLR and Coastal Communities in India

In India, coastal zones have gained importance because of high productivity of its ecosystems, exploitation of renewable and non-renewable natural resources, concentration of population, industrialization, discharge of waste effluents and municipal sewage and growth of recreational activities. Land–ocean interaction in the coastal zone produces a varied set of landforms and organisms having special forms of adaptation. The coastal processes largely operate within 10 m above or below and a few kilometres landward and seaward of the shoreline. The coastal regions are highly vulnerable to storm surges, tsunamis and inland floods, due to their low-lying nature and extremely gentle gradients (ISRO 2012). Coastal zones face the most direct impact of the sea-level rise around the world as they fringe the world oceans and are low lying. The sea-level rise leads to accelerated erosion, shoreline retreat, saltwater intrusion into coastal groundwater, aquifers, inundation of wetlands and estuaries and creates threats to historical and cultural resources as well as infrastructure (Pendleton et al. 2004).

India has a coastline measuring upto 7515 km. Around 35% of Indians live within 100 km of the Indian coast (ISRO 2012). The Indian coast can broadly be divided into two parts: the west coast facing the Arabian Sea and the east coast facing the Bay of Bengal. The coastal areas are in the states of Gujarat, Maharashtra, Goa, Karnataka, Kerala, Tamil Nadu, Andhra Pradesh, Odisha, West Bengal and in the union territories of Daman & Diu, Lakshdweep Islands, Puducherry, Andaman and Nicobar Islands. The western coastal plains are submerged coastal plain; hence, it is a narrow belt and provides natural conditions for the development of ports and harbour. On the other hand, the Eastern coastal plains are emergent coastal plain and their continental shelf extends upto 500 km into the sea; hence, it is difficult to develop good natural harbours and ports in this region. This also makes them more prone to be affected by disasters like tsunami, storm surge and cyclones. The coastal plains of India are covered by fertile soils on which different crops like rice and coconut are grown. In view of its economic importance, the Indian coast has several big and small ports. Fishing is an important occupation in southern India, while low-lying areas of Gujarat are famous for producing salt. Kerala backwaters and Goa beaches are important tourist destinations.

5 Mumbai—A Historical Geography

The area known as Mumbai, then Bombay, consisted of a small archipelago in a harbour on the west coast of India. The number of distinct islands was open to interpretation since large areas were underwater at high tide and during the monsoon season (June to September), when water discharged from Thane and Panvel Creeks to raise the harbour's sea level. At other times, it was possible to cross between the islands on foot. This space contained a range of ecosystems: While some areas were entirely submerged, others consisted of mangrove forests, tidal flats and artificial salt pans. In 1661, Bombay was transferred by Portuguese to the English Crown in the treaty accompanying the marriage of Charles II and Catherine of Braganza. In 1668, the crown transferred the islands to the East India Company. The modern assumption is that seven islands existed: Colaba, Old Woman's Island, Bombay, Mazagaon, Parel, Worli and Mahim. The area today, after extensive reclamation, covers sixty-five square kilometres (Riding 2018) (Fig. 2).

Maharashtra State, barring a few centres, is as poorly developed as many industrially backward states in the country. This concentration is the result of historical forces. The British developed Bombay as the centre of their trade and also developed port in its natural harbour. By the middle of the nineteenth century, Bombay was the unrivalled trading centre in the whole India. Even traders from older centres, like Surat, migrated to Bombay in this period. As in the case of all cities developed to foster colonial trade, the transport network, specifically the railways, was developed with Bombay as the centre, and the city thrived at the expense of its hinterland. Agglomerative forces came into play leading to the growth of industries and diversified it in and around the city. This process continued unchecked till the 1960s, when the State Government intervened to limit further growth of industry in Bombay. Government policy, together with dispersing forces like soaring rents and factor prices, and labour trouble, led to the spill over of industry towards the suburbs at first and, later, to the nearest large urban centre of Pune.

6 Impact of SLR on Mumbai

As the population in developing countries continued growing and many people were living in high-risk, coastal locations, sea-level rise can have significant physical,

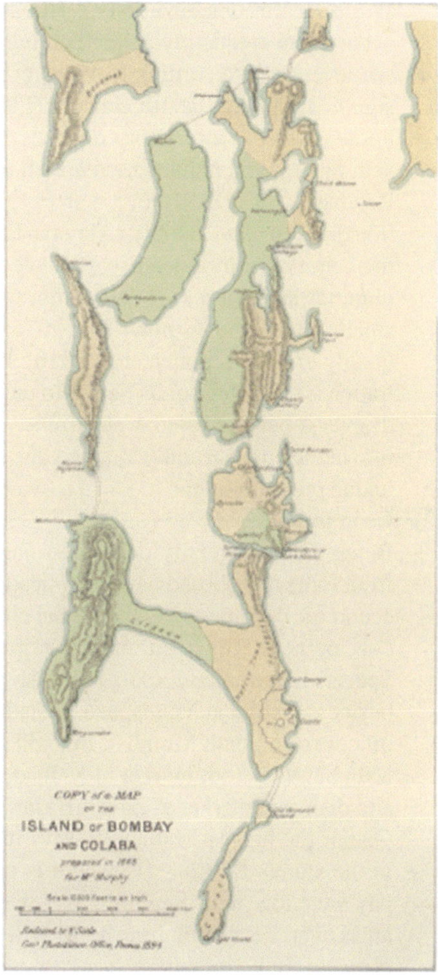

Fig. 2 Islands of Bombay. *Source* Riding (2018)

health, social, economic and environmental consequences on both individuals and communities (Alderman et al. 2012).

Mumbai is termed as the economic capital of India. It is home to major financial institutions such as Reserve Bank of India, the National Stock Exchange of India, Bombay Stock Exchange and also to many of the corporate headquarters and commercial establishments. As per Census (2011), the population of Mumbai is 1.2 million, a large section which still resides in slums. As most of the Bombay city is reclaimed, the water stagnation is a major problem during monsoons. During high tides, the sea water is many a time above the sea level, and hence the storm water drainage is not able to function to its full capacity leading to flood-like situation in the city.

(1) **Health**

World Health Organization (WHO 2009) reports that flooding is expected to occur more frequently in the coming decades because of climate change and sea-level rise. According to Watts et al. (2015), the sea-level rise represents a significant public health concern. Flooding and health issues, such as water-borne diseases; respiratory diseases, such as asthma; vector-borne infections; skin rashes; and malnutrition are all interlinked (Rose and Akpinar-Elci 2015; Reponen et al. 2010), along with psychological issues such as depression, emotional trauma and anxiety (Akpinar-Elci et al. 2018; Bei et al. 2013). Increased incidence of respiratory diseases have been associated with floods, as many communities and houses are left damp, which often results in moulds and mouldy smells (Rose and Akpinar-Elci 2015; Mendell et al. 2011) and a reduction in quality of life (Reponen et al. 2010). It is often observed that the local administration is criticized for its failure to predict as well as respond to flood situations, despite it being a regular occurrence during monsoons. The health infrastructure in the city is usually strained due to excessive population, such floods exert undue pressure on the medical services.

(2) **Economic**

Being a commercially developed megalopolis, many of the corporate houses from India and abroad have their headquarters in the city. Post liberalization the economic development of Mumbai city is supported by a massive infrastructure development. The creation of Bandra Kurla Complex (BKC), restructuring of Special Electronics Export Processing Zone (SEEPZ) at Andheri, creation of business district at Mindspace, Malad was channelled to decentralize corporate offices from South Mumbai. But still several trading zones are located in South Mumbai like cloth market at Kalbadevi, electronic market at Grant road, gold and diamond market at Zaveri Bazar, utensils and tools market at Loharchawl. One of the biggest leather industry in located at Dharavi as a cottage industry in the slums. Shifting elsewhere is not commercially viable to them, as they pay very low rent. Sea level rise will lead to permanent inundation of these areas over a very long time horizon, leaving billions of rupees worth investment obsolete. In a city where currently a large workforce is involved in secondary to quaternary economic activities, this incurs a great economic cost.

(3) **Displacement**

Coastal communities are more densely populated and rapidly growing in population (Neumann et al. 2015). Koli community, which is one of the original inhabitants of Bombay Islands, live in a settlement called "koliwada" near the sea. Versova, Madh, Marve, Colaba, Trombay and Worli are some of the larger koliwadas in Mumbai. Fishing is a major source of livelihood among the members of this community. As their boats are on the beach, the settlements have formed right next to the sea and often during high tide the waves crash on the walls of the dwellings. Over the years, regular flooding during monsoon has started to occur in the koliwadas, as storm water from land and sea water merge in these areas. As these koliwadas are right next to the sea, they fall under Coastal Regulatory Zone (CRZ) due to which construction or repair becomes difficult. This has lead to several unauthorised constructions mushrooming in the koliwadas. But as the sea level rises, the settlements are shrinking in size and leading to relocation of the population (Kotak 2018).

(4) **Infrastructure**

To create multi-modal transport linkages construction of multiple flyovers, express ways, metro & mono railways, skywalks have been carried out in Mumbai as support to the economic boom post liberalization. East-West suburbs connecting link roads like Jogeshwari-Vikroli Link Road (JVLR), Ghatkopar-Mankhurd Link Road (GMLR), Santacruz-Chembur Link Road (SCLR), under construction Mulund-Goregaon Link Road (MGLR) and Sion-Bandra Link Road (SBLR) have been built over the years at a great cost and new residential areas have mushroomed along these link roads. As sea level rise has an impact which will only be visible in the long term, no specific measures are in place to take care of the same. Over the years, Municipal Corporation of Greater Mumbai has installed five pumping stations are Irla, Haji Ali, Cleveland, Lovegrove and Britannia (Reay Road). The work on Gazdarbadh, Mahul and Mogra pumping stations are at various stages of completion (Pinto 2017). But these are only effective in short-term disruptions like flooding in monsoon and are not sufficient to mitigate sea level rise. Further, the government has planned coastal road on the entire west coast from Colaba (South Mumbai) to Virar (Extended Western Suburb) and statue construction in the sea, which will require reclamation of land thereby advancing sea level rise in other parts of the city. Mumbai's southern part was once habited by colonial officials to a great extent which led to construction of several heritage structures and art deco style building viz. the Royal Opera House, Asiatic Building, Horniman Circle and the famed Gateway of India. Sea level rise can lead to irreparable damage to this glorious historical inheritance.

Sea-level rise also impacts hazardous waste management. Sea-level rise would place waste management sites in the floodplain, exacerbating the impacts of a storm or landfall tropical storm. Exposure to toxic waste can increase the risks of toxicity of the liver and kidney as well as cause cancers (EPA 2018).

(5) **Biodiversity**

Seventy percent of the world's wetlands are in coastal regions which include salt swamplands, mangrove woodlands and sludge flats, which are highly vulnerable and very susceptible to changes in the environment (Spencer et al. 2016). A model was developed Spencer et al. (2016) by to assess the biophysical and socio-economic impacts of sea-level rise and wetland environments. The study estimated that if the sea level rose 50 cm up to 59 per cent of all coastal wetlands would be inundated, while in case of a higher sea-level rise scenario (110 cm), the statistic would increase to 78 per cent. Chowdhury and Behera (2015) has studied the changes in mean sea level with the help of tidal gauge stations located at Mumbai, Kandla, Cochin and Hiron Point (Bangladesh). The Mumbai station has a data for over 100 years, which has provided some significant findings. Figure 3 clearly shows that over the past century the average sea level rise has been more than 100 mm at Mumbai station, which is also consistent with the global estimate of 1–2 mm/ year (IPCC 2013).

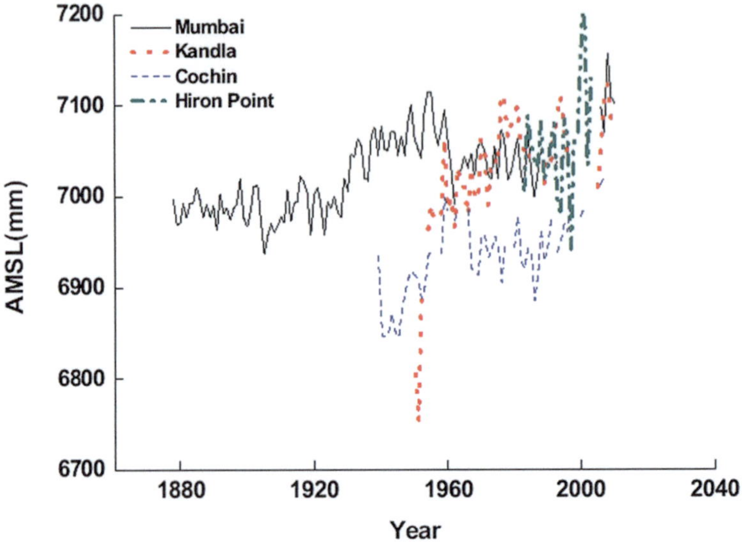

Fig. 3 Average Mean Sea Level (AMSL in mm) at master stations

Mumbai, due to its peculiar peninsular shape, has mangroves on all the sides. These mangroves prevent erosion of soil and inundation by sea water by reducing wave action. They also absorb carbon dioxide. However, in recent years due to reasons like unauthorised reclamation of land, high salinity of water, blockage of intertidal region, coastal pollution, the mangrove cover is depleting (Chatterjee 2018).

7 Adaptation and Mitigation

Sea-level rise (SLR) is a a by-product of Climate Change on a global scale. It is a reality which has to be accepted by the world and adaptive measures need to be placed in this regard. Due to lack of space in urban areas, the damage of SLR will be to a greater extent in urban area, while in rural areas shifting of settlement to higher grounds will be required. As land availability is comparatively more in rural areas, the land inundated by sea water can be used for aquaculture, desalination of water, salt pans.

In terms of mitigation, building of sea walls or barriers is a practice that has been done for a long time for the both areas, below and above the sea level. In Mumbai, such examples of creating barriers with the help of tetrapods can be seen at Colaba, Nariman Point and Marine Drive. But these are rather expensive and may not be economically feasible option for a developing and underdeveloped countries (TNN 2001). The sea walls can be constructed in height for the immediate future and

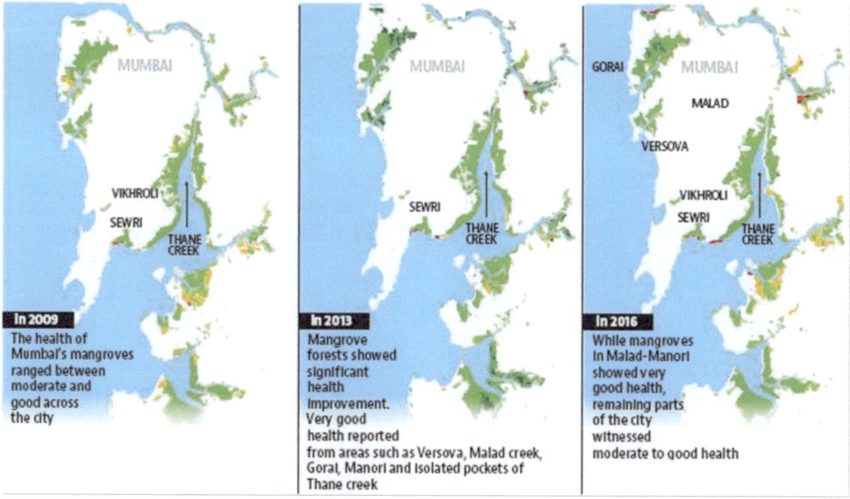

Fig. 4 Changes in Mangrove cover *Source* Chatterjee (2018)

designed to accommodate future increases in height. Dredging, creating off-shore barrier islands and reshaping ocean floors are some of the measures which can also be considered. Extending the wetlands by increasing the mangrove cover is also a good way to mitigate the SLR as mangroves can prevent further encroachment by the sea and at the same time keep alive the biodiversity in the area (Wong 2015).

Physical and digital model simulation and analysis should be carried out in order to study the segment of the ocean basin adjacent to the low-lying coastal area threatened by sea-level rise. An initial mapping of the regional seabed with historical data regarding currents and tides should be carried out with the objective of designing and engineering a hydrological system that would allow an increased volume of flow through the shallow water region, while limiting sea-level rise in the low-lying coastal area. Oceanographers, marine scientists, engineers and regional planners can collectively engage in limiting the adverse effects of sea-level rise (Wong 2015). With the technological advancement of the current world, tools like geographical information system and remote sensing can be used to generate scientific models and data, respectively. E.g. the figures (Figs. 4 and 5) depict the Landsat remote sensing data regarding changing mangrove cover at Thane creek, which is part of the east coast of Mumbai.

8 Conclusion

Research suggests that continued sea-level rise in the coming century is very likely, despite attempts to reduce greenhouse gas (GHG) emissions through agreements such as the Kyoto Protocol or more recently the Paris Climate Accord (Levermann

Fig. 5 Changes in Mangrove cover at Thane creek 1988 and 2017 *Source* NASA (2018)

et al. 2013). The concerns and impacts of climate change and sea-level rise on coastal communities around the world vary to a great extent (Schulte et al. 2015). Geisler and Currens (2017) estimated that 1.4 billion people could be displaced globally as a result of climate change by 2060.

Global commerce and trade rely heavily on historic marine transportation routes and port cities. In the era of globalization, sea-level rise becomes a significant concern at individual, local government as well as to global economy levels. Hence, both worldwide collaboration and localized strategies are very much required to address the challenges of sea-level rise. Following are some of the measures which could be taken into account for the same:

- The rate of sea-level rise may be slowed down by reducing the concentration of heat-trapping, greenhouse gases (GHG) in the atmosphere.
- Retrofitting or reconstruction of infrastructural facilities like roadways, drainage systems, homes and businesses may need to be carried out to reduce the impacts of sea-level rise.
- Successful implementation of counter-measures to sea-level rise will require cooperation, trusting partnerships and locally focused adaptation strategies like investment in green infrastructure and nature-based solutions, revising zoning strategies, construction of dikes and sea walls, improved drainage infrastructure, and inward migration.
- Capacity building of coastal communities and reducing the impacts of sea-level rise on global level will also require international research and worldwide organizational support.

Acknowledgements Thanks are due to Dr. Hemantkumar Pednekar, Principal, ASC College, Onde, Palghar for his critical inputs on the draft paper.

References

Akpinar-Elci, M., Rose, S., & Kekeh, M. (2018). Well-being and mental health impact of household flooding in Guyana, the Caribbean. *Marine Technology Society Journal, 52*(2), 18–22.

Alderman, K., Turner, L. R., & Tong, S. (2012). Floods and human health: A systematic review. *Environment International, 15*(47), 37–47.

Bei, B., Bryanta, C., Gilsona, K., Koh, J., Gibsona, P., Komiti, A. et al. (2013). A prospective study of the impact of floods on the mental and physical health of older adults. *Aging and Mental Health, 17*(8), 992–1002.

Cazenave, A., et al. (2014). The rate of sea-level rise. *Nature Climate Change, 4*(5), 358–361.

Census. (2011). Government of India, New Delhi.

Chatterjee, B. (2018). In a first, a health update on mangroves' in The Hindustan Times Nov. 18, 2018 cited Indian Institute of Space Sciences and Technology report on Mangroves 2009–2016.

Chowdhury, P., & Behera, M. R. (2015). A study on regional sea level variation along the Indian coast. *Procedia Engineering, 116*, 1078–1084.

Church, J. A., Clark, P. U., Cazenave, A., Gregory, J. M., Jevrejeva, S., Levermann, A., ... & Payne, A. J. (2013). Sea level change. Climate change 2013: the physical science basis. *Contribution of working group I to the fifth assessment report of the intergovernmental panel on climate change* (pp. 1137–1216). Cambridge, United Kingdom and New York, NY, USA: Cambridge University Press.

Eggleston, J., & Pope, J. (2013). Land subsidence and relative sea-level rise in the southern Chesapeake Bay region (No. 1392). US Geological Survey. Retrieved from https://pubs.usgs.gov/circ/1392/pdf/circ1392.pdf. Accessed November 10, 2018.

Environmental Protection Agency. (2018). *Potential well water contaminants and their impacts.* Retrieved from http://www.epa.gov/privatewells/potential-well-water-contaminants-and-their-impacts. Accessed November 7, 2018.

Fasullo, J. T., Nerem, R. S., & Hamlington, B. (2016). Is the detection of accelerated sea level rise imminent? *Scientific Reports, 6*(August), 1–7.

Geisler, C., & Currens, B. (2017). Impediments to inland resettlement under conditions of accelerated sea level rise. *Land Use Policy, 66,* 322–330.

Intergovernmental Panel on Climate Change. (2007). Climate change 2007: Impacts, adaptation and vulnerability. In M. L. Parry, O. F. Canziani, J. P. Palutikof, P. J. van der Linden & C. E. Hanson (Eds.), *Contribution of working group II to the fourth assessment report of the intergovernmental panel on climate change*. Cambridge, UK: Cambridge University Press, 967 pp.

Intergovernmental Panel on Climate Change. (2013). Climate change: The physical science basis. Contribution of working group I to the fifth assessment report of the intergovernmental panel on climate change. Cambridge University Press, Cambridge, United Kingdom and New York, NY, USA. https://doi.org/10.1017/cbo9781107415324.

Intergovernmental Panel on Climate Change. (2018). Summary for policymakers. In: Global warming of 1.5 °C. An IPCC special report on the impacts of global warming of 1.5 °C above pre-industrial levels and related global greenhouse gas emission pathways, in the context of strengthening the global response to the threat of climate change, sustainable development, and efforts to eradicate poverty [V. Masson—Delmotte, P. Zhai, H. O. Pörtner, D. Roberts, J. Skea, P. R. Shukla, A. Pirani, W. Moufouma—Okia, C. Péan, R. Pidcock, S. Connors, J. B. R. Matthews, Y. Chen, X. Zhou, M. I. Gomis, E. Lonnoy, T. Maycock, M. Tignor, T. Waterfield (eds.)] (in press).

ISRO. (2012). *Coastal zones of India.* Ahmedabad: Space Application Centre.

Khan, A. A. (2018). Why would sea-level rise for global warming and polar ice-melt? *Geoscience Frontiers,* xxx 1–15.

Kotak, Y. (2018). 'Koliwadas not marked in Mumbai DP as government wants our land, allege fisherfolk'. Retrieved from https://www.hindustantimes.com/mumbai-news/koliwadas-not-marked-in-mumbai-dp-as-government-wants-our-land-allege-fisherfolk/story-4JXFgJGYuvsC5GkmKRdFtO.html. Accessed November 12, 2018.

Levermann, A., Church, J. A., Clark, P. U., Cazenave, A., Gregory, J. M., Jevrejeva, S., et al. (2013). Sea-level rise by 2100. *Science, 342*(6165), 1445.

Mendell, M. J., Mirer, A. G., Cheung, K., Tong, M., & Douwes, J. (2011). Respiratory and allergic health effects of dampness, mold, and dampness-related agents: a review of the epidemiologic evidence. *EHP., 119*(6), 748–756.

Mimura, N. (2013). Sea-level rise caused by climate change and its implications for society. *Proceedings of the Japan Academy. Series B, Physical and Biological Sciences*, 89(7), 281–301. https://doi.org/10.2183/pjab.89.281.

National Aeronautics and Space Administration. (2018). Monitoring Mumbai's Mangroves. Retrieved from https://earthobservatory.nasa.gov/images/91333/monitoring-mumbais-mangroves. Accessed December 5, 2018.

National Climate Assessment. (2014). Climate change impacts in the United States. Full report. Retrieved from https://nca2014.globalchange.gov/. Accessed November 10, 2018.

National Oceanic and Atmospheric Administration. (2015). Climate change: Ocean heat content. Retrieved from https://www.climate.gov/news-features/understanding-climate/climate-change-ocean-heat-content. Accessed November 10, 2018.

National Oceanic and Atmospheric Administration. (2016). Extended reconstructed sea surface temperature (ERSST.v4). National Centers for Environmental Information. https://www.ncdc.noaa.gov/data-access/marineocean-data/extended-reconstructed-sea-surface-temperature-ersst. Accessed November 10, 2018.

National Oceanic and Atmospheric Administration. (2017a). Climate change: Global sea level. Retrieved from https://www.climate.gov/news-features/understanding-climate/climate-change-global-sea-level. Accessed November 10, 2018.

National Oceanic and Atmospheric Administration. (2017b). How is sea level rise related to climate change? Retrieved from https://oceanservice.noaa.gov/facts/sealevelclimate.html. Accessed November 10, 2018.

National Oceanic and Atmospheric Administration. (2017c). What is the difference between local sea level and global sea level? Retrieved from https://oceanservice.noaa.gov/facts/sealevel-global-local.html. Accessed November 10, 2018.

National Oceanic and Atmospheric Administration. (2017d). Examining sea level rise exposure for future populations. Retrieved from https://coast.noaa.gov/digitalcoast/stories/population-risk.html. Accessed November 10, 2018.

National Snow and Ice Data Center. (2018). Quick facts on ice sheets. Retrieved from https://nsidc.org/cryosphere/quickfacts/icesheets.html. Accessed November 10, 2018.

Neumann, B., Vafeidis, A. T., Zimmermann, J., & Nicholls, R. J. (2015). Future coastal population growth and exposure to sea-level rise and coastal flooding-a global assessment. *PloS one, 10*(3), e0118571.

Parris, A., Bromirski, P., Burkett, V., Cayan, D., Culver, M., J. Hall, J. … Weiss, J. (2012). Global sea level rise scenarios for the US National Climate Assessment. NOAA Tech Memo OAR CPO-1. 37pp.

Pendleton, E. A., Thieler, E. R., & Williams, S. J. (2004). Coastal vulnerability assessment of Cape Hettaras National Seashore (CAHA) to sea level rise. USGS Open File Report 2004-1064.

Pinto, R. (2017). Mumbai has 400cr pumping stations, but still uses pumps'. Retrieved from https://timesofindia.indiatimes.com/city/mumbai/mumbai-has-rs-400cr-pumping-stations-but-still-uses-pumps/articleshow/59062324.cms. Accessed November 15, 2018.

Rao, K. N., Subraelu, P., Rao, T. V., Malini, B. H., Ratheesh, R., Bhattacharya, S., & Rajawat, A. S. (2008). Sea-level rise and coastal vulnerability: An assessment of Andhra Pradesh coast, India through remote sensing and GIS. *Journal of Coastal Conservation, 12*(4), 195–207.

Reponen, T., Singh, U., Schaffer, C., Vesper, S., Johansson, E., Adhikari, A., et al. (2010). Visually observed mold and moldy odor versus quantitatively measured microbial exposure in homes. *Science of the Total Environment, 408*(22), 5565–5574.

Riding, Tim. (2018). 'Making Bombay Island': Land reclamation and geographical conceptions of Bombay, 1661–1728. *Published in Journal of Historical Geography, 59,* 27–39.

Rose, S., & Akpinar-Elci, M. (2015). An evaluation of the prevalence of respiratory and general symptoms among occupants of households affected by flooding in Guyana: The case of Cove & John. JARE.

Schulte, D. M., Dridge, K. M., & Hudgins, M. H. (2015). Climate change and the evolution and fate of the Tangier Islands of Chesapeake Bay, USA. *Scientific reports, 5,* 17890.

Slangen, A. B. A., et al. (2016). Anthropogenic forcing dominates global mean sea-level rise since 1970. *Nature Climate Change, 6*(7), 701–705.

Spencer, T., Schuerch, M., Nicholls, R. J., Hinkel, J., Lincke, D., Vafeidid, A. T., & Brown, S. (2016). Global coastal wetland change under sea-level rise and related stresses: The DIVA wetland change model.' Global and planetary change. https://doi.org/10.1016/j.gloplacha.2015.12.018; https://www.sciencedirect.com/science/article/pii/S0921818115301879.

Sweet, W. V., Kopp, R. E., Weaver, C. P., Obeysekera, J., Horton, R. M., Thieler, E. R., & Zervas, C. (2017). Global and Regional sea level rise scenarios for the United States. NOAA Technical Report NOS CO-OPS 083. NOAA/NOS Center for Operational Oceanographic Products and Services.

Times News Network. (2001, September 3). Tetrapods too expensive, says BMC. Retrived from https://timesofindia.indiatimes.com/bombay-times/Tetrapods-too-expensive-says-BMC/articleshow/1238484810.cms. Accessed November 7, 2018.

UN ATLAS OF THE OCEANS. (2016). Human settlements on the coast. Retrieved from http://www.oceansatlas.org/subtopic/en/c/114/. Accessed November 7, 2018.

U.S. Geological Survey. (2017). Louisiana's rate of coastal wetland loss continues to slow. Retrieved from https://www.usgs.gov/news/usgs-louisiana-s-rate-coastal-wetland-loss-continues-slow. Accessed October 18, 2018.

Watson, C. S., et al. (2015). Unabated global mean sea-level rise over the satellite altimeter era. *Nature Climate Change, 5*(6), 565–568.

Watts, N., Adger, W. N., Agnolucci, P., Blackstock, J., Byass, P., Cai, W., … Cox, P. M. (2015). Health and climate change: policy responses to protect public health. *The Lancet, 386*(10006), 1861–1914.

Wong, K. V. (2015). Mitigation and adaptation responses to sea level rise. *The Open Hydrology Journal, 9*(Suppl-1 M4), 24–27.

World Health Organization. (2009). *Protecting health from climate change, connecting science, policy and people* (p. 28). Geneva: World Health Organization.

Sea Level Rise and Coastal Communities

Michele Kekeh, Muge Akpinar-Elci and Michael J. Allen

Abstract Globally, sea level is rising and it represents a serious concern. The trends are indisputable, and the predictions indicate that sea level will continue to rise for the coming centuries. Sea level rise is a consequence of human-induced climate change and has significant implications for low-lying coastal communities around the world. The impacts are felt with increased coastal flooding during storms and higher frequency of tidal flooding. In many countries, local drainage infrastructures were not designed to consider the mean of sea level rise observed in the past couple of years. Consequently, outlet pipes are often inundated and covered at high tide, reducing the efficiency of drainage in the event of heavy rain. With continued population growth and many people living in high-risk, coastal locations, the potential exists for significant physical, health, social, economic, and environmental consequences on both individuals and communities especially in developing countries.

Keywords Sea level rise · Coastal communities · Health outcomes · Flooding · Monitoring · Adaptation

Globally, sea level rise is a serious concern. It is a consequence of human-induced climate change and has significant implications for low-lying coastal communities. With mean sea level projected to increase in the coming decades (Allen et al. 2018), many coastal communities will experience increased probability of extreme events such as heavy rainfall and tropical storms. These hazards may also interact with geophysical processes and modify erosion rates or landslide frequency. Thus, addressing climate change and sea level rise is integral to the success of countries' sustainable development plans. Food security, access to clean water, sustainable innovation, economic security, and preservation of marine ecosystems are all relevant in the context of a changing climate and sustainability (IPCC 2014a, b, c). It is crucial for policymakers and citizens, especially those in coastal locations, to understand the causes

M. Kekeh (✉) · M. Akpinar-Elci
Center for Global Health, Old Dominion University, Health Sciences Bldg, Room 3132, Norfolk, VA, USA
e-mail: mkekeh@odu.edu

M. J. Allen
Department of Political Science & Geography, Old Dominion University, Norfolk, VA, USA

R. Akhtar (ed.), *Extreme Weather Events and Human Health*,
https://doi.org/10.1007/978-3-030-23773-8_12

and long-term impacts of sea level rise. This chapter provides an examination of the regions that are currently experiencing sea level rise, the threats it poses, and highlights some strategies that coastal communities are developing to improve resilience to the impacts of sea level rise.

1 Regional Causes of Sea Level Rise

Global mean sea level is computed over the entire ocean, while local sea level refers to the relative height of the sea at a specific location (Mimura 2013). According to the National Climate Assessment, global sea level has increased by about 8 inches (203.2 mm) since 1880 (NCA 2014). While this increase represents a global average, local factors play a significant role in regional observations. According to the National Oceanic and Atmospheric Administration (NOAA), some coastal communities experience a greater increase in sea level compared to the global average (NOAA 2017a, b, c, d). In the USA, Sweet et al. (2017) highlight regional variability in sea level trends and found that parts of Alaska have experienced negative trends or land uplifts exceeding 10 mm/year. Norfolk, Virginia, has experienced an increase of 1.51 ft (0.47 m) since 1927, nearly twice the global rate. Some of this increase is attributed to complicated geologic processes such as isostatic rebound and groundwater pumping (Eggleston and Pope 2013).

There are two significant causes of global sea level rise: thermal expansion and the melting of land-based ice (Mimura 2013). Since the beginning of the twentieth century, global sea surface temperatures have increased at an average rate of 0.13 °F (−17.70 °C) per decade (1901–2015; NOAA 2016). Carbon dioxide and other greenhouse gases have enhanced the greenhouse effect and led to a rapid increase in global land and sea surface temperature over the past 150 years. While climate change has happened in the past, the rate of this change is unprecedented and not explainable without considering human contribution as a major factor (NCA 2014). The release of heat-trapping, greenhouse gases through anthropogenic sources has warmed both the atmosphere and the oceans; NOAA estimates that 90% of this energy has been stored in the ocean (NOAA 2015).

As water warms, it expands thus requiring more space (NOAA 2017a, b, c, d). This thermal expansion helps explain some of the increase in global sea level—water expanding due to warming. The second main cause of global sea level rise is the melting of land-based ice. As global temperatures increase, the decay of land-based glaciers contributes to increased water volume found in the oceans. Much of the climate change research focuses on the ice sheets found on Greenland and Antarctica. Together, these ice sheets contain more than 99% of the freshwater ice on Earth and, if fully melted, would contribute to dozens of meters in sea level rise (NSIDC 2018). Currently, most of sea level rise can be attributed to thermal expansion and melting of glaciers rather than the decay of these large ice sheets.

While thermal expansion and the decay of glaciers contribute to sea level rise, tropical storms, prolonged rain, fluctuations in ocean currents, and moon phases also play an important role in the spatial–temporal variability in the height of the ocean (Church et al. 2013). Understanding the coupled ocean-atmosphere factors that influence global sea level are important, but so too are the human systems. Land-use policy plays an important role in the development along the coast and, in some cases, may enhance risk to sea level rise and coastal flood events.

Due to improvements in the modeling of land-ice contribution to sea level rise, many projections suggest that it is very likely that the rate of global mean sea level rise observed in the twenty-first century will be much higher compared to that seen during the period from 1971 to 2010 (Church et al. 2013). While global average sea level has increased by an average rate of 3.4 mm per year over this time, studies suggest this rate will increase in the coming decades (IPCC 2013; Parris et al. 2012; Sweet et al. 2017). A recent study was the first to detect an acceleration and suggested that sea level rise could more than double the amount compared to the static rate of 3 mm/year (Nerem et al. 2018). The processes governing the recent trends in sea level rise could be different later in the century. As melting ice sheets become the primary contributor to global sea level, the potential exists for significant sea level rise not experienced in millennia.

Despite attempts to reduce greenhouse gas (GHG) emissions through agreements such as the Kyoto Protocol or more recently the Paris Climate Accordingly, research suggests that continued sea level rise in the coming century is very likely (Levermann et al. 2013). The unmitigated release of heat-trapping gases will have significant ramifications on human health, economic viability, and ecological habitats. In planning for the future, policymakers should consider the evidence of climate change and sea level rise. The future of low-lying, coastal communities may hinge on our ability to mitigate the causes and adapt to the consequences of sea level rise.

2 Sea Level Rise and Coastal Communities: An Overview of Selected Cases

Sea level rise will impact coastal communities around the world. According to the UN Atlas of the Oceans, eight of the ten most populous cities are in coastal regions (UN Atlas of the Oceans 2016). These cities include Tokyo, Mumbai, New York, Shanghai, Lagos, Los Angeles, Calcutta, and Buenos Aires. Of these cities, only Mexico City and Sao Paulo are not considered coastal. In the USA, 123.3 million people or 39% of the entire population live in a coastal shoreline county, which are often impacted by flooding, erosion, and hurricanes (NOAA 2013).

Rising sea levels pose severe problems to the city of New York. At least half of the city's population lives in coastal regions. The level of the sea in New York City has increased by 1.1 ft (0.335 m) since 1900, which is almost twice the observed global rate during a similar period (NASA 2015). By 2100, it is projected that New York's sea level will rise an additional 18–50 in. (5.4864–15.24 m), and this trend is expected to continue for centuries (DEC 2019). The rate of this rise is increasing and will threaten coastal communities and resources with more frequent and severe flooding and storms, which will damage communities, infrastructure, and ecosystems. This new trend was illustrated in 2012, when Hurricane Sandy put part of New York City under 9 ft (2.743 m) of water and caused the deaths of dozens of people (Rice 2017). The peak water level recorded surpassed the highest level documented by the Federal Emergency Management Agency (FEMA) since 1991. The storm caused severe building damage to over 70,000 housing units, placed an unprecedented strain on the healthcare system, and disrupted communication systems (ACEP 2015). The total cost of Sandy was estimated at $70.2 billion (ACEP 2015). The anticipated steady rise of the sea will only continue to erode beaches and inundate low-lying lands over time. New York City may very likely be confronted with compromises to or even collapses of low-lying sewage systems, transportation, and wastewater management. The sea level rise will likely cause catastrophic flooding in the coming decades.

A rising sea level is putting the entire state of Louisiana at risk of flooding because the land is sinking. Currently, there are approximately 955,000 people at risk of coastal flooding in the state (The New York Times 2016). It is predicted that an additional 262,000 more people inland would be at risk due to the rising of the sea. The level of the sea has risen 24 inches (0.609 m) since 1950; however, for the past decade, the speed of this rise has been increasing by 1 in. (0.025 m) every two years. Louisiana has lost 4,900 km^2 since 1932 and could possibly lose an additional 4,600 km^2 due to severe weather and storm surge (The City of New Orleans 2017). In Louisiana, New Orleans is at higher risk because the city is the largest population center in the state and is located below sea level (Campanella 2018). The city has invested in flood defenses along the coast due to recent events. In 2005, Hurricane Katrina flooded over 80% of New Orleans, filling neighborhoods with up to 12 ft (3.657 m) of seawater, killing 1,833 people, and leaving millions more homeless (Zimmermann 2015). An increase in sea level of just 5 feet will cause 88% of the city to flood (The New York Times 2016), which has led city leaders to focus on need-based approaches to flood management.

Second only to New Orleans, Norfolk stands as the second most vulnerable to sea level rise in the USA. The city is situated at the mouth of the Chesapeake Bay and is penetrated by various creeks and rivers. With 144 miles (232 km) of coastline, the port city has thrived in an age of trade and commerce and currently serves as the home to the world's largest naval bases. Before 1971, the city averaged 20 h of minor flooding, but since that time, a ninefold increase in flood hours has been observed (Atkinson et al. 2012). Local drainage infrastructure was also not designed to consider the 1.51 ft (46.02 cm) of sea level rise over the last 90 years. Consequently, outlet pipes are often inundated and covered at high tide, reducing the efficiency of drainage in the event of heavy rain. Flooding cancels school, disrupts transportation infrastructure, and erodes sand dunes designed to reduce the impact of storm surge. The U. S. Department of Defense recently called climate change and sea level rise a threat multiplier (Banusiewicz 2014), drawing attention to the issue in the context of global geopolitics and national security.

In Florida, although the sea level only rose by 8 inches on average (0.20 m) from 1950 to 2017 (NOAA 2018a, b), this increase has cost the state about four billion dollars so far. The speed at which the sea level rises will accelerate over the next 10 years, increasing to 1 in. (0.025 m) every 3 years. The four most vulnerable counties in Florida are Broward, Miami-Dade, Palm Beach, and Monroe due to the low-lying locations. In many of these locations, people who are most susceptible to sea level rise have a high income level. The wealthiest tend to live more closely to the coast compared to the poor. Bloetscher et al. (2016) suggested that areas vulnerable to sea level rise may only increase over time and highlighted the necessity to define incremental strategies and planning actions to reduce socioeconomic impacts and provide better monitoring and reporting of diseases in the region. The challenge with Florida is that many traditional mitigation strategies such as building walls or barriers will not apply to the area due to the porous nature of the ground, which allows saltwater to seep into the drinking water reserve and may compromise sewage plants. In addition, at least 120,000 properties are vulnerable due to frequent flooding. Hauer (2017) speculated that Florida may lose an estimated 2.5 million of its population due to just 1.8 m of sea level rise.

Throughout USA, rising seas pose a significant threat to coastal infrastructure. For example, a report from the California Department of Boating and Waterways highlighted the 1,100 miles of California coastline and estuarine shoreline vulnerable to sea level rise (King et al. 2010). The impacts are already being felt with increased coastal flooding during storms, higher frequency of tidal flooding (e.g., nuisance flooding), and increased erosion. With continued population increase and development which does not often fully consider the future risk of sea level rise and climate change, 480,000 people and $100 billion in property stand to be at risk of flooding by 2100 (PHI 2016).

In the USA, the impacts of sea level rise are well studied (Brown et al. 2011). However, with continued population growth and many people living in high-risk, coastal locations, the potential exists for significant physical, health, social, economic, and environmental consequences on both individuals and communities especially in coastal areas countries (Alderman et al. 2012).

3 Sea Level Rise and Health

Sea level rise represents a significant public health concern (Watts et al. 2015). Globally, low-income communities are often most vulnerable to the impacts of hazards as recovery is dependent upon coping capacity, institutional power relations, and availability of resources (Rufat et al. 2015). In the USA, low-cost flood insurance that does not fully consider the rising sea incentivizes risky land development along the coast (Cleetus 2013). Land-use policy often does not adequately account for long-term risk but rather focuses on short-term economic benefit.

With projected changes to sea level, the added volume of water will increase the probability of severe flooding. Compound hazards associated with sea level rise can put community members at risk of water-borne diseases, reduce food security, increase stress-related disorders, and contribute to massive displacement of people. According to the NOAA, flooding is expected to occur more frequently in the coming decades because of climate change and sea level rise (NOAA 2019). Research has linked flooding and health issues, such as water-borne diseases; respiratory diseases, such as asthma; vector-borne infections; skin rashes; malnutrition (Rose and Akpinar-Elci 2015; Reponen et al. 2010); and psychological issues such as depression, emotional trauma, and anxiety (Akpinar-Elci et al. 2018; Bei et al. 2013). After floods, many communities and houses are left damp, which often results in molds and moldy smells. These have been associated with increased incidence of respiratory diseases (Rose and Akpinar-Elci 2015; Mendell et al. 2011) and a reduction in quality of life (Reponen et al. 2010). Furthermore, as sea level rises and intrudes on freshwater aquifers, populations dependent on these reservoirs will experience an increased risk for hypertension and diarrheal diseases (Vineis et al. 2011). Sea level rise also impacts hazardous waste management. Sea level rise would place waste management sites in the floodplain, exacerbating the impacts of a storm or landfall tropical storm. Exposure to toxic waste can increase the risks of toxicity of the liver and kidney as well as cause cancers (EPA 2018).

Sea level rise plays a role in the displacement of people. Following Hurricane Katrina, 800,000 Louisiana residents were displaced (Kent 2006). While Hurricane Katrina was not caused by climate change, the probability of extreme events increases with a warming climate. With higher waters, storms may inundate more inland locations not historically prone to flooding. Additionally, mass migrations from low-lying island communities may result. For example, the island of Salomon has experienced some of the highest rates of sea level rise since 1993—an increase of more than 8 mm per year (Australian Bureau of Meteorology and CSIRO 2011). This has led to an increase in the frequency and intensity of flooding in the villages, forcing people to relocate to other islands. The displacement of individuals has led to disharmony in communities and between families. In assessing the perceptions of residents, Asugeni et al. (2015) found that many people were fearful of the future with respect to rising sea level and flooding. Citizens were worried about food security, the inability to mitigate saltwater intrusion, and the uncertainty of where to live in the future (Asugeni et al. 2015). Salomon is not alone. As coastal communities around the world cope with the impacts of climate change and sea level rise, some are faced with migration concerns (Schulte et al. 2015). Globally, Geisler and Currens (2017) estimated that by 2060, 1.4 billion people could be displaced as a result of climate change. The potential displacement of people is one reason the U.S. Department of Defense considers climate change a threat multiplier (Banusiewicz 2014).

4 Impacts on Coastal Wetlands and Biodiversity

In addition to the health implications of sea level rise, reduced biodiversity is also of concern. In coastal regions, the changes in sea level combined with development and agricultural production are the most critical long-term causes of the changes in biodiversity. With future projections in sea level rise, ecosystems of low-lying islands are at high risk of being submerged, leading to significant biodiversity loss. Species rely on such environments for habitat, food, and migration routes. In total, 70% of the world's wetlands are in coastal regions and include salt swamplands, mangrove woodlands, and sludge flats. Salt marsh wetlands are highly vulnerable and very susceptible to changes in the environment (Spencer 2018).

Spencer et al. (2016) developed a model to assess the biophysical and socioeconomic impacts of sea level rise and wetland environments. The study estimated that up to 59% of all coastal wetlands would be inundated if the sea level rose 50 cm. Under a higher sea level rise scenario (110 cm), the statistic would increase to 78%. Since 1932, Louisiana has lost 5,197 km^2 of land (USGS 2017). While the changes in the landscape are attributed to both climate and non-climate factors, the reduction has compounding impacts on local fisheries and critical infrastructure along the coast. With exponential increases to sea level projected, the landscape changes will also likely intensify.

Additionally, changing the acidity, temperature, or overall composition of water threatens estuarine and coral environments which are highly diverse and important for local identity, food sources, and tourism.

5 Socioeconomic Impacts of Sea Level Rise

Compared to more inland locations, coastal communities are more densely populated and rapidly growing in population (Neumann et al. 2015). Additionally, more people are living in and migrating to urban locations than ever before (UN 2014). The ability for cities to maintain healthy environments requires a proactive approach in planning for climate change and sea level rise impacts. The issue remains also synergies with sustainable development goals.

Individuals live near the coast for a variety of reasons. Marine and estuary environments provide valuable resources for seafood production, and global commerce relies upon transportation routes and historic port cities. Sea level rise projections pose significant costs to individuals, local municipalities, and the interconnected, global economy. One study estimated the cost of storm surge and sea level rise to exceed $990 billion for the USA (Neumann et al. 2015). Infrastructure including roadways, drainage systems, homes, and businesses may need retrofitting or reconstructed to be resilient to the impacts of sea level rise. Some locations may not have the capacity to address such a financial burden. Reinforcing the shoreline—through green or concrete structures—comes at a large economic cost, depending on the type of protection. Often, the price tag is beyond the means of many developing countries. In addition, the cost of insuring and inspecting these structures can be unsustainable. For example, the Caribbean region suffered significant damage in the 1980s and 1990s. As a result, many insurance companies withdrew from providing coverage in the region, which led to the remaining companies raising their prices and incorporating higher deductibles (IPCC 2013).

As populations migrate to urban, coastal locations, increased demand for housing adds stress to development. The increase in the frequency or intensity of flood-related events could adversely affect the built environment of many coastal communities, especially in countries with limited capacity.

6 Adaptation and Mitigation

Many strategies exist designed to reduce risk associated with the impact of sea level rise. These strategies include mitigating the causes (e.g., reducing greenhouse gases (GHG) emissions) and/or adapting to the consequences. This section provides a snapshot of possible solutions to addressing risk and reducing vulnerability related to sea level rise.

Multiple opportunities exist in reducing GHG from diet modifications, improving the efficiency of products, transforming the energy sector from fossil fuels to renewables, or developing a carbon fee or cap-and-trade program. Together, these individual and collective efforts reduce the amount of greenhouse gases emitted into the atmosphere.

Long-term development strategies that utilize scientific evidence are needed to address the challenges of sea level rise. Adaptation measurements included managed retreat, construction of dikes and sea walls, investment in green infrastructure and nature-based solutions, improved drainage infrastructure, and revising zoning strategies. No one solution exists in addressing the issue, and thus, cross-disciplinary, cross-national cooperation is required. McLeman (2018) writes about migration with dignity. Acknowledging that some locations may not be inhabitable, developing and incentivizing managed retreat strategies may be a method by which worst-case chaos may be averted. The management of this across geopolitical boundaries remains an area of future research.

The Multihazard Mitigation Council (2017) showed that for every dollar spent on proactively planning and risk reduction, $4 are saved in the long-term recovery. Local governments can reduce risk by designating non-habitable zones and discouraging development in high-risk areas. Minimizing assets in high-risk zones eliminates the economic cost in recovery.

Efforts to address the challenges of sea level rise require worldwide collaboration and localized strategies. Located in the Chesapeake Bay in Virginia, Tangier Island faces similar challenges to other coastal communities (Gertner 2016). However, the strategies designed to improve resilience for Tangier Island may not work for other coastal communities such as Bangladesh. Comparing the two locations, the political and economic structures are quite different. Locally focused adaptation strategies are important as successful implementation will require cooperation and trusting partnerships. Technical assistance and capacity building should help the most vulnerable prepare for the future impacts of sea level rise. Organizational support and international research will be essential in building capacity and reducing the impact of sea level rise on our coastal communities, well-being, and global economy. In addition to the transformation of the energy sector, the economic and political discourse surrounding the science of sea level rise must also be considered.

7 Monitoring and Impact Assessments

Coastal regions in the USA are main centers for economic, social, ecological and environmental, and cultural development. To address the challenges faced by coastal areas due to sea level rise, it is essential to have a social conversation and political support to determine the future of coastlines (Allen et al. 2018). These conversations include a good understanding of future predictions and regional variabilities, the storm patterns, what type of monitoring measures is economically sound do not put a burden on the local economies. Agreements should involve all stakeholders and their active participation with decision-makers. For example, a significant portion of Florida's tourism depends on its natural resources, and the constant inundation of the shoreline in addition to anthropogenic pressures only increases the likelihood of their degradation. To address these issues, the South Florida Regional Planning Council started a dialogue with decision-makers and city planners. The Council adopted a multi-sector collaboration strategy to monitor and assess the impacts of sea level rise. The work team's focus is to develop a timely health impact assessment to ensure that human health is considered as a component of any adaptation measures (Bloetscher et al. 2016). Some of these solutions include protecting sewage systems, raising roads, improving storm water management, and creating seawalls. To make more informed, evidence-driven decisions to address the wide range of impacts, the Council used data collected through monitoring and surveillance tools. Many tools have been developed and are satellite-based to collect meaningful information that gives us a broader understanding of the issue. Jason-3, a US-European oceanography satellite was launched in 2016 to monitor sea level rise (NASA 2016). Such innovations allow for improved resolution in modeling and are useful for evaluating risk at a finer-scale.

In the USA, NOAA has developed a program called Sea Level Rise Viewer which storehouses data, photos, and maps to visualize and compare sea level rise scenarios along the coast of the USA (NOAA 2018a, b). This viewer is a web-based visualization and planning tool that provides a community-level analysis as to the impacts of sea level rise. In addition, HAZUS-MH is a tool employed by the Federal Emergency Management Agency (FEMA) that provides a risk-based approach to disaster and risk mitigation, emergency preparedness, and recovery by identifying and displaying hazards and vulnerabilities (FEMA 2018). By incorporating demographic and community-level data, HAZUS serves as a data management system and used to support risk decision-making by estimating potential losses from natural hazards.

8 Summary Statement

Coastal regions are directly affected by recurrent flooding, loss of biodiversity, and storm damage exacerbated by sea level rise. To adequately prepare for the impacts, individual citizens, community leaders, and policymakers must address the causes of climate change and sea level rise while identifying the multi-sector risks associated with the consequences. At present, research has established the health and socioeconomic impacts of sea level rise on coastal communities in the USA, yet capacity to address the challenges is still limited. This hinders the assessment of coastal impacts and future predictions. Therefore, further research may continue to synergize across disciplines and geopolitical boundaries to address these deficiencies.

References

Akpinar-Elci, M., Rose, S., & Kekeh, M. (2018). Well-being and mental health impact of household flooding in Guyana, the Caribbean. *Marine Technology Society Journal, 52*(2), 18–22. https://doi.org/10.4031/MTSJ.52.2.3.

Alderman, K., Turner, L. R., & Tong, S. (2012). Floods and human health: A systematic review. *Environment International, 15*(47), 37–47.

Allen, M. R., Dube, O. P., Solecki, W., Aragón-Durand, F., Cramer, W., Humphreys, S. … Zickfeld, K. (2018). Framing and Context. In: Global warming of 1.5°C. An IPCC Special Report on the impacts of global warming of 1.5 °C above pre-industrial levels and related global greenhouse gas emission pathways, in the context of strengthening the global response to the threat of climate change, sustainable development, and efforts to eradicate poverty. Retrieved from https://www.ipcc.ch/site/assets/uploads/sites/2/2018/11/SR15_Chapter1_Low_Res.pdf.

American College of Emergency Physicians. (2015). *Lessons Learned from Hurricane Sandy and recommendations for improved healthcare and public health response and recovery for future catastrophic events*. Final Report. Retrieved from https://www.acep.org/globalassets/uploads/uploaded-files/acep/by-medical-focus/disaster/lessons-learned-from-hurricane-sandy-webpage.pdf.

Asugeni, J., MacLaren, D., Massey, P. D., & Speare, R. (2015). Mental health issues from rising sea level in a remote coastal region of the Solomon Islands: Current and future. *Australasian Psychiatry*; 23(6) Supplement 22–25. https://doi.org/10.1177/1039856215609767.

Atkinson, L. P., Ezer, T., & Smith, E. (2012). Sea level rise and flooding risk in Virginia. *Sea Grant Law & Policy Journal, 5*, 3.

Australian Bureau of Meteorology and CSIRO. (2011). Climate change in the Pacific: Scientific assessment and new research. Volume 2: Country reports. Retrieved from www.pacificclimatechangescience.org/publications/reports/. Accessed May 15, 2008.

Banusiewicz, J. D. (2014). Hagel to address "treat multiplier" of climate change. U.S. Department of Defense. Retrieved from https://www.defense.gov/News/Article/Article/603440/.

Bei, B., Bryanta, C., Gilsona, K., Koh, J., Gibsona, P., Komiti, A, … Judda, F. (2013). A prospective study of the impact of floods on the mental and physical health of older adults. *Aging and Mental Health, 17*(8), 992–1002.

Bloetscher, F., Polsky, C., Bolter, K., Mitsova, D., Garces, K. P., King R., … Hamilton K. (2016). Assessing potential impacts of sea level rise on public health and vulnerable populations in Southeast Florida and providing a framework to improve outcomes. Retrieved from https://www.mdpi.com/2071-1050/8/4/315.

Brown, S., Kebede, A. S., & Nicholls, R. J. (2011). *Sea-level rise and impacts in Africa, 2000–2100*. School of Civil Engineering and the Environment, University of Southampton, Southampton SO17 1BJ, UK. Report.

Campanella, R. (2018). How humans sank New Orleans. The Atlantic. Retrieved from https://www.theatlantic.com/technology/archive/2018/02/how-humans-sank-new-orleans/552323/.

Church, J. A., Clark, P. U., Cazenave, A., Gregory, J. M., Jevrejeva, S., Levermann, A., … Unnikrishnan, A. S. (2013). Sea level change. In: Climate change 2013: The physical science basis. Contribution of working group I to the fifth assessment report of the intergovernmental panel on climate change.

Cleetus, R. (2013). *Overwhelming risk: Rethinking flood insurance in a world of rising seas*. Union of Concerned Scientists. Online at http://www.ucsusa.org/assets/documents/global_warming/Overwhelming-Risk-Full-Report.pdf. Accessed December 23, 2018.

Department of Environmental Conservation. (2019). Sea level rise. What is expected for New York. Fast facts. Retrieved from https://www.dec.ny.gov/energy/45202.html.

Eggleston, J., & Pope, J. (2013). Land subsidence and relative sea-level rise in the southern Chesapeake Bay region (No. 1392). US Geological Survey. Retrieved from https://pubs.usgs.gov/circ/1392/pdf/circ1392.pdf. Accessed May 15, 2018.

Environmental Protection Agency. (2018). *Potential well water contaminants and their impacts*. Retrieved from https://www.epa.gov/privatewells/potential-well-water-contaminants-and-their-impacts. Accessed May 20, 2018.

Federal Emergency Management Agency. (2018). HAZUS Software. Retrieved from https://www.fema.gov/hazus-hardware-and-software-requirements. Accessed May 7, 2018.

Geisler, C., & Currens, B. (2017). Impediments to inland resettlement under conditions of accelerated sea level rise. *Land Use Policy, 66*, 322–330.

Gertner, J. (2016). Should the United States save Tangier Island from oblivion? Retrieved from https://www.nytimes.com/2016/07/10/magazine/should-the-united-states-save-tangier-island-from-oblivion.html.

Hauer, M. (2017). Migration induced by sea level rise could shape the US population landscape. Retrieved from https://www.nature.com/articles/nclimate3271.

Intergovernmental Panel on Climate Change. (2013). *Climate change: The physical science basis. Contribution of working group I to the fifth assessment report of the intergovernmental panel on climate change*. Cambridge University Press, Cambridge, United Kingdom and New York, NY, USA. https://doi.org/10.1017/cbo9781107415324.

Intergovernmental Panel on Climate Change. (2014a). *Climate change 2014. Synthesis Report*. Retrieved from https://www.ipcc.ch/pdf/assessmentreport/ar5/syr/SYR_AR5_FINAL_full_wcover.pdf. Accessed May 15, 2018.

Intergovernmental Panel on Climate Change. (2014b). *Small Islands*. Retrieved from http://www.ipcc.ch/pdf/assessment-report/ar5/wg2/WGIIAR5-Chap29_FINAL.pdf. Accessed June 15, 2018.

Intergovernmental Panel on Climate Change. (2014c). *Europe*. Retrieved from http://www.ipcc.ch/pdf/assessment-report/ar5/wg2/WGIIAR5-Chap23_FINAL.pdf. Accessed June 17, 2018.

Kent, J. D. (2006). *Louisiana Hurricane impact Atlas* (p. 39). Louisiana Geographic Information Center, Baton Rouge, LA.

King, P. G., McGregor, A. R., & Whittet, J. D. (2010). *The economic cost of sea-level rise to California Beach communities*. Retrieved from https://dbw.parks.ca.gov/pages/28702/files/CalifSeaLevelRise.pdf. Accessed May 11, 2018.

Levermann, A., Clark, P. U., Marzeion, B., Milne, G. A., Pollard, D., Radic, V., et al. (2013). The multimillenial sea level commitment of global warming. *Proceedings of the National Academy of Sciences of the United States of America, 110*(3), 13745–13750. https://doi.org/10.1073/pnas.1312464110.

McLeman, R. (2018). Migration and displacement risks due to mean sea-level rise. *Bulletin of the Atomic Scientists, 74*(3), 148–154.

Mendell, M. J., Mirer, A. G., Cheung, K., Tong, M., & Douwes, J. (2011). Respiratory and allergic health effects of dampness, mold, and dampness-related agents: A review of the epidemiologic evidence. *EHP, 119*(6), 748–756.

Mimura, N. (2013). Sea-level rise caused by climate change and its implications for society. Proceedings of the Japan Academy. *Series B, Physical and Biological Sciences*, 89(7), 281–301. http://doi.org/10.2183/pjab.89.281.

Multihazard Mitigation Council. (2017). Natural hazard mitigation saves: 2017 interim report. Retrieved from https://www.fema.gov/media-library-data/1516812817859-9f866330bd6a1a93f54cdc61088f310a/MS2_2017InterimReport.pdf. Accessed May 16, 2018.

National Aeronautics Space Administration. (2015). NASA Science Leads New York City Change 2015 Report. Retrieved from https://www.giss.nasa.gov/research/news/20150217/.

National Aeronautics and Space Administration. (2016). Jason-3 launches to monitor global sea level rise. Retrieved from https://www.nasa.gov/press-release/jason-3-launches-to-monitor-global-sea-level-rise. Accessed May 16, 2018.

National Climate Assessment. (2014). Climate change impacts in the United States. Full Report. Retrieved from https://nca2014.globalchange.gov/. Accessed May 15, 2018.

National Oceanic and Atmospheric Administration. (2013). *National Coastal population report. Population trends from 1970 to 2020.* Retrieved from https://aamboceanservice.blob.core.windows.net/oceanservice-prod/facts/coastal-population-report.pdf. Accessed May 12, 2018.

National Oceanic and Atmospheric Administration. (2015). Climate change: Ocean heat content. Retrieved from https://www.climate.gov/news-features/understanding-climate/climate-change-ocean-heat-content. Accessed May 15, 2018.

National Oceanic and Atmospheric Administration. (2016). Extended reconstructed sea surface temperature (ERSST.v4). National Centers for Environmental Information. www.ncdc.noaa.gov/data-access/marineocean-data/extended-reconstructed-sea-surface-temperature-ersst. Accessed May 15, 2018.

National Oceanic and Atmospheric Administration. (2017a). Climate change: Global sea level. Retrieved from https://www.climate.gov/news-features/understanding-climate/climate-change-global-sea-level. Accessed April 23, 2018.

National Oceanic and Atmospheric Administration. (2017b). How is sea level rise related to climate change? Retrieved from https://oceanservice.noaa.gov/facts/sealevelclimate.html. Accessed May 10, 2018.

National Oceanic and Atmospheric Administration. (2017c). What is the difference between local sea level and global sea level? Retrieved from https://oceanservice.noaa.gov/facts/sealevel-global-local.html. Accessed April 23, 2018.

National Oceanic and Atmospheric Administration. (2017d). Examining sea level rise exposure for future populations. Retrieved from https://coast.noaa.gov/digitalcoast/stories/population-risk.html. Accessed April 29, 2018.

National Oceanic and Atmospheric Administration. (2018a). Tidal Station Data Key West. Retrieved from https://tidesandcurrents.noaa.gov/waterlevels.html?id=8723214&units=standard&bdate=19500101&edate=20171231&timezone=GMT&datum=MSL&interval=m&action=data.

National Oceanic and Atmospheric Administration. (2018b). Understanding and planning for sea level rise in California. Retrieved from https://coast.noaa.gov/digitalcoast/stories/ca-slr.html. Accessed April 28, 2018.

National Oceanic and Atmospheric Administration. (2019). Is sea level rising? Retrieved from https://oceanservice.noaa.gov/facts/sealevel.html.

National Snow and Ice Data Center. (2018). Quick facts on ice sheets. Retrieved from https://nsidc.org/cryosphere/quickfacts/icesheets.html. Accessed May 15, 2018.

Nerem, R. S., Beckley, B. D., Fasullo, J. T., Hamlington, B. D., Masters, D., Mitchum, G. T. (2018). Climate-change–driven accelerated sea-level rise detected in the altimeter era. Proceedings of the National Academy of Sciences. https://doi.org/10.1073/pnas.1717312115.

Neumann, B., Vafeidis, A. T., Zimmermann J., & Nicholls, R. J. (2015). Future coastal population growth and exposure to sea-level rise and coastal flooding-a global assessment. *PloS one, 10*(3), e0118571.

Parris, A., Bromirski, P., Burkett, V., Cayan, D., Culver, M., J. Hall, J., … Weiss, J. (2012). Global Sea Level Rise Scenarios for the US National Climate Assessment. NOAA Tech Memo OAR CPO-1. 37pp.

Public Health Institute. (2016). *Sea level rise, climate change and health*. Retrieved from http://climatehealthconnect.org/wp-content/uploads/2016/09/SeaLevelRise.pdf. Accessed April 24, 2018.

Reponen, T., Singh, U., Schaffer, C., Vesper, S., Johansson, E., Adhikari, A., et al. (2010). Visually observed mold and moldy odor versus quantitatively measured microbial exposure in homes. *Science of the Total Environment, 408*(22), 5565–5574.

Rice, D. (2017). Catastrophic, Sandy-like floods could hit New York City every 5 years due to sea-level rise. *USA Today*. Retrieved from https://www.usatoday.com/story/weather/2017/10/23/catastrophic-sandy-like-floods-could-hit-nyc-every-5-years-due-sea-level-rise/791581001/.

Rose, S., & Akpinar-Elci, M. (2015). An evaluation of the prevalence of respiratory and general symptoms among occupants of households affected by flooding in Guyana: The case of Cove & John. JARE.

Rufat, S., Tate, E., Burton, C. G., & Maroof, A. S. (2015). Social vulnerability to floods: Review of case studies and implications for measurement. *International Journal of Disaster Risk Reduction, 14,* 470–486.

Schulte, D. M., Dridge, K. M., & Hudgins, M. H. (2015). Climate change and the evolution and fate of the Tangier Islands of Chesapeake Bay, USA. *Scientific reports, 5,* 17890.

Spencer, T. (2018). Up to four-fifths of wetlands worldwide could be at risk from sea level rise. University of Cambridge. Retrieved from https://www.cam.ac.uk/research/news/up-to-four-fifths-of-wetlands-worldwide-could-be-at-risk-from-sea-level-rise. Accessed May 20, 2018.

Spencer, T., Schuerch, M., Nicholls, R. J., Hinkel, J., Lincke, D., Vafeidid, A. T., … Brown, S. (2016). Global coastal wetland change under sea-level rise and related stresses: The DIVA Wetland Change Model.' Global and Planetary Change (2016). https://doi.org/10.1016/j.gloplacha.2015.12.018; https://www.sciencedirect.com/science/article/pii/S0921818115301879.

Sweet, W. V., Kopp, R. E., Weaver, C. P., Obeysekera, J., Horton, R. M., Thieler, E. R., & Zervas, C. (2017). *Global and Regional Sea Level Rise Scenarios for the United States*. NOAA Technical Report NOS CO-OPS 083. NOAA/NOS Center for Operational Oceanographic Products and Services.

The City of New Orleans. (2017). Climate action for a resilient New Orleans. Retrieved from https://nola.gov/nola/media/Climate-Action/Climate-Action-for-a-Resilient-New-Orleans.pdf.

The City of New York. (2017). Climate action for a resilience New Orleans. Retrieved from https://nola.gov/nola/media/Climate-Action/Climate-Action-for-a-Resilient-New-Orleans.pdf.

The New York Times. (2016). What could disappear. Retrieved from https://archive.nytimes.com/www.nytimes.com/interactive/2012/11/24/opinion/sunday/what-could-disappear.html.

United Nations. (2014). World's population increasingly urban with more than half living in urban areas. Retrieved from http://www.un.org/en/development/desa/news/population/world-urbanization-prospects-2014.html. Accesses May 18, 2018.

UN Atlas of the Oceans. (2016). Human settlements on the Coast. Retrieved from http://www.oceansatlas.org/subtopic/en/c/114/. Accessed May 18, 2018.

U.S. Geological Survey. (2017). Louisiana's rate of Coastal wetland loss continues to slow. Retrieved from https://www.usgs.gov/news/usgs-louisiana-s-rate-coastal-wetland-loss-continues-slow. Accessed May 11, 2018.

Vineis, P., Chan, Q., & Khan, A. (2011). Climate change impacts on water salinity and health. *Journal of Epidemiology and Global Health, 1*(1), 5–10. https://doi.org/10.1016/j.jegh.2011.09.001.

Watts, N., Adger, W. N., Agnolucci, P., Blackstock, J., Byass, P., Cai, W., … Cox, P. M. (2015). Health and climate change: Policy responses to protect public health. *The Lancet, 386*(10006), 1861–1914.

Zimmermann, K. A. (2015). Hurricane Katrina: Facts, Damage & Aftermath. LiveScience. Retrieved from https://www.livescience.com/22522-hurricane-katrina-facts.html.

Storm Surges, Heavy Rain and Strong Wind: Impacts of Tropical Cyclone Winston in Fiji—Focus on Health

Helene Jacot Des Combes

Abstract On February 20th, 2016, tropical cyclone (TC) Winston, a category 5 cyclone, hit Fiji, affecting more than half the population of the country and leaving behind damages worth more than 30% of the GDP of the country. The health sector alone suffered damages and losses amounting to FJ$13.9 million, corresponding to the damage or destruction of 88 health facilities out of 214. One characteristic of this cyclone was the importance of damages caused by the storm surge associated with the cyclone in some areas. This is extremely worrying in a changing climate and more specifically the context of sea-level rise, and studies indicate that climate change has an influence on cyclones and storm surges although it is still difficult to precisely quantify it. TC Winston had important impacts on health and on the health system in Fiji. A total of 44 casualties were recorded, and more than 120 people were injured, including 45 who were hospitalized. Cases of diarrheal diseases, leptospirosis and typhoid were observed and monitored, but actions were taken and there was no large outbreak after the cyclone. This chapter provides general information about TC Winston, then presents the specific impacts of the storm surge associated with TC Winston and discusses the influence of climate change, and the last sections details and discusses the impacts of TC Winston on health in Fiji.

1 Introduction—Tropical Cyclone Winston in Fiji

The tropical disturbance TD09F, which would become tropical cyclone (TC) Winston, was first observed between Fiji and Vanuatu on February 7. The system evolved into a category 1 cyclone and was named Winston on February 11, 2016, by the Regional Specialised Meteorological Centre (RSMC) based in Nadi, Fiji. The category 1 cyclone was located west of Fiji and was moving southwards. On February 14th, TC Winston turned northeast and stayed north of Tonga on the 17th. At that stage, most of the forecasts indicated it would continue northeast, but it turned back

H. Jacot Des Combes (✉)
Pacific Centre for Environment and Sustainable Development, The University of the South Pacific, Suva, Fiji
e-mail: helenejdc@hotmail.com

R. Akhtar (ed.), *Extreme Weather Events and Human Health*,
https://doi.org/10.1007/978-3-030-23773-8_13

and started to move westward and reached category 5 on both the Australian Tropical Cyclone and Saffir-Simpson intensity scales by February 19th. TC Winston made a landfall in Fiji on February 20th (Fig. 1), with maximum sustained 10-min wind speed estimated at 204 km/h (110 kt) near its center (Fiji Meteorological Services 2016a, b) but maximum gust wind speed was recorded at 303 km/h (Government of Fiji 2016). TC Winston also brought heavy rain to Fiji, with accumulated rainfall over February 20th and 21st reaching 479 mm at Nadarivatu, in the north of Viti Levu, close to the cyclone's path, compared to the long-term average rainfall for February based on the 1971–2000 period which stands below 300 mm. This heavy rain over a short time period generated flooding in several locations in Fiji although the main urban centers were not affected (Fiji Meteorological Services 2016b).

Not only was TC Winston a very intense cyclone but it made a direct landfall to Fiji, the eye passing just between the two main islands of Vanua Levu (north) and Viti Levu (south) where the capital city Suva is located. This led a large part of Fiji to sustain the maximum wind speed over 100 km/h and in some areas up to over 150 km/h (Fig. 2). This resulted in a significant part of the population, infrastructure and environment of Fiji exposed to these strong winds (Fig. 3).

The consequences of the direct landfall of a category 5 cyclone in Fiji were disastrous for the country. Forty-four people were killed, twenty-seven hospitalized, one hundred and seventeen injured, four hundred eighty-five thousand and thirty lost the livelihood of the family breadwinner, and a total of five hundred forty thousand four hundred and fourteen were affected. In May 2016, damages were estimated at more than FJ$1.5 million and the losses at more than FJ$1.3 billion. The total of over FJ$2.85 billion corresponded to one-fifth of Fiji's 2014 GDP (Government of

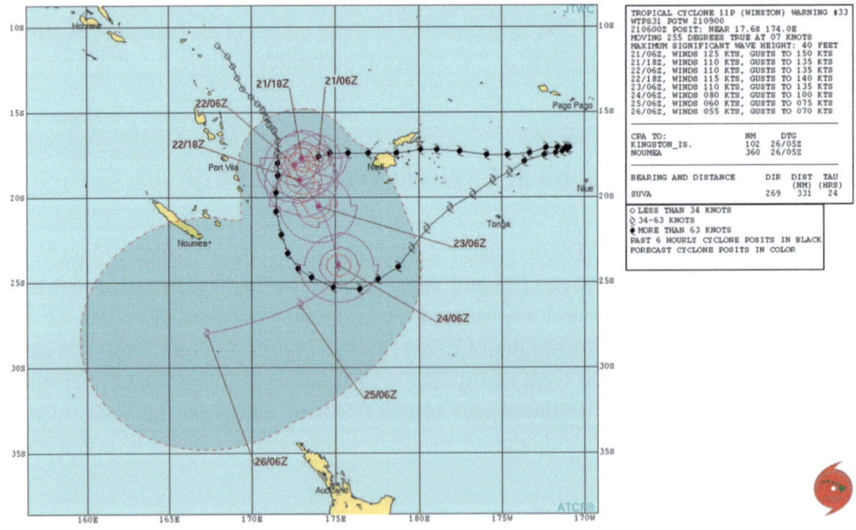

Fig. 1 Tropical cyclone Winston observed (black spirals) and projected (pink spirals) path. *Source* Joint Typhoon Warning Centre (2016)

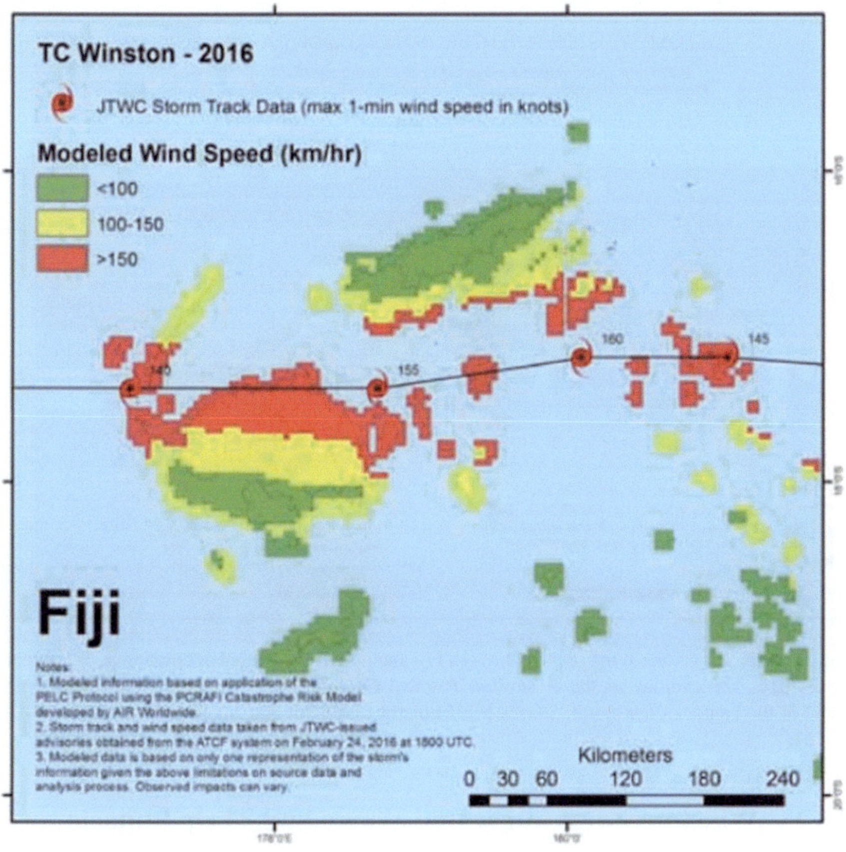

Fig. 2 Map of modeled wind speed during tropical cyclone Winston path through Fiji between February 19 and 22, 2016. *Source* Government of Fiji (2016)

Fiji 2016). All economic sectors were affected although the agriculture sector alone suffered damages and loss accounting for FJ$542 million, while damages and loss in the health sector amounted to FJ$13.9 million (Government of Fiji 2016).

The government declared the state of emergency for all four divisions on February 20th. The nine national humanitarian clusters coordinated the government in its response to the event and different national, regional and international organization and governments supported Fiji in the response and reconstruction phases (Government of Fiji 2016).

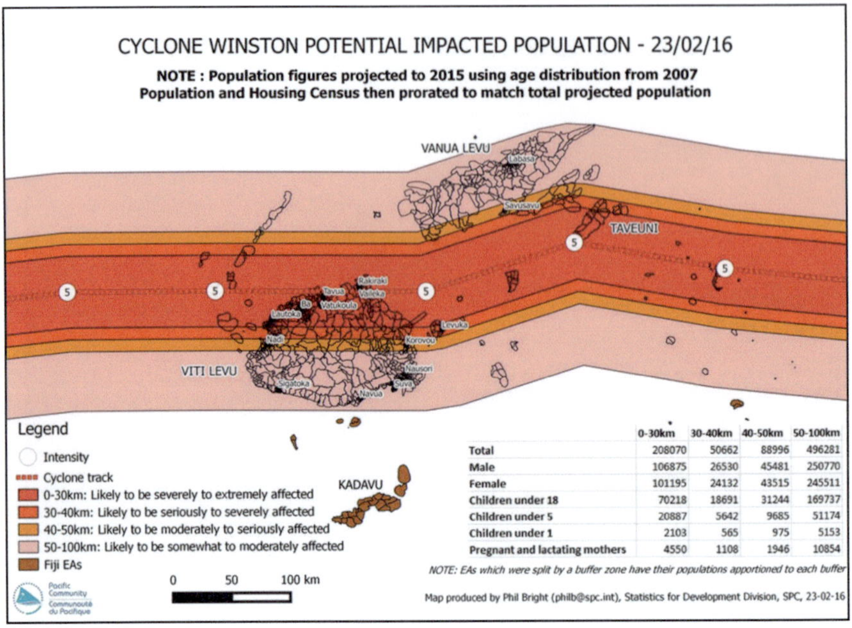

Fig. 3 Tropical cyclone Winston path through Fiji and potential affected population as of February 23rd, 2016. The cyclone on the Australian Tropical Cyclone Intensity Scale is indicated by the figure at the center of the spiral. *Source* Fiji Ministry of Health and Medical Services and WHO (2016)

2 TC Winston Storm Surge and Waves and Their Impacts

With an atmospheric pressure of 930 hPa measured at the center of the cyclone on February 19, a storm surge was expected to affect the coastal areas of Fiji during TC Winston (Fiji Meteorological Services 2016a). Contrarily to the Australian Tropical Cyclone intensity scale, the Saffir-Simpson Hurricane scale indicates a range of height for storm surges associated with tropical cyclones of different categories. For category 5, when winds are stronger than 249 km/h, the US National Weather Service from the National Oceanic and Atmospheric Administration indicates that the storm surge can be higher than 5.5 m (18 ft). However, the actual height of a storm surge is controlled by local parameters such as bathymetry and flow bottom friction (Bastidas et al. 2016). Another important parameter is the influence of storm size (in particular its radius) on the height of storm surges. When this parameter is included in the model to predict storm surges, there is a better correlation with the observed surges in the USA (Irish et al. 2008). There is no such study conducted in the Pacific so far but it could be useful for early warning systems to get a more precise prediction of storm surges.

Reports on the characteristics of TC Winston include precise information on the wind speed and amount of rainfall; however, there is no precise information on the height of the associated storm surge, although it has been recognized that the storm surge was responsible for a significant part of the damages observed in different locations around Fiji; in particular Koro and Vanua Balavu Islands and the southern coast of Vanua Levu where the seawater reached 200 m inland in some locations (Government of Fiji 2016). This significant storm surge generated damages on infrastructure and coastal vegetation and plantations (Fig. 4). Marine ecosystems such as coral reef which are known to protect the coast were also damaged by the storm surge (Government of Fiji 2016).

The intensity of TC Winston and its occurrence, one year after TC Pam, another category 5 cyclone which affected Vanuatu, Tuvalu and Kiribati in 2015 led to the question of the influence of climate change on the intensity and/or frequency of these "super cyclones" and the attribution of their characteristics (e.g. Emanuel 2005, 2017; Knutson et al. 2010; Terry and Lau 2018; Tukayabu et al. 2015). This question was very relevant in the case of TC Winston due to the regional context. Climate change has been recognized since the mid-1990s by all leaders of the Pacific Small Islands Developing States (PSIDS) as a very serious threat for their development and even, in the case of the low lying atoll countries like Tuvalu, Kiribati or the Republic of the Marshall Islands, as a threat to their mere existence, leading to the inclusion of climate change adaptation and mitigation in two major frameworks for the region: the Framework for Pacific Regionalism in 2014 and the Framework for Resilient Development in the Pacific in 2016 (Pacific Community et al. 2016; Pacific Islands Forum Secretariat 2014). There is thus an intense activity of observation and monitoring of climate change impacts in the region, with a special focus on the impact of sea-level rise. The question was also raised in response to the succession

Fig. 4 Picture of Vatulele Village, north-eastern extremity of Koro Islands after TC Winston. Koro was one of the most affected small islands in Fiji, where most of the damages were caused by strong wind (185 km/h) and the storm surge. *Source* Fiji National Disaster Risk Management Office (n.d.)

of intense tropical cyclones (categories 4–5) in the region before 2016, including (but not limited to): TC Pam (cat. 5 in 2015 in Vanuatu), TC Ian (cat. 4, in 2014 in Tonga), TC Evan (cat. 4 in 2012 in Samoa and Fiji) and TC Tomas (category 4 in 2010 in Fiji). In this context, the two questions on TC Winston and climate change were: Was the frequency of these intense cyclones the result of climate change? Was sea-level rise partly responsible for the height of the storm surge associated with TC Winston?

In response to the first question, although recent studies have shown that model results indicate and increase in the frequency of intense cyclone due to climate change (e.g., Sugi et al. 2016; Backmeister et al. 2018), globally by an average of 25–30% per °C of global warming (Holland and Bruyère 2013), there is a large regional variability and the frequency is expected to decrease in the South Pacific Basin (Sugi et al. 2016).

With regards to the second question, a recent study of super typhoon Haiyan in the Philippines, comparing the models of the storm surge associated with this cyclone in a changing climate with the storm surge associated with a similar cyclone in a pre-industrial environment, indicated that up to 20% of the height of the storm surge could be attributed to climate change (Takayabu et al. 2015). This supported the findings of a study conducted in Bangladesh which concluded that sea-level rise of 0.3 m and increased sea surface temperature by 2 °C would increase the area inundated by a storm surge by about 15% (Karim and Mimura 2008). However, a study on the strength of the waves associated with TC Winston based on the displacement of coral boulders on the coasts indicated that the waves for TC Winston reached up to 10 m high with an associated flow around 14 m/s but were not unprecedented and were, in fact, comparable to those generated by TC Tomas (cat. 4) in 2010 (Terry and Lau 2018). It seems thus likely that climate change, and in particular sea-level rise, will increase the height and the intensity of storm surges associated with tropical cyclones, widening the inundated areas and generating more damages. Even if comparable events were observed in the previous 400 years, the extent of coastal development since then, both in terms of population increase and infrastructure building, would significantly increase the risks in the coastal region of Fiji.

3 Impacts on Health

In terms of direct impacts, the number of casualties was 43 on March 10th, 2016 but increased to 44 on March 18th (Fiji Health and Nutrition Cluster 2016a, b). On March 10th, 126 people were reported injured, 45 were hospitalized as a direct consequence of TC Winston, probably due to trauma caused by collapsed structures, debris carried by strong winds or near drowning. These cases were treated in the different health structures, and tetanus shots were also provided to limit the risk of contracting the disease (Fiji Health and Nutrition Cluster 2016a). However, tetanus being part of the national childhood immunization program, coverage in 2013 was estimated to be above 95% and 74% for maternal tetanus toxoid; a part of the population, including elderlies who were not included in the childhood program, remains at risk though (Fiji Ministry of Health and Medical Services and WHO 2016). A total of 19,812 people were displaced (Fiji Health and Nutrition Cluster 2016a). This figure decreased to 4,299 on March 18th, and 361 on March 29th (Fiji Health and Nutrition Cluster 2016b, c).

The high number of displaced people, sheltered in 557 evacuation centers, caused worry on the higher risk of secondary impacts due to living in overcrowded conditions. In such situation, the risk of transmission of communicable diseases, such as acute respiratory infections, measles and rubella, diarrheal disease (bacterial and viral), hepatitis A, Leptospirosis, meningococcal disease and typhoid, increases (Fiji Ministry of Health and Medical Services and WHO 2016). Children under 5 are particularly at risk, especially from acute respiratory infection and diarrheal disease, as was observed after the floods in Honiara and Guadalcanal, Solomon Islands, in 2014 which were followed by a diarrheal epidemic affecting more widely children under 5 who had 20 more chances to contract the disease than older children (Fiji Ministry of Health and Medical Services and WHO 2016). One of the priorities of the Fijian authorities and of the Health and Nutrition Cluster was to limit the secondary impacts as much as possible. These secondary impacts are generally attributed to three main causes: lack of proper nutrition, in particular for children below 5, lack of access to clean water, and life in overcrowded conditions in shelters (Fiji Ministry of Health and Medical Services and WHO 2016). A few case suspicions of measles were recorded after TC Winston (Fiji Health and Nutrition Cluster 2016b) and several cases of conjunctivitis were recorded in Suva (Fiji Health and Nutrition Cluster 2016c).

The reduction in the public health risks associated with TC Winston comprised different actions which built on public health programs in place before the cyclone. A specific Early Warning Alert and Response System (EWARS) was put in place to monitor risks of transmission of communicable diseases with 33 stations participating on March 18(Fiji Health and Nutrition Cluster 2016b). Among the diseases monitored by the EWARS was acute or bloody diarrhea, expected to be caused by bacterial since the other potential cause, rotavirus, vaccination, has been part of the national childhood immunization program since October 2012, thus limiting the risks of a large outbreak (Fiji Ministry of Health and Medical Services and WHO 2016). Some clusters of diarrhea were observed in Keiyasi and other badly affected areas (Fiji Health and Nutrition Cluster 2016b; Fiji Ministry of Health and Medical Services and WHO 2016). Priorities identified to be addressed by the first appeal to international aid included: supporting the delivery of health services to all people, procurement and distribution of urgent and essential medical products such as treatment and vaccines and provide children and maternal care as well as address the risk of sexually transmissible diseases (Fiji Ministry of Health and Medical Services and WHO 2016).

In addition to diarrheal disease, leptospirosis and typhoid were monitored by EWARS. Leptospirosis had been observed after the floods in Fiji in 2012 when 576 cases were reported with 7% fatalities (Lau et al. 2016). Although there were some concerns for the Navua area on March 19th, no cases were observed after TC Winston (Fiji Health and Nutrition Cluster 2016b). A typhoid outbreak of 35 cases was observed after TC Winston in the villages of Qelekuro and Nabulini on the north-east part of Viti Levu, the main island of Fiji but no fatalities were observed. This outbreak was explained by a worsening of the environmental conditions due to the cyclone in an area of high seroprevalence of antibodies against *Salmonella enterica*, one of the bacteria responsible for typhoid fever (de Alwis et al. 2018). To address this outbreak, twenty thousand doses of typhoid vaccine have arrived in Fiji for outbreak prevention and control, if indicated in line with the typhoid fever vaccination strategy that has been finalized by the Public Health Intervention group (Fiji Health and Nutrition Cluster 2016d).

Vector-borne diseases such as dengue, zika and chikungunya were also monitored after TC Winston since several of these diseases are present in Fiji. On March 19th, 71 cases of dengue-like disease (fever, rash, edema of the hands and feet, joint and muscle pain) were reported from Lautoka, west of Viti Levu (Fiji Health and Nutrition Cluster 2016b) and on April 9th, 15 cases of Zika had been confirmed (Fiji Health and Nutrition Cluster 2016d). Prevention measures to limit the proliferation of mosquitoes were implemented to limit further spreading of these diseases.

Malnutrition was a serious problem in some areas after TC Winston, in particular for children under 5. On March 9th, a total of 15 malnutrition cases were recorded, 12 mild and 3 severe ones which were treated at divisional hospitals (Fiji Health and Nutrition Cluster 2016a). The number of cases increased to 26 on March 29th (Fiji Health and Nutrition Cluster 2016c). To address this issue, food ration contents were reviewed, micronutriment powder was distributed and a nutrition specialist joined the staff of the Fiji Ministry of Health and Medical Services for three months after TC Winston (Fiji Health and Nutrition Cluster 2016c, d). The cyclone had a significant impact on local fisheries which provide fish to schools, limiting the supply of proteins to the children (World Conservation Society 2016). Several cases of malnutrition were observed on Koro Island. Because of the damages caused by the cyclone on this island, the terrestrial food sources were damaged or destroyed by strong wind, heavy rain and storm surge which also led to a contamination of the soil by saltwater (Government of Fiji 2016). In addition, coastal fisheries in Koro were very badly affected by the cyclone, for example around 96% of the fishing boats in Koro were destroyed, affecting the food supply of the households, 92% of which depends on local fisheries for subsistence and leading to a decrease in the amount of fish delivered to the local school to 0% (World Conservation Society 2016). The cyclone also had a longer impact on livelihoods in Koro since 50% of their income came from Kava farming which was completely destroyed after the cyclone (World Conservation Society 2016). The situation regarding food supply was difficult and there was some evidence of young girls offering sexual services in exchange for food, thus increasing the risk of sexually transmitted diseases and of violence. A system of food vouchers was established by the government to facilitate food supply for all affected people (Government of Fiji 2016).

A total of 8,466 people received psychological first aid after TC Winston. Training of trainers workshops was organized by the Ministry of Health and Medical Services to support the deployment of staff in all areas of Fiji, and more than 300 nurses were trained to provide psychological help (Fiji Health and Nutrition Cluster 2016b, d).

Persons with disabilities were also affected by TC Winston. A survey was conducted by UNICEF and the Pacific Disability Forum on a sample of 963 people. The findings indicate that 13.6% of children with disabilities were in need of psychological aid, and that specific first aid kits and dignity kits were needed to ensure a level of hygiene preventing the development of diseases. It was also noted that a significant number of persons with disability did not go to the shelter during Winston, either because there were accessibility issues or because of the lack of security (Pacific Disability Forum and UNICEF 2016).

4 Impacts on the Health System

The damages and loss faced by the health sector in Fiji amounted to FJ$13.9 million and corresponded to the damage or destruction of 88 health facilities out of 214, including clinics (Table 1). The damages included damage to the infrastructure themselves, loss of medical supplies, interruption of telecommunications, road and sea transportation and disruption of water and energy supplies. For example, parts of the main hospital in Fiji were flooded (Canyon 2017). Priority requests to partners included the procurement of vaccine cold chains (Fiji Health and Nutrition Cluster 2016a). Despite these damages, the priorities were the establishment of medical care centers to treat the different medical emergencies and to ensure the continuous care for patients with chronic diseases or infections (Fiji Health and Nutrition Cluster 2016a). However, some of the main health facilities were able to reopen and function normally relatively rapidly after the cyclone (Canyon 2017).

To face the needs to treat the population after the cyclone, retired nurses were re-engaged to support the medical teams deployed in the country (Government of Fiji 2016). Despite the support for the extra personnel, it was sometimes difficult to send medical personnel in affected areas because of the lack of means of transportation and because the safety of the first responders and other personnel on the ground needed to be ensured (Fiji Health and Nutrition Cluster 2016a). However, one public health outreach program was organized with the support of the extra nurses and another one was carried in mid-March. While the Eastern Division of Fiji was fully covered, 85% of the villages and 84% of the settlements were visited in the Northern Division and 96% of the villages and 82% of the settlements were visited in the Western Division (Fiji Health and Nutrition Cluster 2016b).

The reconstruction of the health infrastructures was expected to cost around FJ$31 million to ensure that all health facilities are fully operational. The most important costs for the reconstruction covers the relocation of some facilities in less exposed areas, the reconstruction of damaged facilities and the provision of drug and supplies (Government of Fiji 2016). In addition to these costs, FJ$59 million are needed to make the sector more resilient to climate change and disasters in the long term. Costs planned for this building-back-better include the retrofitting of existing facilities, build back better the destroyed facilities and the costs associated with the relocation of facilities (Government of Fiji 2016).

Table 1 Number of facilities damaged or with some functionality lost by division

Type of facility	Number of facilities reported				Total	Facilities that lost some form of Functionality				Total	Facilities that were damaged				Total
	Central	West	East	North		Central	West	East	North		Central	West	East	North	
Hospitals	8	7	5	4	24	–	–	–	–	–	7	7	1	3	18
Health centers	20	27	15	18	80	–	2	1	–	3	9	21	2	3	35
Nursing stations	22	28	36	21	107	4	–	–	–	4	12	18	1	3	34
Old people's home	1	1		1	3	–	–	–	–	–	1	–	–	–	1
Total	51	56	63	44	214	4	2	1	0	7	29	46	4	9	88

Source Fiji Health and Nutrition Cluster (2016a)

5 Conclusions

TC Winston caused devastation in Fiji in February 2016, leading to an important loss of lives and catastrophic socio-economic consequences. The health sector faced important damages and losses and struggled to continue to care for the health of Fijians in the aftermath of the cyclone. In some places, such as Koro island, strong winds, heavy rain and high storm surges combined to destroy infrastructures and food sources for coastal villagers. This destruction had long-term impacts and affected food security and several malnutrition cases were recorded on the island. The magnitude of the cyclone and its associated impacts raised the question of the influence of climate change and sea-level rise in the frequency and intensity of these extreme events. Although new studies focus on the attribution of cyclones to climate change, it is still difficult to precisely quantify the influence anthropogenic climate change has on cyclones. The high number of people seeking shelter in the evacuation centers increased the risk of secondary outbreak of diarrheal, acute respiratory diseases and vector-borne diseases that were already present in Fiji such as typhoid, dengue or leptospirosis. Although some cases were observed in the most affected areas or in urban environment, the monitoring and treatment of these cases prevented the generation of large outbreaks of these diseases. Malnutrition was closely monitored as well as vector-borne disease and campaigns of reduction of mosquito populations were conducted in the months after the cyclone. A total of 88 health facilities were damaged or destroyed, out of the 214 located in Fiji, for a total of FJ$13.9 million. Infrastructures damages, issues with power and clean water supply were the biggest problems, and there was a need to procure system to ensure the continuity of cold chain for vaccines against the secondary diseases. However, these difficulties were addressed so the health of Fijian could be supported as best as possible under the circumstances. The reconstruction of the health infrastructure, including building back better for a more resilient health sector to climate change and extreme events, is expected to cost FJ$ 90 million, but it is considered an important investment due to the vulnerability of the country to these issues.

References

Backmeister, J. T., Reed, K. A., Hannay, C., Lawrence, P., Bates, S., Truesdale, J. E., et al. (2018). Projected changes in tropical cyclone activity under future warming scenarios using a high-resolution climate model. *Clim Change, 146*(3–4), 547–560.

Bastidas, L. A., Knighton, J., & Kline, S. W. (2016). Parameter sensitivity and uncertainty analysis for a storm surge and wave model. *Natural Hazards and Earth System Sciences, 16,* 2195–2210.

Canyon, D. V. (2017). Fiji's cyclone Winston. In J. Ear, A.D.B. Cook, D.V. Canyon, K. Daniel (Eds.), *Disaster response regional architectures—assessing future possibilities* (74 p). Inouye Asia-Pacific Center for Security Studies, Honolulu, Hawaii, USA.

de Alwis, R., Watson, C., Nikolay, B., Lowry, J. H., Thieu, N. T. V., Van, T. T., et al. (2018). Role of environmental factors in shaping spatial distribution of Salmonella enterica serovar Typhi, Fiji. *Emerging Infectious Diseases, 24*(2), 284–293. https://doi.org/10.3201/eid2402.170704.

Emanuel, K. (2005). Increasing destructiveness of tropical cyclones over the past 30 years. *Nature, 436,* 686–688.

Emanuel, K. (2017). Assessing the present and future probability of Hurricane Harvey's rainfall. *Proceedings on the National Academy of Sciences, 114*(48), 12681–12684.
Fiji Health and Nutrition Cluster. (2016a). *Fiji tropical cyclone bulletin 1* (p. 16). Suva, Fiji: Fiji Health and Nutrition Cluster.
Fiji Health and Nutrition Cluster. (2016b). *Fiji tropical cyclone bulletin 2* (p. 5). Suva, Fiji: Fiji Health and Nutrition Cluster.
Fiji Health and Nutrition Cluster. (2016c). *Fiji tropical cyclone bulletin 3* (p. 5). Suva, Fiji: Fiji Health and Nutrition Cluster.
Fiji Health and Nutrition Cluster. (2016d). *Fiji tropical cyclone bulletin 4* (p. 6). Suva, Fiji: Fiji Health and Nutrition Cluster.
Fiji Meteorological Services. (2016a). *Tropical disturbance advisory number A37*. Online source, accessible at http://weather.noaa.gov/pub/data/raw/wt/wtps11.nffn.txt.
Fiji Meteorological Services. (2016b). *Fiji climate summary February 2016* (Vol. 37(2), 11 p).
Fiji National Disaster Management Office (n.d.), *Tropical cyclones—action guide*, Fiji National Disaster Management Office, Suva, Fiji.
Government of Fiji. (2016). *Fiji post-disaster needs assessment, tropical cyclone Winston*. February 20, 2016, Government of Fiji, Suva, Fiji, 160 p.
Government of Fiji. (2016b). *Fiji post-disaster needs assessment, tropical cyclone Winston, February 20, 2016* (p. 160). Suva, Fiji: Government of Fiji.
Holland, G., & Bruyère, C. L. (2013). Recent intense hurricane response to global climate change. *Climate Dynamics, 42,* 617–627.
Irish, J.L., Resio, D.T., Ratcliff, J.J. (2008). The Influence of storm size on Hurricane Surge. *Journal of Physical Oceanography.* https://doi.org/10.1175/2008jpo3727.1.
Joint Typhoon Warning Centre. (2016). *Tropical cyclone Winston track. Online source*. Accessible at http://oceanweatherservices.com/blog/2016/02/19/tc-winston-bears-down-on-fiji/.
Karim, M. F., & Mimura, N. (2008). Impacts of climate change and sea-level rise on cyclonic storm surge floods in Bangladesh. *Global Environmental Change, 18,* 490–500.
Knutson, T. R., McBride, J. L., Chan, J., Emanuel, K., Holland, G., Landsea, C., et al. (2010). Tropical cyclones and climate change. *Nature Geoscience, 3,* 157–163.
Lau, C. L., Watson, C. H., Lowry, J. H., David, M. C., Craig, S. B., Wynwood, S. J., et al. (2016). Human leptospirosis infection in Fiji: an eco-epidemiological approach to identifying risk factors and environmental drivers for transmission. *PLoS Neglected Tropical Disease, 10*(1), e0004405. https://doi.org/10.1371/journal.pntd.0004405.
Pacific Community (SPC), Secretariat of the Pacific Regional Environment Programme (SPREP), Pacific Islands Forum Secretariat (PIFS), United Nations Development Programme (UNDP), United Nations Office for Disaster Risk Reduction (UNISDR), the University of the South Pacific (USP) (2016), *Framework for Resilient Development in the Pacific: An Integrated Approach to Address Climate Change and Disaster Risk Management (FRDP) 2017–2030*, The Pacific Community, Geoscience Division, Suva, Fiji, 40 p.
Pacific Disability Forum & UNICEF (2016), *TC Winston Disability Needs Assessment Fiji Islands* (22 p), UNICEF, Suva, Fiji.
Pacific Islands Forum Secretariat. (2014). *The framework for Pacific regionalism* (p. 16). Suva, Fiji: Pacific Islands Forum Secretariat.
Sugi, M., Murakami, H., Yoshida, K. (2016). Projection of future changes in the frequency of intense tropical cyclones. *Climate Dynamics* https://doi.org/10.1007/s00382-016-3361-7.
Terry, J. P., & Lau, A. Y. A. (2018). Magnitudes of nearshore waves generated by tropical cyclone Winston, the strongest landfalling cyclone in South Pacific records. Unprecedented or unremarkable? *Sedimentary Geology, 364,* 276–285.
Tukayabu, I., Hibino, K., Sasaki, H., Shiogama, H., Mori, N., Shibutani, Y., et al. (2015). Climate change effects on the worst-case storm surge: a case study of Typhoon Haiyan. *Environmental Research Letters, 10,* 064011.
World Conservation Society. (2016). *Impact of tropical cyclone Winston on fisheries-dependent communities in Fiji* (p. 80). Suva, Fiji: World Conservation Society.

Extreme Weather Events and Health Responses in Taiwan

Li-San Hung and Mei-Hui Li

Abstract In this study, epidemiological studies on the effects of temperature-related and water-related extreme events on health in Taiwan were reviewed. Results of studies on the effects of temperate-related extreme events on health have varied because of the different indices used and diseases targeted; however, overall, these studies suggest that extreme low-temperature events generate greater mortality risk than do extreme high-temperature events, although health consequences from extreme high-temperature events cannot be ignored. With regard to health consequences from water-related extreme events, most related studies have focused on the association between floods and excessive rainfall, caused mainly by typhoons, with morbidity and mortality. Results of numerous studies have suggested that floods and excessive rainfall increase the risk of various diseases, especially water-borne, vector-borne and food-borne diseases.

Keywords Climate change · Urban areas · Cardiovascular diseases · Kidney disease · Infectious diseases · Cold surges

1 Introduction

Taiwan is located at the border that delineates tropical and subtropical regions; the island's annual average ambient temperature is above 20 °C, and its annual rainfall is approximately 3000 mm (Huang et al. 2016). Average temperatures in Taiwan have risen by 1.4 °C between 1911 and 2009, and it is approximately twice the increase in global surface temperature (0.74 °C; Hsu et al. 2011). The median near-surface temperature change has been estimated at between 2.1 and 2.4 °C based on the projected 2080–2099 average minus the 1980–1999 average (Hsu et al. 2011). Additionally, in the future, Taiwan may experience increased rainfall in the wet season and less rainfall in the dry season (Hsu et al. 2011). Studies have suggested

L.-S. Hung
Department of Geography, National Taiwan Normal University, Taipei, Taiwan

M.-H. Li (✉)
Department of Geography, National Taiwan University, Taipei, Taiwan
e-mail: meihuili@ntu.edu.tw

R. Akhtar (ed.), *Extreme Weather Events and Human Health*,
https://doi.org/10.1007/978-3-030-23773-8_14

that extreme temperature and water events seriously affect morbidity and mortality (Cann et al. 2013; McMichael et al. 2006). Accordingly, the effects of temperature-related and water-related extreme events on health in Taiwan warrant examination. This chapter summarizes the effects of temperature-related and water-related extreme events on health based on the findings of epidemiological studies of Taiwanese cases conducted over the past 25 years and particularly since 1994.

2 Temperature-Related Extreme Events

No universal definitions of heat waves or cold surges have been established. According to the World Meteorological Organization, a heat wave is defined as a period of more than 5 consecutive days where the daily maximum temperature exceeds the average maximum temperature by at least 5 °C. However, this definition is not suitable in Taiwan because conditions where the daily maximum temperature exceeds the average maximum temperature by 5 °C occur infrequently. An official definition of a heat wave has still not been determined for Taiwan. The most common method to establish specific thresholds involves selecting the tail of distributions for extreme temperatures. Diurnal temperature range (DTR—the difference between maximum and minimum temperatures) is one of the derived variables applied in studies of climate change or extremes.

Lu et al. (2012) analyzed climate variation and change based on records gathered from 1911 to 2009 from six meteorological stations (Taipei, Taichung, Tainan, Hengchun, Taitung and Hualien) (Fig. 1). The threshold for an extremely hot day from June to August was defined as one with temperatures reaching the 90th percentile of the probability distribution of the 99-year daily temperatures recorded at each station. The annual numbers of extremely hot days have increased at all six stations over the preceding 99 years (Lu et al. 2012). Among the aforementioned six stations, the Taipei station is located in the most urbanized area and the records from this station exhibited the steepest trend in temperature increase, with 1.4 extremely hot days per decade over the preceding 100 years and 4 extremely hot days per decade over the preceding 30 years. Moreover, the number of extremely hot days between 2000 and 2009 was more than 10 greater than that for 1911–1920 at the Taipei station. By contrast, the annual number of extremely hot days was highest for the periods of 1950–1960 at the Tainan station and 1955–1965 at the Hengchun station (Lu et al. 2012). In addition, data from these six stations revealed that the frequency and intensity of cold surges decreased because of the effects of warming from 1911 to 2009 (Lu et al. 2012). Cold surges were defined in relation to the following temperature changes: (1) the rank of the daily minimum temperature within a 48-h temperature drop was higher than the 90th percentile, or (2) the percentile rank of the daily minimum temperature was lower than the 10th percentile at each station (Lu and Lee 2009).

Shiu et al. (2009) analyzed hourly meteorological data collected at 26 meteorological stations in Taiwan from 1961 to 2005. The decreasing trends in DTR were

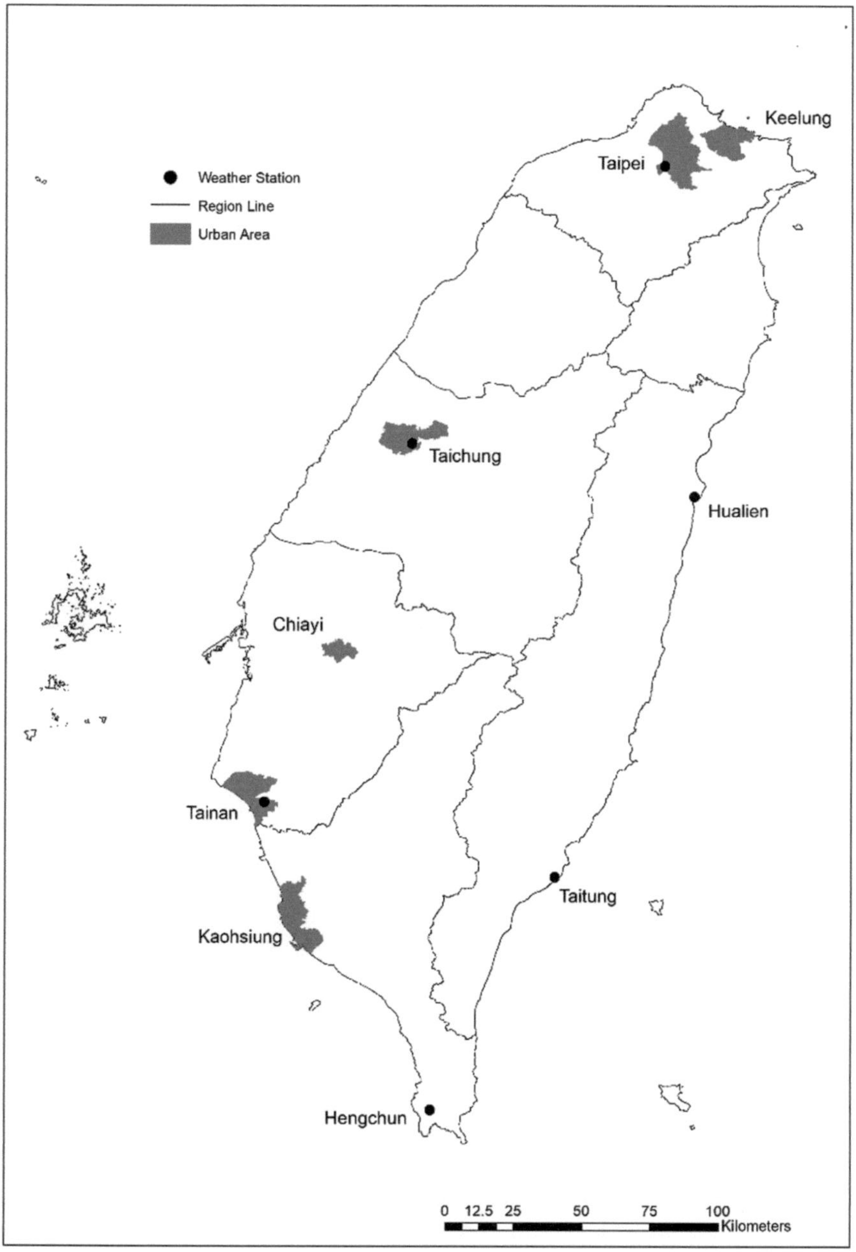

Fig. 1 Locations of mentioned meteorological stations, urban areas and regions in this study

significant at most stations. The decrease of DTR was largest in major urban areas and smaller in smaller cities. In the study of Kueh et al (2017), a 95th percentile threshold was used as the standard to identify hot extremes over a period of consecutive days. These authors examined heat waves in Taiwan and their relevance to the anomalous western North Pacific subtropical high by performing dynamically downscaled simulations based on observations from 1979 to 2003 as well as simulations for 2075–2099. Heat waves are projected to be more severe in the future; in addition, daily minimum temperatures are expected to be high, and the DTR is expected to be relatively large. Lin et al. (2017) studied climate change based on heat waves that occurred from 1971 to 2010 in three major cities of Taiwan—Taipei, Taichung and Kaohsiung (Fig. 1). In that study, a heat wave was defined as daily maximum air temperature in the 95th percentile for more than 3 consecutive days during the study period; the results revealed that from 2001 to 2010, heat wave temperatures in Taipei and Kaohsiung significantly increased by 0.6 °C, and heat wave events or numbers of days and duration also increased. Moreover, the researchers calculated the wet-bulb globe temperature heat stress for the aforementioned three cities from 2003 to 2012 and simulated the warming scenario for 2075–2099 by using ECHAM5/MPIOM-WRF (ECW) with dynamic downscaling and 5-km resolution. Their simulation results indicated that heat stress in the three cities will exceed or approach the danger level by the end of the current century. In summary, the aforementioned studies have indicated that the effects of dangerous heat stress on public health are expected to increase and require more attention in Taiwan.

3 Temperature-Related Mortality and Morbidity

Increased mortality and morbidity due to extreme temperature events are recognized as a major global public health challenge. Cardiac failure, stroke, and respiratory infections are expected to increase with exposure to cold extremes. Additionally, elevated temperatures and more frequent and prolonged hot days may induce health complications such as heat stroke, heat exhaustion and exacerbation of health conditions in people with obesity, cardiovascular disease, respiratory disease and diabetes mellitus (Turner et al. 2012; Bunker et al. 2016). In Taiwan, many island-wide epidemiological studies have been conducted to identify vulnerable regions that exhibit links between socioeconomic factors and susceptibility to extreme temperature events. Urban areas also exhibit critical vulnerabilities under extreme weather events. Therefore, some studies have assessed extreme temperature events and health complications in urban areas. The following subsections are organized around the two types of studies, namely urban-area-focused and island-wide studies, in their presentations of epidemiological evidence regarding the health effects of extreme temperature events in Taiwan.

3.1 Health Responses to Extreme Temperature Events in Urban Areas

Several studies have investigated health responses associated with extreme temperature events in four major cities (Taipei, Taichung, Tainan and Kaohsiung) in Taiwan (Fig. 1). Lin et al. (2011) revealed that elderly individuals exhibited the lowest overall mortality when the temperature was 26 °C from 1994 to 2003; this finding suggests that low temperatures may have caused greater mortality than high temperatures through the aggravation of circulatory diseases. Lin et al. also discovered that mortality risk slightly increased under more intense and prolonged heat extremes but did not increase under more intense or prolonged cold extremes. In another study by the same research team, low temperatures were associated with greater risk of mortality from cardiovascular diseases than were high temperatures from 1994 to 2007 (Lin et al. 2013a). Wang et al. (2012) evaluated emergency room visit (ERV) risk for all causes and independently for cardiopulmonary disease associated with long-lasting extreme temperatures from 2000 to 2009. Low temperatures were associated with a slightly higher risk for circulatory-disease-related ERVs than were high temperatures. After accounting for the 4-day cumulative temperature effect, the ERV risks for all causes and respiratory diseases independently were associated with extreme cold in the 5th percentile that lasted for > 8 days and that in the 1st percentile that lasted for > 3 days. In addition, the first extreme heat event reaching the 99th percentile temperature for each year was associated with higher ERV for all causes and circulatory diseases.

Sung et al. (2013) adopted a daily mean heat index that combined temperature and relative humidity to predict the risk of mortality in populations from six major cities (Keelung, Taipei, Taichung, Tainan, Chiayi and Kaohsiung) in Taiwan from 1994 to 2008 (Fig. 1). In all six cities, significantly higher risk ratios of daily mortality were found when the daily mean heat indices were at and above the 95th percentile compared with the risk ratios associated with the lowest percentile. The increased risks were similar among individuals aged 65 years and older in the five largest cities but not among those in the least urbanized area, namely Chiayi.

3.2 Factors of Vulnerability to Extreme Temperature Events in Island-Wide Studies

In Taiwan, all deaths are reported to the township and district household registry offices. Socioeconomic and demographic variables for each township may be estimated using data from the Department of Health and the Taiwanese government's census report. Chen et al. (2010) examined the nonstationary effects of social determinants on cardiovascular mortality between 1997 and 2003 for 349 townships in Taiwan (the nine townships from the four offshore islands were excluded). Their results revealed that an immediate increase in cardiovascular mortality occurred after each cold surge. Five social determinants, namely social disadvantage, lack of economic opportunity, stability, vulnerable groups and rurality, exhibited spatial nonstationary effects on cardiovascular mortality rates after cold surges. This finding provided an empirical basis for developing public health programs with local emphasis on the effects of extreme cold. The same research team (Wu et al. 2011) used spatial regression models to examine the relationships among the spatial characteristics of temperature, extracted factors from demographic and socioeconomic parameters, and mean cardiovascular mortality in all 358 of Taiwan's townships 2 weeks before and after cold and heat events. The study results indicated that urban areas were associated with substantially lower mortality than were rural areas after cold and heat events that occurred from 1994 to 2003. Compared with heat events, cold events exerted greater effects on mortality ratios in most townships. Furthermore, a negative association was found between mortality after extreme cold and heat events and urbanization or medical resource availability. High percentages of older people, vulnerable groups and aborigines in certain township populations might have contributed to the high vulnerability of township populations to extreme cold and heat events.

Based on geographical and socioeconomic characteristics, several studies have examined the association of health effects and extreme weather in seven regions of Taiwan (Fig. 1) from 2000 to 2008. Lin et al. (2013b) examined the association of hospital admissions for kidney disease in relation to extreme and prolonged heat and cold events. The effect of hot temperatures on kidney-disease-related morbidity leading to hospital admission was more significant than that of cold temperatures but the additional adverse effects were weak for prolonged extreme heat events. Wang and Lin (2014) evaluated ambient temperatures and extreme temperature events' associations with risk of outpatient visits for asthma, and chronic airway obstruction not elsewhere classified (CAO). The results indicated that elderly patients with respiratory disease and CAO were vulnerable to temperatures of 30 °C and higher, whereas younger patients associated with all types of respiratory diseases were vulnerable to temperatures of 18 °C and below. Elevated numbers of outpatient visits for all respiratory diseases and asthma were associated with events of extreme heat lasting for 6–8 days. Conversely, events of extreme cold lasting for more than 8 days were significantly negatively associated with outpatient visits for all respiratory diseases. Lin et al. (2015b) analyzed the all-cause and cardiovascular disease mortality risks associated with regional ambient temperatures for the same period. The mortality risks were greater during low temperatures than high temperatures, especially that from cardiovascular disease. These results are in agreement with the results of studies from other subtropical (Ou et al. 2013) or tropical area (Xuan et al. 2014). People living in tropical or subtropical regions are probably experienced a higher mortality risk or health effects to cold temperature than those living in cold regions. It suggests cold-related mortality is also an important public health problem and should not be underestimated on temperature-attributable mortality and morbidity.

4 Water-Related Extreme Events

Global climate change is expected to influence the global frequency, intensity and duration of extreme water-related events such as droughts, storm surges, floods, and excessive rainfall (Cann et al. 2013). Extreme rainfall events may increase the risks of pathogen transmission in the environment through bodies of water (Chen et al. 2012). Such events can increase water turbidity and lead to gastrointestinal illness (Hunter 2003). Exposure to contaminated water, moist soil, and vegetation after typhoon or flood events may lead to various infectious diseases such as leptospirosis and melioidosis (Chiu et al. 2009; Ko et al. 2007). Extreme water-related events are likely to be more severe in the future (Hsu et al. 2011), and this will likely increase the number and intensity of public health challenges that the public and private sectors must address.

Several studies have examined the historical patterns of precipitations in Taiwan and have used climate models to predict future precipitation scenarios. Tung et al. (2016) used a probability index (PI) to evaluate changes in extreme rainfall between 1960 and 2009 in Taiwan. For the PI, the cumulative density function of a generalized extreme value distribution was used to fit and standardize the annual daily maximum precipitation (RX1 day). Although the RX1 day is a popular index for representing extreme rainfall events, it is sensitive to outliers and difficult to apply to large areas. In their study, Tung et al. divided Taiwan into four regions: north, center, south and east. Their study results indicated that no distinct linear trend occurred for decadal change in any of the four regions before the year 2000. That study also projected future extreme rainfall events based on RCP4.5 and RCP8.5. The RCP8.5 scenario predicted that the PI will increase 10% by 2065 and 14% by 2100. Huang et al. (2012) investigated the effects of climate change on rainfall frequency in Taiwan based on five Intergovernmental Panel on Climate Change climate models under the Special Report on Emission Scenario A1B. The change in the frequency of events of maximum consecutive dry days and events of maximum 1-, 2- and 3-day rainfall for 1980–1999 and 2080–2099 were observed. The results indicated that the risk of droughts and floods in Taiwan is likely to increase in 2080–2099. Additionally, water resources are likely to be more unevenly distributed in the future. Chen et al. (2009) investigated the historical trends of meteorological drought in Taiwan based on data obtained from 22 climate stations. Their study suggested that since 1960, incidence of meteorological drought has decreased in northeastern Taiwan but increased in central and southern Taiwan. In addition, the results revealed that the number of dry days has increased.

5 Water-Related Mortality and Morbidity

Water-related climate events include droughts, storm surges, floods and excessive rainfall. Few studies have examined the relationships between drought events and mortality and morbidity in Taiwan, especially with regard to the influence of climate change. In the present study, we focused on the available literature concerning the associations of floods with excessive rainfall, mortality and morbidity in Taiwan. Many annual flood events and excessive rainfall events in Taiwan are caused by typhoons during summer and fall. Between 1958 and 2016, approximately 4,388 people were reported as missing or dead and approximately 15,010 people were reported as wounded because of typhoons (Central Weather Bureau 2018). A preliminary study on the 1,556 deaths attributed to typhoons in Taiwan between 2000 and 2014 suggested that the work-related deaths therein were mostly related to outdoor activities such as farming and fishing, and the non-work-related deaths were mostly associated with inundation in people's homes and landslides (Chang 2016).

At least two studies have investigated the relationships between water-related extreme events and incidences of several infectious diseases in Taiwan. Chen et al. (2012) studied the effects of extreme precipitation on the distribution of eight infectious diseases, both water-borne and vector-borne, across the main island of Taiwan from 1994 to 2008. The infectious diseases studied were hepatitis A, enteroviruses, bacillary dysentery, leptospirosis, melioidosis, scrub typhus, dengue fever and Japanese encephalitis. The results indicated that daily precipitation levels were significantly correlated with incidences of all of these infectious diseases in Taiwan. For water-borne diseases, extreme torrential precipitation, defined as > 350 mm/day, was correlated with higher relative risks for bacillary dysentery and enterovirus infections compared with an ordinary precipitation level, defined as < 100 mm/day. Regarding vector-borne diseases, the relative risk of dengue fever and Japanese encephalitis increased with greater precipitation only up to 350 mm/day. Huang et al. (2016) investigated the incidences of flood-related diseases of the eyes, skin and gastrointestinal tract in Taiwan between 1998 and 2008; the study results suggested that the ratios of disease incidence within 10 days after flood events to 10 days before the events were 1.15 for eyes, 1.08 for skin and 1.11 for the gastrointestinal tract after covariates had been controlled for.

Several studies have investigated the relationships between extreme rainfall events and the incidence of intestinal infections such as those involving *Vibrio parahaemolyticus* and diarrhea-associated diseases. Hsiao et al. (2016) studied the effects of climate variation on *Vibrio parahaemolyticus* infection outbreaks in Taiwan between 2000 and 2011. The study results indicated that the level of maximum daily rainfall (current and one month ago) significantly and negatively affected the incidence of *Vibrio parahaemolyticus* infection. Pathogen concentrations in estuaries may have been diluted during heaving rainfall, thereby reducing levels of contamination in fish and other seafood; this may explain the study results. However, a study by Chou et al. (2010) suggested that heavy rainfall was positively correlated with the incidence of diarrhea-associated diseases in Taiwan. The authors used the number of days of extreme rainfall as the indicator of heavy rainfall, and their results suggested that extreme rainfall days were strongly related to diarrhea-associated morbidity in Taiwan between 1996 and 2007. Specifically, the authors discovered that extreme rainfall days were significantly correlated with diarrhea-associated morbidity in young adults (15–39-year-olds) but not in children (0–14-year-olds) or older adults (40–64-year-olds).

Health consequences after the strike of typhoons were widely discussed. Chiu et al. (2009) reported that a hospital in Taiwan admitted six patients with serologically confirmed cases of leptospirosis between 2004 and 2008. All six patients had been in contact with contaminated soil or water, likely from typhoon-associated floods and heavy rainfall. Thus, the authors suggested that people must be educated about how not to come into contact with contaminated water or soil and to wear protective clothing and footwear in areas of suspected contamination during typhoon season. Ko et al. (2007) reported an outbreak of 40 cases of melioidosis that followed widespread flooding in southern Taiwan caused by Typhoon Haitung, which hit on July 16, 2005. That study also confirmed that melioidosis is endemic in Taiwan.

Several diseases were reported after the struck of Typhoon Morakot, which had record-breaking rainfall over southern Taiwan in the past 50 years (Chien and Kuo 2011), in August 2009. These diseases included leptospirosis (Lin et al. 2012; Su et al. 2011) and melioidosis (Su et al. 2011). Lin et al. (2013c) discussed the epidemiological characteristics of lower extreme cellulitis occurrence after floods caused by Typhoon Morakot in 2009. The researchers found that the number of patients with cellulitis increased from 183 to 344 during the 30-day period after the typhoon. In another study, Lin et al. (2015a) investigated the risk factors for post-flood infectious diseases among people displaced by Typhoon Morakot in Taiwan. Among the 288 studied samples, seroconversion of Entamoeba histolytica was observed in 128 participants (44.4%). A significantly higher rate of seroconversion was identified in the shelter group (56.1%) than in the community group (36.8%). A few cases of vaccine-preventable diseases of measles, mumps and rubella were also found in both the community group and shelter group. These results suggest that a clean water supply is crucial for the management of post-disaster situations, and vaccination programs should be extended to populations at high risk of post-disaster displacement.

6 Conclusion

This chapter reviews the health consequences of extreme weather-related and water-related events in Taiwan for the past 25 years. The results suggested that overall, in Taiwan, extremely low-temperature events have resulted in greater mortality risk than have extreme high-temperature events, while extreme high-temperature events have increased in higher risk for several kinds of diseases, such as kidney-related diseases and cardiovascular diseases. Besides, floods and excessive rainfall are associated with increased incidences of various diseases, especially water-borne, vector-borne and food-borne diseases. Mitigation of these negative effects remains a critical topic for Taiwanese society. Taiwan's temperature is likely to increase. In addition to focusing on the health effects of extreme temperature events, the public and private sectors in Taiwan may be required to increase their attention on the health consequences associated with water-related extreme events.

References

Bunker, A., Wildenhain, J., Vandenbergh, A., Henschke, N., Rocklöv, J., Hajat, S., et al. (2016). Effects of air temperature on climate-sensitive mortality and morbidity outcomes in the elderly; a systematic review and meta-analysis of epidemiological evidence. *EBioMedicine, 6,* 258–268.

Cann, K. F., Thomas, D. R., Salmon, R. L., Wyn-Jone, A. P., & Kay, D. (2013). Extreme water-related weather events and waterborne disease. *Epidemiology Infection, 141*(4), 671–686.

Chang, C. T. (2016). A preliminary study on mortality attributed to tropical cyclones in Taiwan 2000–2014. *Environmental and Hazards, 15*(2), 178–187.

Chen, S. T., Kuo, C. C., & Yu, P. S. (2009). Historical trends and variability of meteorological droughts in Taiwan. *Hydrological Sciences Journal, 54*(3), 430–441.

Chen, V. Y. J., Wu, P. C., Yang, T. C., & Su, H. J. (2010). Examining non-stationary effects of social determinants on cardiovascular mortality after cold surges in Taiwan. *Science of the Total Environment, 408*(9), 2042–2049.

Chen, M. J., Lin, C. Y., Wu, Y. T., Wu, P. C., Lung, S. C., & Su, H. J. (2012). Effects of extreme precipitation to the distribution of infectious diseases in Taiwan, 1994–2008. *PLoS ONE, 7*(6), e34651.

Chien, F. C., & Kuo, H. C. (2011). On the extreme rainfall of Typhoon Morakot (2009). *Journal of Geophysical Research, 116,* D05104.

Chiu, C. H., Wang, Y. C., Yang, Y. S., & Chang, F. Y. (2009). Leptospirosis after typhoon in Taiwan. *Journal of Medical Sciences, 29*(3), 131–134.

Chou, W. C., Wu, J. L., Wang, Y. C., Huang, H., Sung, F. C., & Chuang, C. Y. (2010). Modeling the impact of climate variability on diarrhea-associated diseases in Taiwan (1996–2007). *Science of the Total Environment, 409*(1), 43–51.

Central Weather Bureau. (2018). *How many losses were caused by typhoons?* https://www.cwb.gov.tw/V7/knowledge/encyclopedia/ty055.htm. Accessed 6 June 2018.

Hsiao, H. I., Jan, M. S., & Chi, H. J. (2016). Impacts of climatic variability on *Vibrio parahaemolyticus* outbreaks in Taiwan. *International Journal of Environmental Research and Public Health, 13*(2), 188.

Hsu, H. H., Chou, C., Wu, Y. C., Lu, M. M., Chen, C. T., & Chen, Y. M. (2011). *Climate change in Taiwan: Scientific report 2011 (summary).* Taipei: National Science Council.

Huang, W. C., Chiang, Y., Wu, R. Y., Lee, J. L., & Lin, S. H. (2012). The impact of climate change on rainfall frequency in Taiwan. *Terrestrial, Atmospheric and Oceanic Sciences, 23*(5), 553–564.

Huang, L. Y., Wang, Y. C., Wu, C. C., Chen, Y. C., & Huang, Y. L. (2016). Risk of flood-related diseases of eyes, skin and gastrointestinal tract in Taiwan: A retrospective cohort study. *PLoS ONE, 11*(5), e0155166.

Hunter, P. R. (2003). Climate change and waterborne and vector-borne disease. *Journal of Applied Microbiology, 94*(s1), 37–46.

Ko, W. C., Cheung, B. M. H., Tang, H. J., Shih, H. I., Lau, Y. J., Wang, L. R., et al. (2007). Melioidosis outbreak after typhoon, southern Taiwan. *Emerging Infectious Diseases, 13*(6), 896–898.

Kueh, M. T., Lin, C. Y., Chuang, Y. J., Sheng, Y. F., & Chien, Y. Y. (2017). Climate variability of heat waves and their associated diurnal temperature range variations in Taiwan. *Environmental Research Letters, 12*(7), 074017.

Lin, Y. K., Ho, T. J., & Wang, Y. C. (2011). Mortality risk associated with temperature and prolonged temperature extremes in elderly populations in Taiwan. *Environmental Research, 111*(8), 1156–1163.

Lin, C. Y., Chiu, N. C., & Lee, C. M. (2012). Leptospirosis after typhoon. *The American Journal of Tropical Medicine and Hygiene, 86*(2), 187–188.

Lin, Y. K., Chang, C. K., Wang, Y. C., & Ho, T. J. (2013a). Acute and prolonged adverse effects of temperature on mortality from cardiovascular diseases. *PLoS ONE, 8*(12), e82678.

Lin, Y. K., Wang, Y. C., Ho, T. J., & Lu, C. A. (2013b). Temperature effects on hospital admissions for kidney morbidity in Taiwan. *Science of the Total Environment, 443,* 812–820.

Lin, P. C., Lin, H. J., Guo, H. R., & Chen, K. T. (2013c). Epidemiological characteristics of lower extremity cellulitis after a typhoon flood. *PLoS ONE, 8*(6), e65655.

Lin, C. Y., Chen, T. C., Dai, C. Y., Yu, M. L., Lu, P. L., Yen, J. H., et al. (2015a). Serological investigation to identify risk factors for post-flood infectious diseases: a longitudinal survey among people displaced by Typhoon Morakot in Taiwan. *British Medical Journal Open, 5*(5), e007008.

Lin, Y. K., Yu, S. W., Lin, P. L., & Wang, Y. C. (2015b). Ambient temperature and all-cause and cardiovascular diseases mortality risk by areas in Taiwan. *Taiwan Gong Gong Wei Sheng Za Zhi, 34,* 103–114. [in Chinese with English abstract].

Lin, C. Y., Chien, Y. Y., Su, C. J., Kueh, M. T., & Lung, S. C. (2017). Climate variability of heat wave and projection of warming scenario in Taiwan. *Climate Change, 145*(3–4), 305–320.

Lu, M. M., & Lee, S. Y. (2009). Cold surges in Taiwan—A critique of using Taipei station temperature only to define the events. *Atmospheric Science, 37,* 1–10. [in Chinese with English abstract].

Lu, M. M., Cho, Y. M., Lee, S. Y., Lee, C. T., & Lin, Y. C. (2012). Climate variations in Taiwan during 1911–2009. *Atmospheric Science, 40,* 297–317. [in Chinese with English abstract].

McMichael, A. J., Woodruff, R. E., & Hales, S. (2006). Climate change and human health: present and future risks. *The Lancet, 367*(9513), 859–869.

Ou, C. Q., Song, Y. F., Yang, J., Chau, P. Y., Yang, L., Chen, P. Y., et al. (2013). Excess winter mortality and cold temperatures in a subtropical city, Guangzhou. *China. PLoS One, 8*(10), e77150.

Shiu, C. J., Liu, S. C., & Chen, J. P. (2009). Diurnally asymmetric trends of temperature, humidity, and precipitation in Taiwan. *Journal of Climate, 22*(21), 5635–5649.

Su, H. P., Chan, T. C., & Chang, C. C. (2011). Typhoon-related leptospirosis and melioidosis, Taiwan, 2009. *Emerging Infectious Diseases, 17*(7), 1322–1324.

Sung, T. I., Wu, P. C., Lung, S. C., Lin, C. Y., Chen, M. J., & Su, H. J. (2013). Relationship between heat index and mortality of 6 major cities in Taiwan. *Science of the Total Environment, 442,* 275–281.

Tung, Y. S., Chen, C. T., Min, S. K., & Lin, L. Y. (2016). Evaluating extreme rainfall changes over Taiwan using a standardized index. *Terrestrial, Atmospheric & Oceanic Sciences, 27*(5), 705–715.

Turner, L. R., Barnett, A. G., Connell, D., & Tonga, S. (2012). Ambient temperature and cardiorespiratory morbidity: a systematic review and meta-analysis. *Epidemiology, 23,* 594–606.

Wang, Y. C., & Lin, Y. K. (2014). Association between temperature and emergency room visits for cardiorespiratory diseases, metabolic syndrome-related diseases, and accidents in metropolitan Taipei. *PLoS ONE, 9*(6), e99599.

Wang, Y. C., Lin, Y. K., Chuang, C. Y., Li, M. H., Chou, C. H., Liao, C. H., et al. (2012). Associating emergency room visits with first and prolonged extreme temperature event in Taiwan: a population-based cohort study. *Science of the Total Environment, 416,* 97–104.

Wu, P. C., Lin, C. Y., Lung, S. C., Guo, H. R., Chou, C. H., & Su, H. J. (2011). Cardiovascular mortality during heat and cold events: determinants of regional vulnerability in Taiwan. *Occupational and Environmental Medicine, 68,* 525–530.

Xuan, L. T., Egondi, T., Ngoan, L. T., Toan, D. T., & Huong, L. T. (2014). Seasonality in mortality and its relationship to temperature among the older population in Hanoi, Vietnam. *Global Health Action., 7*(1), 23115.

Extreme Weather and Climate Events and Occupational Health in Thailand

Uma Langkulsen and Desire Rwodzi

Abstract *Background* Extreme weather and climate events impact public health in multiple ways and extreme heat is one of the major environmental challenges for public health in Thailand. *Objective* We assessed the effects of heat stress on occupational health among agricultural workers who are exposed to the heat at different levels for both males and females in different age groups. We further investigated the responses to extreme heat among the government agencies. *Methods* A cross-sectional study was used to identify heat exposure situations that create heat stress-related health risks. The question guidelines have been developed to assess occupational heat impacts in the study site and used semi-structured interviews to describe how agricultural workers experience heat exposures where they work. The Wet Bulb Globe Temperature (WBGT) index and temperatures were measured by WBGT monitors in both winter and summer. Twenty-nine participants aged 24–76 were recruited to the study. *Results* All participants had one or more signs and symptoms of heat stress; however, they had concerns about work-related heat stress prevention. Outdoor WBGT was found to be highest between 11:00 AM and 11:59 AM, and 13:00 PM and 13:59 PM in winter and summer, respectively. Our findings revealed that study sites had heat indices necessitating extreme caution, where sunstroke, muscle cramps, and/or heat exhaustion were possible with prolonged exposure. *Conclusions* A heat health warning system is essential to reduce the negative impacts of extreme weather.

Keywords Climate change · Extreme weather · Heat exposure · Heat stress · Thailand

U. Langkulsen (✉)
Faculty of Public Health, Thammasat University, Rangsit Campus, Pathum Thani 12121, Thailand
e-mail: u.langkulsen@gmail.com

D. Rwodzi
Paediatric HIV, Clinton Health Access Initiative, Harare, Zimbabwe

R. Akhtar (ed.), *Extreme Weather Events and Human Health*,
https://doi.org/10.1007/978-3-030-23773-8_15

1 Introduction

A clear manifestation that the global climate has changed over the past half-century is the increased frequency and duration of extreme weather events such as heat waves around the world (McMichael et al. 2006). Extreme weather events have made the risks associated with exposure to high ambient temperatures a major public health concern. Several epidemiological investigations have linked exposure to high ambient temperatures with heat stress in humans, culminating in morbidity and mortality. Heat stress occurs when a person's surrounding environment, clothing and activity interact in a manner that forces the body temperature to rise (Health and Safety Professionals Alliance: HaSPA 2012; Parsons 2014). The rate of metabolic heat production leads to a rise in body core temperature (Jay and Brotherhood 2016). Heat stress has had its toll on human populations in a variety of settings across the world. There is enough evidence to characterize the temperature on morbidity and mortality relationships (Bai et al. 2014; Gasparrini et al. 2015; Hajat and Kosatky 2010; Jackson et al. 2010; Metzger et al. 2010; Tawatsupa et al. 2014; Ye et al. 2012). The heterogeneity in heat-related morbidity and mortality between populations could also be explained by population density, gross domestic product (GDP) as well as age distribution (Hajat and Kosatky 2010).

Both acute and chronic effects of heat stress on human health are well-documented (Xiang et al. 2014). Acute effects include deterioration of concentration and fine motor skills, and heat exhaustion may start to manifest within a rise in core body temperature of 2 °C (HaSPA 2012). Prolonged heat exposure causes chronic diseases affecting the kidney, liver, heart, digestive system, nervous system, skin and respiratory system, as well as sleep disorders, and alterations in gestational length (Baccini et al. 2008; HaSPA 2012; Tawatsupa et al. 2012). Furthermore, prolonged exposure to heat threatens the thermoregulatory system of outdoor workers and causes significant productivity loss in occupational settings (Langkulsen et al. 2010). HaSPA (2012) revealed that dehydration of 1–2% of body weight results in a 6–7% reduction in physical work rate. In Thailand, working under heat stress was associated with both worse overall health and psychological distress (Tawatsupa et al. 2012). In addition, vector-borne diseases are also prevalent among workers who work at dawn or dusk to avoid heat stress (Bennett and McMichael 2010).

Most of the evidence from epidemiological studies of extreme temperatures has been done in North American and European populations (Hajat and Kosatky 2010; McMichael et al. 2006). These populations are in temperate-developed countries, whereas little has been studied about the effects of hot and humid weather in tropical countries such as Thailand, where temperatures are expected to rise even further under global climate change (Tawatsupa et al. 2014). Because of its geography and stage of economic development, Thailand is one of the countries that is experiencing a substantial increase in temperature over recent years. Thailand's annual mean maximum temperature increased significantly from 1951 to 2017 by approximately 1 °C (Thai Meteorological Department 2018a). Chronic illnesses such as diseases of the circulatory system and diseases of the respiratory system, which have been linked to excessive heat exposure, have also become a serious public health issue as they are among the top five leading categories of cause of death in Thailand (Strategy and Planning Division 2018). Morbidity rates induced by the effects of heat and sunlight exposure (International Statistical Classification of Diseases and Related Health Problems 10th Revision: ICD 10 T67) of the Thai population for the whole kingdom (excluding Bangkok Metropolis) rose from 2.42 per 100,000 population in 2013 to 4.61 in 2016, and decreased by 0.22 per 100,000 population in 2017. Children aged 10–14 and the elderly (aged 60 years and above) are the most affected by heat and sunlight; based on occupational groups, agricultural and farm workers are disproportionately affected, as substantiated by their morbidity rates (Ministry of Public Health 2018).

2 Climate Variability and Extreme Heat in Thailand

The map in Fig. 1 shows the annual average maximum temperature across Thailand from 2012 to 2016. The dataset was produced by the Weather Forecast Bureau of the Thai Meteorological Department. During March to May 2013, the daily maximum temperatures exceeded 40 °C in ten stations are shown in Table 1. The maximum temperature of the year 2014 was 41.8 °C at Thoen in Lampang province and at Mueang in Mae Hong Son province on March 31 and April 24, respectively (Thai Meteorological Department 2018b).

In 2015, the maximum temperature in several areas of Thailand reached a new record. Based on the annual mean temperature data, there was an increase of 0.8 °C

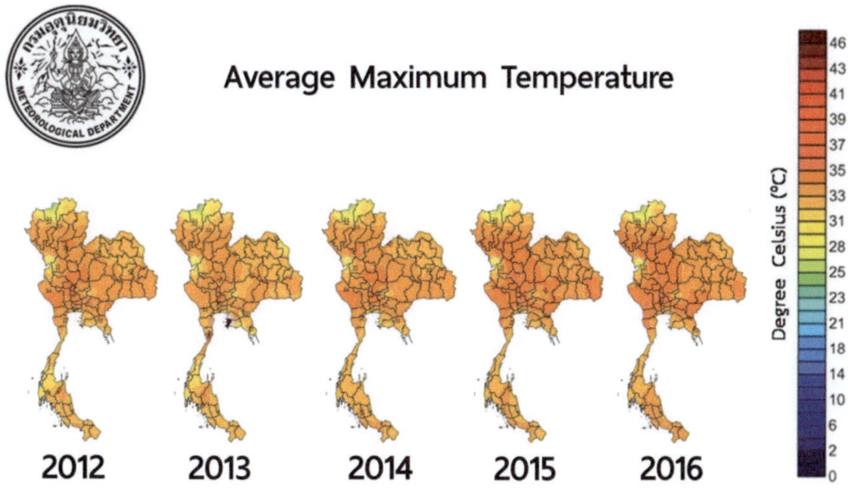

Fig. 1 Annual average maximum temperature in Thailand, 2012–2016. *Source* Thai Meteorological Department, 2017

Table 1 Description and location of meteorological stations

No.	Station name	Province	Region	Latitude	Longitude	Temperature (°C)
1	Sakon Nakhon Agromet	Sakon Nakhon	Northeastern	17.7.0	104.3.0	41.0
2	Burirum	Burirum	Northeastern	15.13.0	103.14.0	41.2
3	Bangkok Metropolis	Bangkok Metropolis	Central	13.43.35	100.33.36	40.1
4	Aranyaprathet	Sa Kaeo	Eastern	13.42.0	102.35.0	40.3
5	Nong Phlub Agromet	Prachuap Khiri Khan	Southern East Coast	12.35.0	99.44.0	40.8
6	Phichit Agromet	Phichit	Northern	16.26.17	100.17.33	40.7
7	Tha Phra Agromet	Khon Kaen	Northeastern	16.20.0	102.49.0	42.6
8	Kamalasai	Kalasin	Northeastern	16.19.57	103.35.18	41.8
9	Roi Et Agromet	Roi Et	Northeastern	16.4.0	103.37.0	41.2
10	Chai Nat Agromet	Chai Nat	Central	15.9.0	100.11.0	41.6

Source Thai Meteorological Department

above normal for all months, with the annual mean temperature of 27.9 °C. The Thai Meteorological Department reported that 2015 was the second warmest year in 65 years of record-keeping, following only 1998. Similarly, the annual mean temperature for 2016 was 28.0 °C, and the mean temperature was above normal by more than 1 °C. The mean temperatures over Thailand were 2.2 °C above normal during April and 1.8 °C above in May. Several areas of Thailand have the new highest recorded maximum temperature (Thai Meteorological Department 2018b).

3 Health Impacts of Extreme Heat

Given the paucity of studies on the subject, particularly in populations residing in hot and humid environments, we aimed to investigate the effects of heat stress on agricultural workers in Thailand.

3.1 Methods

3.1.1 Study Design

A cross-sectional study design was used to examine the relationship between climate variables and health status in a natural occupational setting exposed to heat.

3.1.2 Study Area

Pathum Thani province is located in the central part of Thailand and occupies 1525.856 km^2 (Pathum Thani 2018) as shown in Fig. 2. The province is consisting of seven districts include Mueang Pathum Thani, Thanyaburi, Khlong Luang, Lam Luk Ka, Lat Lum Kaeo, Sam Khok, and Nong Suea. Lat Lum Kaeo district had the highest number of rice farming households and a rice planting area of 84,896 rai (Pathum Thani Agricultural Provincial Extension Office 2018). The study area of paddy in Lat Lum Kaeo district in Pathum Thani province was selected purposively.

This province was selected for the study because the average rice yield is 758 and 754 kg/rai in crop year 2011/2012 and 2012/2013, respectively (Office of Agricultural Economics 2016). In addition, the provinces in Thailand are officially grouped into four levels, based on rice production potential; Pathum Thani province was considered to be R1 as the rice production potential is more than 550 kg/rai (Rice Department 2018).

The climate of Pathum Thani province is governed by two monsoon winds, which originate from the northeast between mid-October and mid-February, and the southwest from mid-May to mid-October. In 2011, the average daily temperature ranges from 20.8 to 33.6 °C, a mean annual temperature of 28.6 °C, average daily humidity of 75.1%, and mean annual rainfall of about 1530 mm (Pathum Thani Provincial Office 2018).

3.1.3 Study Participants

Sample size calculation is not required for this qualitative research, 29 rice farmers of Lat Lum Kaeo district were included in the study. The standard approach for in-depth interviews as well as standard practice for estimates of interviews was applied. Inclusion criteria were agricultural workers aged 20 and 80. The exclusion criteria were a refusal to give informed consent. All participants provided written informed consent prior to their participation in the study.

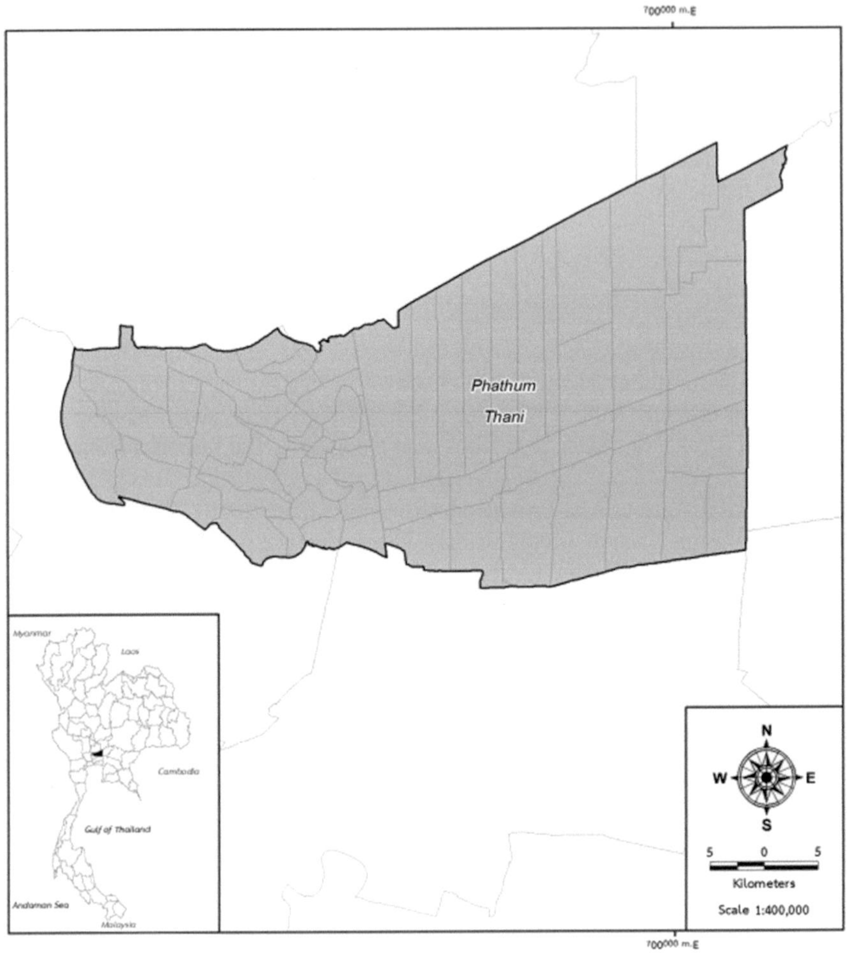

Fig. 2 Location of the study area

3.1.4 Interviews

The study on occupational health was conducted on 29 participants in Pathum Thani province. The question guidelines were first developed in English by the Hothaps team (High Occupational Temperature: Health and Productivity Suppression) under the Hothaps program for climate change impact assessment and prevention (Kjellstrom 2012). The guidelines were then translated into Thai and tested to ensure they are phrased the right way. We formulated 7 parts regarding (1) general information; (2) type of work; (3) heat exposure at work; (4) impacts of heat on health; (5) impacts on work activities and productivity; (6) heat prevention approaches; and (7) heat exposure outside of work.

Fig. 3 WBGT monitor in paddy at Lat Lum Kaeo district

3.1.5 WBGT Measurements

The data collection took place between January and May 2012. Wet Bulb Globe Temperature (WBGT) heat stress index is used in the assessment of the effects of a wide range of outdoor weather conditions on human responses. Heat exposures were monitored at three locations in a paddy where rice farmers normally worked for 10 consecutive days, using QUESTemp° 34 Monitor (Quest Technologies Inc.) made in the USA as shown in Fig. 3. These locations represent the heat exposure that working rice farmers experience outdoors.

3.1.6 Data Analysis

The hourly data were presented as time trend graphs based on WBGT measurements. Comparisons of dry season (winter and summer) measurements were made in this study. The collected data were analyzed using SPSS 23.0 for Windows. Descriptive statistics are used in the study to describe the characteristics of participants.

3.1.7 Ethical Approval

This study was approved by the Human Research Ethics Committee of Thammasat University (No. 2) (Number of COA 010/2554).

Table 2 Description of the participants

Variables[a]	Age group of participant	
	< 60 years of age ($n = 25$)	≥ 60 years of age ($n = 4$)
Male, n (%)	22 (88)	1 (25)
Mean age, years ± SD	42.4 ± 11.1	68.2 ± 5.8
Work experience, year average	24.6	55.0
Signs and symptoms of heat stress, n (%)		
Excess sweating	25 (100)	4 (100)
Feeling thirsty	25 (100)	4 (100)
Heat rash	10 (40)	4 (100)
Tiredness and exhaustion	25 (100)	4 (100)
Feeling faint	5 (20)	4 (100)
Moody	17 (68)	4 (100)
Insomnia	13 (52)	2 (50)
Nausea	11 (44)	3 (75)
Dizziness	15 (60)	3 (75)
Headache	20 (80)	3 (75)
Muscle cramps	14 (56)	3 (75)
Heat exposure reduce work output, n (%)	14 (56)	2 (50)
Aware of any safety standards regarding heat, n (%)	14 (56)	2 (50)

[a]SD = Standard deviation

3.2 *Results*

3.2.1 Characteristics of Study Participants

A total of 29 participants were enrolled in the study and the mean age was 46.0 years (SD = 13.8). The group of participants was comprised of 23 males (79.3%) and 6 females (20.7%). The mean work experience was 28.8 years (SD = 17.3; range 3–60). Rice cultivation activities include preparation of fields, transplantation, field maintenance, and harvesting. Participants are exposed to heat and work in outdoor environments of a hot climate. A description of the participants is presented in Table 2.

3.2.2 Impacts of Heat on Health

In terms of signs and symptoms of heat stress, both age groups reported experiencing excess sweating, feeling thirsty, tiredness and exhaustion while working or spending time outside. Headaches were reported by 75–80% of those interviewed in both age groups, as shown in Table 2.

3.2.3 Impacts on Work Activities and Productivity

In the age group 60 years and older, all participants reported heat exposure impacts on work activities during work on hot days. Among them, only half reported work output and productivity decreases for each crop.

Among participants less than 60 years of age, most participants (76%) reported heat exposure impacts on work activities, and 11 (57.9%) of the 19 participants reported work output and productivity decreases as shown in Table 2.

3.2.4 Heat Prevention Approaches

Participants less than 60 years of age were more aware of any safety standards regarding heat than the participants aged 60 years and older, as shown in Table 2. Participants reported that methods to manage heat stress conditions included increasing the number of rice farmers over summer, longer workdays, earlier starts, split shifts, regular breaks, lighter protective clothing, active rehydration, and more time in the shade.

3.2.5 Heat Exposure Outside of Work

Participants were also asked about heat exposure outside of work. Only 11 participants less than 60 years of age (44%) reported heat exposure at home or during travel to and from work affecting their health, while participants aged 60 years and older reported that heat exposure outside of work did not affect their health.

3.2.6 Exposure Results

In winter, the outdoor WBGT data for 10 consecutive days from 6:00 AM on January 10 to 5:59 AM on January 20, varied from 19.2 to 36.7 °C, with an average of 25.5 °C. Relative humidity (RH) ranged from 28.0 to 100%, with an average of 74.9%. The ambient temperature (Ta) ranged from 20.7 to 42.5 °C, with an average of 27.2 °C as shown in Figs. 4 and 5.

In summer, the outdoor WBGT data for 10 consecutive days from 6:00 AM on May 11 to 5.59 AM on May 21, varied from 20.9 to 36.3 °C, with an average of 28.2 °C. RH ranged from 28.0 to 100%, with an average of 76.8%. The ambient temperature ranged from 24.2 to 41.4 °C, with an average of 30.2 °C as shown in Figs. 6 and 7.

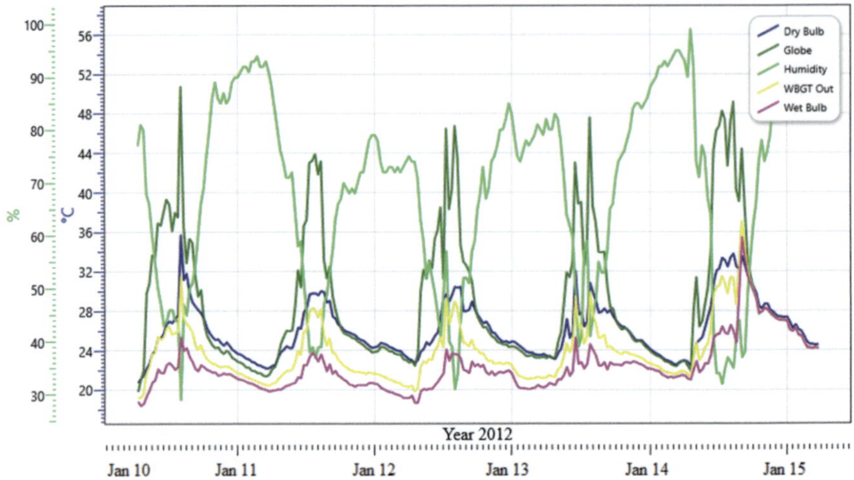

Fig. 4 WBGT, temperatures, and RH variation in winter 2012, January 10–15. (Dry bulb temperature = T_a, globe temperature = T_g, natural wet bulb temperature = Tnwb)

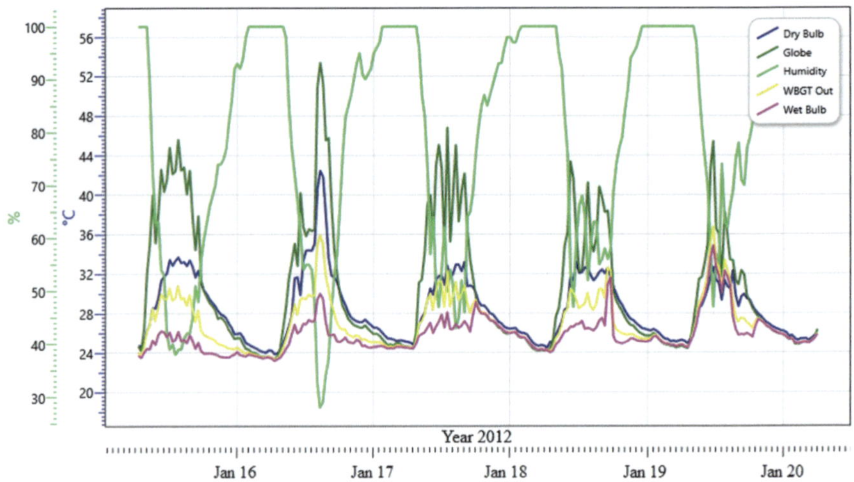

Fig. 5 WBGT, temperatures, and RH variation in winter 2012, January 16–20. (Dry bulb temperature = T_a, globe temperature = T_g, natural wet bulb temperature = Tnwb)

The outdoor WBGT was found to be highest during 11:00 AM and 11:59 AM in winter and during 13:00 PM and 13:59 PM in summer. RH was high at midnight and in the early morning and decreased after sunrise, as shown in Figs. 4, 5, 6, and 7.

The number of days in which the heat index categorized as extreme caution (32–41 °C) (NOAA 2018) is 4 days in winter and 10 days in summer as shown in Table 3.

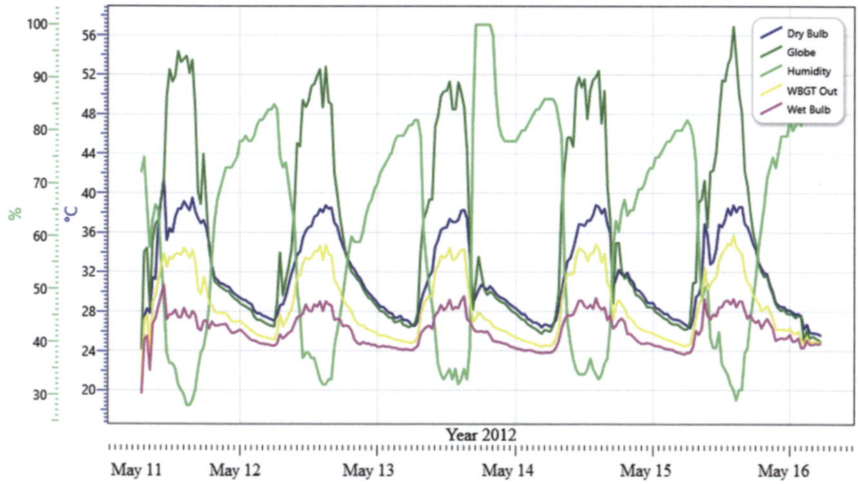

Fig. 6 WBGT, temperatures, and RH variation in summer 2012, May 11–16. (Dry bulb temperature $= T_a$, globe temperature $= T_g$, natural wet bulb temperature = Tnwb)

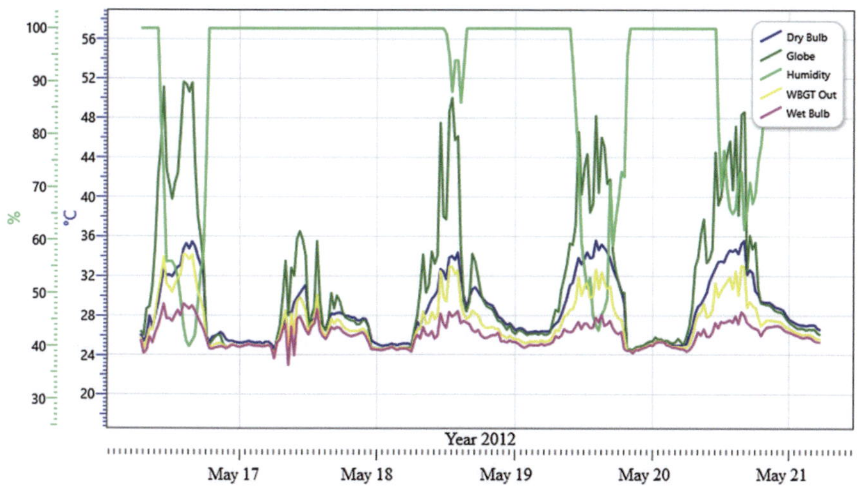

Fig. 7 WBGT, temperatures, and RH variation in summer 2012, May 17–21. (Dry bulb temperature $= T_a$, globe temperature $= T_g$, natural wet bulb temperature = Tnwb)

Table 3 Heat indices in paddy in dry season, 2012

Winter	Heat index (°C)	Summer	Heat index (°C)
January 10	26.4	May 11	38.1
January 11	26.8	May 12	36.6
January 12	27.1	May 13	36.4
January 13	27.4	May 14	35.9
January 14	31.3	May 15	36.0
January 15	32.0	May 16	33.6
January 16	34.1	May 17	35.1
January 17	33.3	May 18	39.5
January 18	33.7	May 19	35.4
January 19	32.9	May 20	39.5

4 Responses to Temperature Extremes

4.1 Response from Ministry of Natural Resources and Environment

The Department of Environmental Quality Promotion (DEQP) collaborated with the Thai Meteorological Department for developing an interpretation of heat index (HI) program for Thailand. This program was developed on Microsoft IIS (Internet Information Server) and uses temperature and relative humidity data that display by province in Google Maps application, as shown in Fig. 8. This program covered 50 meteorological stations in Thailand that users can choose to display as both map and satellite images.

Four categories for the heat index included caution, extreme caution, danger, and extreme danger as shown in Table 4. Heat index exceeding 54.5 °C can lead to extreme danger, including heat stroke.

4.2 Response from Ministry of Public Health

In 2017, the Health Impact Assessment Division under the Department of Health developed a heat health warning system for Thailand based on threshold temperatures. Morbidity data used in this project from 2010 to 2015 were sourced from the Strategy and Planning Division under the Office of the Permanent Secretary Ministry of Public Health. This project obtained meteorological data from 2006 to 2015 from the Thai Meteorological Department. Threshold temperatures by region of Thailand range from 38.23 to 39.97 °C, which are the lowest temperatures that can affect

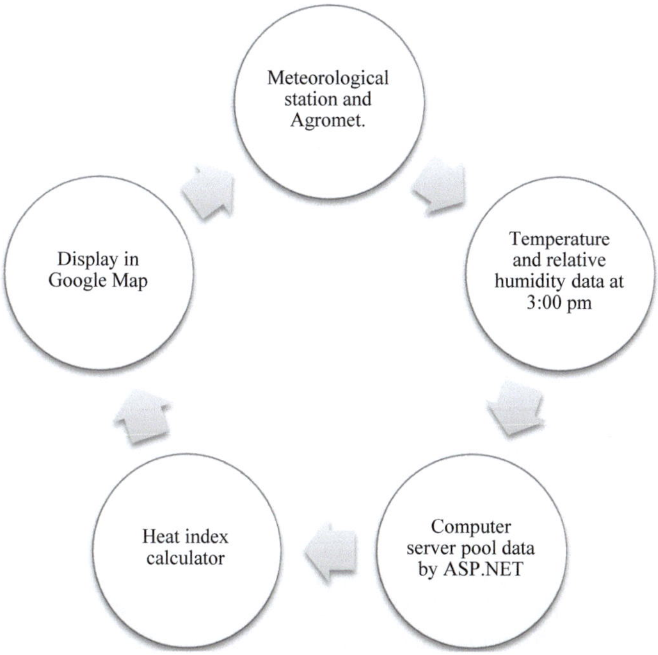

Fig. 8 Flow of interpretation of heat index program

Table 4 Heat index category

	Category	Heat index (°C)
	Caution	26.7–32.2
	Extreme caution	32.2–40.6
	Danger	40.6–54.5
	Extreme danger	Over 54.5

human health in Thailand. The heat health warning system for Thailand has 4 risk levels: surveillance, alert, warning, and danger, as shown in Table 5.

Table 5 Criteria of heat health warning system for Thailand

Level	Risk level	Temperature (°C)
1	Surveillance	Less than 38.1
2	Alert	38.1–40.0
3	Warning	40.1–43.0
4	Danger	Over 43 for 3 consecutive days or over 45

5 Discussion

5.1 Most at-Risk Populations

For assessment of an individual's thermal environment, six basic factors are essential. These include environmental (ambient air temperature, radiant temperature, air velocity, humidity), and personal factors (clothing insulation and metabolic heat) (HaSPA 2012). Previous studies in Europe and North America have reported positive associations between heat waves and mortality, with elderly persons especially women being the most affected (McMichael et al. 2006). This is particularly true among the elderly with diminished physiological capacity for thermoregulation. Children and homeless persons are also vulnerable to heat stress, particularly because of the greater time spent outdoors (IPCC 2014). Outdoor workers and those who work in hot environments, such as farmers, firefighters, construction workers, miners, and factory workers, also have elevated the risk of heat-stress-induced morbidity and mortality (HaSPA 2012). Due to urban heat island (UHI) effects, people in urban environments are at much greater risk of heat stress than those in non-urban regions (McMichael et al. 2006). Heat-related mortality rates differ from place to place, and a number of factors could be responsible. These may include the unexpected duration and intensity of a heat wave, lack of readiness of health care and social systems for such an event, the non-existence of intervention plans and the absence of effective technical solutions (Koppe et al. 2004).

5.2 Physiological Responses to Heat Exposure

According to the homeostatic mechanism, the maintenance of a constant internal environment is the body's natural response to any heat impact on the body, whether from external environment conditions or because of work performed by the individual. HaSPA (2012) reported that the most effective way of regulating body temperature during an episode of intense heat is via evaporation of sweat. Prolonged exposure to heat stress eventually leads to heat acclimatization, which basically refers to the physiological changes that make the body accustomed to a hot environment (HaSPA 2012).

5.3 Temperature Warning Thresholds

Questions regarding the best meteorological indicators to predict heat-related excess mortality and the best warning thresholds remain unanswered; hence, setting up of heat warning systems is beset with difficulties (Pascal et al. 2013). Nevertheless, heat warning systems remain a key component of any heat prevention plan, and the choice of indicators and thresholds is critical as they determine the timelines and effectiveness of the warnings. Evaluating the effectiveness of heat warning systems is problematic in that the systems can only go as far as educating the public, yet it is the decisions that individuals make that largely determine their vulnerability. Thus it is challenging to predict with accuracy the number of lives saved (Sheridan and Kalkstein 2004). Heat warnings, however, should always strive to target the whole population, especially vulnerable groups and institutions that are responsible for their welfare. It is unfortunate that most governments do not perceive heat as a major problem, mainly because of the predominantly simple measures people can take to mitigate heat effects (Koppe et al. 2004).

5.4 Intervention Strategies to Prevent Heat-Related Illness

Several long-term intervention strategies have been proposed. These include the conversion of energy by using solar systems for heating and cooling, reducing the number of vehicles in cities to lessen the urban heat island effect, and the maintenance of high levels of heat acclimatization by adopting an active lifestyle with properly adjusted climatic exposure (Koppe et al. 2004). Options to address heat-related morbidity and mortality may include warning systems linked to response strategies, and also urban planning and improvements to the built environment to reduce heat stress (IPCC 2014).

6 Conclusions

In conclusion, heat health warning systems are important for reducing the harmful effects of extreme weather and climate events. Government and stakeholders require appropriate mechanism to effectively improve health and address the social determinants of health equity in Thailand.

Acknowledgements This study was supported by the National Research University Project of Thailand Office of Higher Education Commission and School of Global Studies at Thammasat University. We would like to express our sincere thanks and appreciation to Dr. Kamol Promsakha na Sakolnakhon from the Thai Meteorological Department, Ministry of Digital Economy and Society, for annual average maximum temperature and location of the study area maps. Our special thanks go to participants for their patience and contribution to this study.

Declaration of Interests We declare no competing interests.

References

Baccini, M., Biggeri, A., Accetta, G., Kosatsky, T., Katsouyanni, K., Analitis, A., et al. (2008). Heat effects on mortality in 15 European cities. *Epidemiology, 19*(5), 711–719.

Bai, L., Woodward, A., & Liu, Q. (2014). Temperature and mortality on the roof of the world: A time-series analysis in three Tibetan counties, China. *Science of the Total Environment, 485*, 41–48.

Bennett, C. M., & McMichael, A. J. (2010). Non-heat related impacts of climate change on working populations. *Global Health Action, 3*, 5640. https://doi.org/10.3402/gha.v3i0.5640.

Gasparrini, A., Guo, Y., Hashizume, M., Lavigne, E., Zanobetti, A., Schwartz, J., et al. (2015). Mortality risk attributable to high and low ambient temperature: A multicountry observational study. *Lancet, 386*(9991), 369–375.

Hajat, S., & Kosatky, T. (2010). Heat-related mortality: A review and exploration of heterogeneity. *Journal of Epidemiology & Community Health, 64*(9), 753–760.

HaSPA (Health and Safety Professionals Alliance). (2012). *The Core Body of Knowledge for Generalist OHS Professionals*. Tullamarine, VIC: Safety Institute of Australia.

IPCC. (2014). Climate change 2014: Synthesis report. In Core Writing Team, R. K. Pachauri, & L. A. Meyer (Eds.), *Contribution of working groups I, II and III to the fifth assessment report of the intergovernmental panel on climate change* (151 pp.). Geneva, Switzerland: IPCC.

Jackson, J. E., Yost, M. G., Karr, C., Fitzpatrick, C., Lamb, B. K., Chung, S. H., et al.. (2010). Public health impacts of climate change in Washington State: projected mortality risks due to heat events and air pollution. *Climate Change, 102*(1–2), 159–186.

Jay, O., & Brotherhood, J. R. (2016). Occupational heat stress in Australian workplaces. Temperature: Multidisciplinary Biomedical Journal, *3*(3), 394–411.

Kjellstrom, T. (2012). *The "Hothaps" program for climate change impact assessment and prevention: Background document and proposed protocols for field studies*. Australia: Umea University, Sweden and Australian National University, Canberra.

Koppe, C., Kovats, S., Jendritzky, G., & Menne, B. (2004). *Heat-waves: Risks and responses. Health and Global Environmental Change Series, No.2*. World Health Organization.

Langkulsen, U., Vichit-Vadakan, N., & Taptagaporn, S. (2010). Health impact of climate change on occupational health and productivity in Thailand. *Global Health Action, 3*, 5607. https://doi.org/10.3402/gha.v3i0.5607.

McMichael, A. J., Woodruff, R. E., & Hales, S. (2006). Climate change and human health: Present and future risks. *Lancet, 367*(9513), 859–869.

Metzger, K. B., Ito, K., & Matte, T. D. (2010). Summer heat and mortality in New York City: How hot is too hot? *Environmental Health Perspectives, 118*(1), 80–86.

Ministry of Public Health. (2018). Occupational and environmental diseases. Retrieved June 10, 2018, from: https://hdcservice.moph.go.th/hdc/reports/page.php?cat_id=f16421e617aed29602f9f09d951cce68. (in Thai).

National Oceanic and Atmospheric Administration (NOAA), National Weather Service, U.S. Department of Commerce. (2018). Heat index information. Retrieved June 10, 2018, from: https://www.weather.gov/gjt/heat_index.

Office of Agricultural Economics, Ministry of Agriculture and Cooperatives. (2016). Statistics of rice cultivation, Pathum Thani province: crops year 2005/06-2014/2015. Retrieved June 10, 2018, from: http://service.nso.go.th/nso/web/statseries/statseries14.html.

Parsons, K. (2014). *Human thermal environments: the effects of hot, moderate, and cold environments on human health, comfort, and performance* (3rd ed.). New York: CRC Press.

Pascal, M., Wagner, V., Le Tertre, A., Laaidi, K., Honoré, C., Bénichou, F., et al. (2013). Definition of temperature thresholds: the example of the French heat wave warning system. *International Journal of Biometeorology, 57*(1), 21–29.

Pathum Thani. (2018). Area: The geography of Pathum Thani province. Retrieved June 10, 2018, from: http://eng.pathumthani.go.th/index.php?option=com_content&view=article&id=9:point&catid=8&Itemid=153.

Pathum Thani Agricultural Provincial Extension Office, Department of Agricultural Extension, Ministry of Agriculture and Cooperatives. (2018). Agricultural database: summary of agricultural area, year 2011/2012. Retrieved June 10, 2018, from: http://www.pathumthani.doae.go.th/datakaset/Datapathum54-55.pdf.

Pathum Thani Provincial Office, Ministry of Interior. (2018). Provincial development plan for Pathum Thani province, year 2014–2017. Retrieved August 25, 2018, from: http://www2.pathumthani.go.th/551017.pdf. (in Thai).

Rice Department, Ministry of Agriculture and Cooperatives. (2018). Rice production potential map: Pathum Thani. Retrieved June 10, 2018, from: http://brrd.ricethailand.go.th/ricemap/. (in Thai).

Sheridan, S. C., & Kalkstein, L. S. (2004). Progress in heat watch-warning system technology. *Bulletin of the American Meteorological Society, 85*(12), 1931–1941.

Strategy and Planning Division, Office of the Permanent Secretary Ministry of Public health, Ministry of Public health. (2018). *Statistical Thailand 2017: rate per 100,000 population of first 10 leading cause groups of death (according to ICD mortality tabulation list 1, 10th revision), 2014–2016* (p. 142). Nontaburi: Ministry of Public health.

Tawatsupa, B., Dear, K., Kjellstrom, T., & Sleigh, A. (2014). The association between temperature and mortality in tropical middle income Thailand from 1999 to 2008. *International Journal of Biometeorology, 58*(2), 203–215.

Tawatsupa, B., Lim, L. L., Kjellstrom, T., Seubsman, S. A., Sleigh, A., & The Thai Cohort Study Team. (2012). Association between occupational heat stress and kidney disease among 37 816 workers in the Thai Cohort Study (TCS). *Journal of Epidemiology, 22*(3), 251–260.

Thai Meteorological Department, Ministry of Digital Economy and Society. (2018a). Annual mean maximum temperature in Thailand from 1951 to 2017. Retrieved June 10, 2018, from: https://www.tmd.go.th/climate/climate.php?FileID=7.

Thai Meteorological Department, Ministry of Digital Economy and Society. (2018b). Thailand annual weather summary, 2013–2016. Retrieved June 10, 2018, from: https://www.tmd.go.th/en/climate.php?FileID=5.

Xiang, J., Bi, P., Pisaniello, D., & Hansen, A. (2014). Health impacts of workplace heat exposure: An epidemiological review. *Ind Health, 52*(2), 91–101.

Ye, X., Wolff, R., Yu, W., Vaneckova, P., Pan, X., & Tong, S. (2012). Ambient temperature and morbidity: A review of epidemiological evidence. *Environmental Health Perspectives, 120*(1), 19–28.

Extreme Events, Disasters, and Health Impacts in Indonesia

Budi Haryanto, Fatma Lestari and Triarko Nurlambang

Abstract Extreme weather events caused by climate change may destroy many components of environment and facilities, direct physical harm, loss of income, psychological stress, and other direct and indirect human health. In Indonesia, 80% of disasters due to climate change during 1998–2018 which were dominated by flooding (39%), heavy wind/storm (26%), landslides (22%), and drought (8%). In 2017, there were 2263 events, and 198 of these events are considered as health crisis. There were 305,837 person impacted and 198 deaths, major injuries 2314 person, minor injury 63,578 person and refugee 243,691 person in 2017. The estimated impacts cost to the economy is about IDR 132 trillion (approximately US $ 8.8 billion) in 2050 as the consequences.

Keywords Climate change · Extreme events · Disasters · Health impacts · Economic costs

1 Background

Some studies claimed that the occurrence of extreme weather events has a significant connection with climate change (Rosenzweig et al. 2001; Scott et al. 2013), such as heavy rainfall, heat waves, storm and many others leading to severe disasters such as flooding and drought. Heavy rainfall frequently happened in tropical climate as a mechanism of water evaporation from the land and sea, and eventually returns to Earth

---

B. Haryanto (✉)
Department of Environmental Health, Faculty of Public Health, Universitas Indonesia, Jawa Barat, Indonesia
e-mail: bharyanto@ui.ac.id

B. Haryanto · T. Nurlambang
Research Center for Climate Change, Universitas Indonesia, Jawa Barat, Indonesia

F. Lestari
Department of Occupational Health, Faculty of Public Health, Universitas Indonesia, Jawa Barat, Indonesia

T. Nurlambang
Department of Geography, Universitas Indonesia, Jawa Barat, Indonesia

R. Akhtar (ed.), *Extreme Weather Events and Human Health*,
https://doi.org/10.1007/978-3-030-23773-8_16

as rain and snow. Climate change intensifies this cycle because as air temperatures increase, more water evaporates into the air. In nature, warmer air holds more water vapor which is contributing to an increase in the average annual amount of rain and snow in some places and creating more intense rainstorms in others. The result is major problems like extreme flooding in communities around the world. Additionally, at the same time, sea levels are rising faster than at any time in almost 3000 years, and its worsening coastal flooding globally.

The United Nations Environmental Program estimates that half of the world's population lives within 60 km (about 37 miles) of a coast—and three-quarters of all major cities are on a shoreline. At the same time that some areas are experiencing more intense precipitation, others are experiencing more drying and even drought. In urban, suburban, and agricultural areas, this runoff can pick up pollutants from the landscape—including sewage from overwhelmed single-pipe systems—and carry them to nearby rivers and other waterways, including reservoirs. On the other end of this same spectrum, periods of drought, enhanced evaporation, and decreases in overall annual rainfall result in reduced water levels in streams, rivers, and lakes. This leaves less water to dilute even relatively common pollutants—and eventually less water to irrigate crops or drink.

A major drought can have serious consequences for people's livelihoods, affecting everything from agriculture and transportation to public health. And of course, all of this has some major implications for food security and supplies of drinkable water. Then there are consequences of severe drought in worsening forest fires. Many deaths from extreme events and natural disasters occur in developing countries. For Indonesia, climate change increases the risk of both heavy rains and extreme droughts and causes massive health impact (BNPB 2018a, b). The character and severity of health impacts from climate extremes depend not only from the extremes themselves but also on exposure and vulnerability (IPCC 2012).

2 Indonesia's Vulnerability

Indonesia is amongst the countries with the highest disaster risk globally. This risk is driven by the country's high exposure to a range of geophysical and hydro-meteorological hazards, combined with grave vulnerabilities resulting from population growth, unequal economic development, urbanization, a lack of social and environmental considerations within development processes, and other drivers. Disasters caused by environmental hazards are becoming increasingly costly and severe in Indonesia (Djalante and Thomalla 2012).

Indonesia is regarded as the second-ranking country after Bangladesh for highest disaster risks in the world (Malpecroft 2010) and is known for its vulnerability against disasters. The country's reputation for disaster risk is intricately related to its unique geographic characteristics. Indonesia is an archipelago consisting of around 17 thousand large and small islands that stretches along the equator (RBI-BNPB 2016). The country has 1.95 million km^2 of land mass and 3.26 million km^2 of water (5.21 million km^2 in total). Indonesia's regional geographic characteristics have an intrinsic functional tie to its geological and climatological characteristics. These characteristics in turn magnify the country's vulnerability towards hydro-climatological disaster events or extreme weathers. These events will likely rise in occurrence in the future, along with increasing population and centralized migration patterns to potentially vulnerable areas, mainly along coastlines (Lestari et al. 2017a, b, c).

While climate-related disasters occurred more frequently and affected more people, geological disasters were the deadliest. Recent publications on climate change risks indices such as the World Risk Index (Buendnis Entwicklung Hilft and UNU-EHS 2011) or the Global Climate Risk Index (Germanwatch 2011) and vulnerability indices such as the Climate Change Vulnerability Map (Maplecroft 2011) and the Multiple Climate Hazard Index (Yusuf and Francisco 2009), show that Indonesia is at the top-end of the spectrum of most vulnerable country to natural hazards. In addition, through its hydro-climatological characteristics, Indonesia has a climactic zone that is influenced by its geographic position located along the equator. Because of its equatorial position, the country's climate is also affected by the Inter-Tropical Convergence Zone (ITCZ) (Fig. 1).

The climate patterns of the regions of Indonesia is categorized into three types (Aldrian 2003: 97–100): the monsoonal region of south Indonesia, the semi-monsoonal region of northwest Indonesia, and the anti-monsoonal region of Molucca. The existence of the three different regions also affects how extreme events are identified and located.

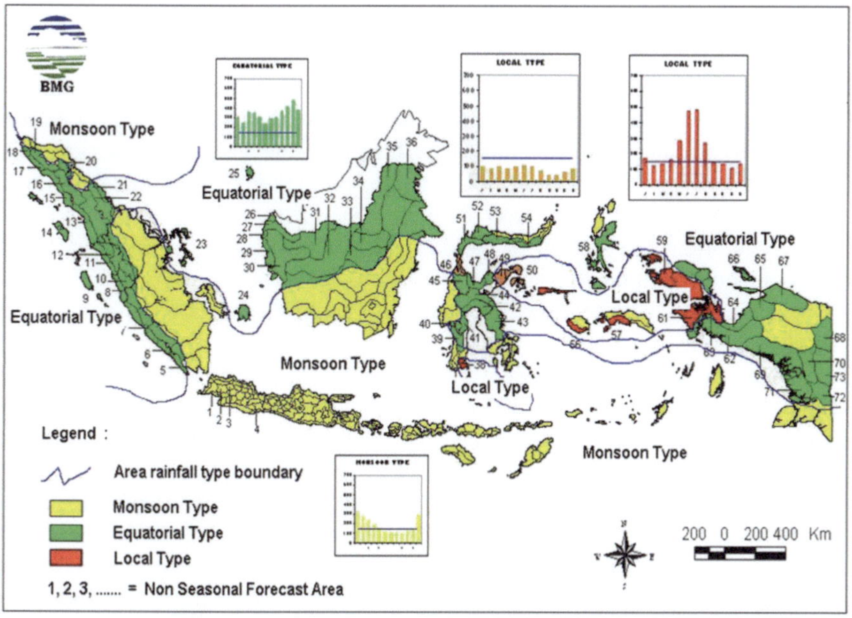

Fig. 1 Types of Precipitation in Indonesia. *Source* https://www.bmkg.go.id/

3 Climate Disasters and Extreme Event: Occurrences and Impacts

In Indonesia, 80% of disasters reported due to climate change during 1998–2018 which were dominated by flooding (39%), heavy wind/storm (26%), landslides (22%), and drought (8%). Disaster such as flooding occurs due to extreme rainfall in a short time, due to weather disruption. The climate change phenomenon also raises the drought longer. Data derived from CRED-EMDAT (2011a, b) show that in the last 100 years 400 disasters occurred in Indonesia killing almost 241,000 people, affecting almost 28 million people, and costing US$ 24 billion. These events were predominantly geological and climate-related disasters.

The mechanism of connection between global temperature rise and wildfires is pretty simple science: Droughts dry out the land, killing plant life—which then also dries out itself, becoming far easier to ignite. And, with less predictable rains, it is harder to stop these fires once they begin. Wildfires can leave communities and governments with billions of dollars in damages, not to mention the incalculable costs of lost plant, animal, and even human life. National Wildlife Federation (NWF) offers a bit more detail on how increases in average annual temperatures create conditions that dramatically elevate the risk and severity of forest fires. Below-average rainfall in some areas—a result of the enhanced evaporation rates associated with warming and the climate crisis—naturally increases the probability and duration of a fire. One major impact of climate change is increasing the potential for severe storms. Severe storms carry a lot of energy, and NWF estimates that "lightning in the [American West] could increase by 12–30% by mid-century."

The terminology of weather and climate refer to different patterns which shares similar components, such as air temperature, rain, wind, humidity, evaporation, sunlight radiation, and so on. Conceptually, weather refers to atmospheric conditions that constantly vary due to earth's rotation, having absolute value, lasting for a short duration, and having a more direct impact. Climate, on the other hand, refers to relatively consistent atmospheric conditions, affected directly from a latitudinal position, having a more lasting duration, and indirect impact. Meteorological processes and weather conditions are studied in the domain of meteorology, while climatology studies climate conditions (Fig. 2).

The differences between extreme weather conditions and extreme climate are difficult to be precisely defined, but they are more easily differentiated in terms of its time scale. An extreme weather event is typically associated with patterns lasting for one day up to a few weeks. Extreme climate events, such as droughts, on the other hand, lasts for much longer and is an accumulation of weather events, be they extreme or non-extreme.

The National Oceanic and Atmospheric Administration (NOAA) have also differentiated weathers and climates based on the duration of the atmospheric conditions or events. NOAA specifically differentiated the time dimension of weathers and climates for its Seamless Prediction of Weather and Climate framework which is used for various meteorological or climatological applications. We can predict basic

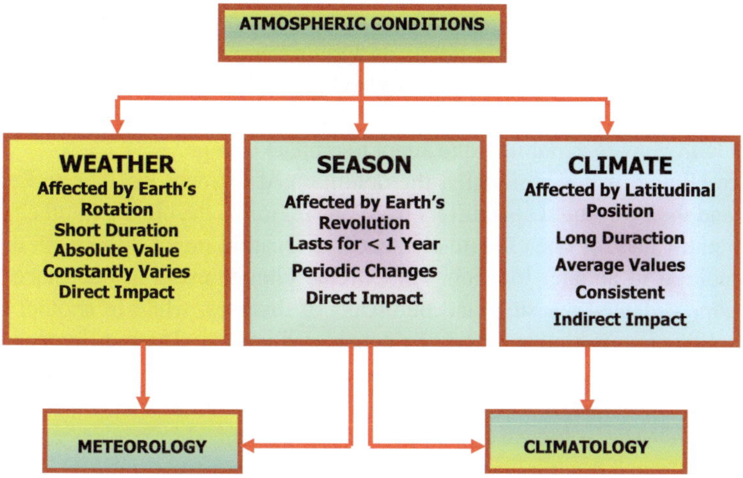

Fig. 2 Conceptual differences between weather and climate. *Source* The Geography Department of Universitas Indonesia and BNPB (2013)

weather conditions in the duration of minutes, hours, and days, and only up to a period of two weeks in the future (Fig. 3). It is fundamental for the meteorological community to gain further understanding of the differences between extreme cli-

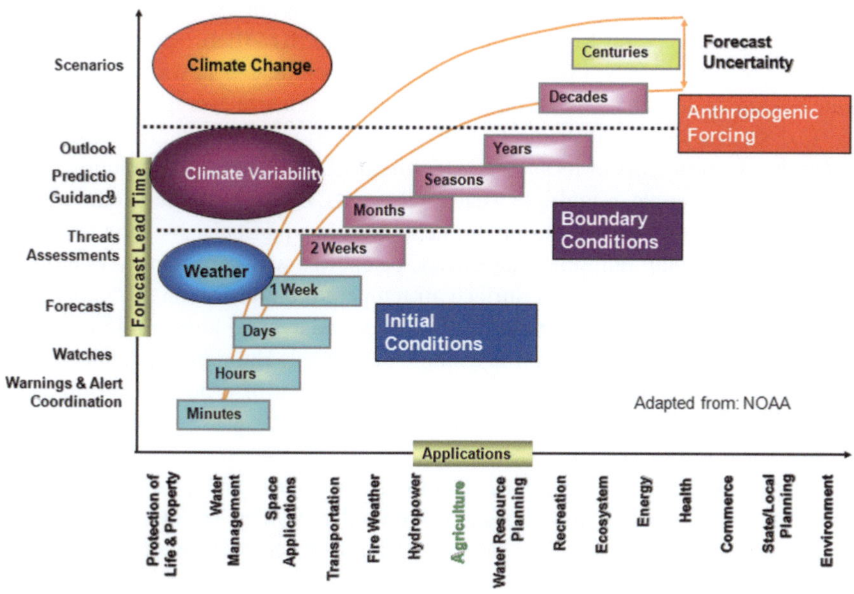

Fig. 3 Seamless prediction of weather and climate. *Source* NOAA, 2009 in The Geography Department of Universitas Indonesia and BNPB (2013)

mate and weather in their spatial and temporal context through a consistent regional or global methodology. This is based on the definition and the calculation of its threshold, so that an efficient monitoring and predictive system could be deployed to increase public resilience. These developments will improve the public's ability to cope with climate variability and adapt to climate change.

Since climate varies regionally, the definition of extreme weather and climate events and their threshold also differs from one region to the others. In other words, the extreme value for one climate element in a location might differ with the normal condition in another location. An extreme climate event in one place might cause unprecedented disasters and outbreaks of diseases, while in another region that extreme climate element might be considered normal. These differences naturally cause the impact of extreme climate events to the people they affect to also vary.

The World Meteorological Organization (WMO) defined several key characteristics to identify and describe extreme weather and climate events (WMO 2018). The first characteristic is magnitude, a measure of the threshold and a clear reflection of extreme events which could show whether the current value is over or under a certain threshold. The second characteristic is duration, the time in which an event begins and ends. Severity is another indicator which measures the destructive potential of extreme events. An example of the usage of severity is to measure the loss experienced by farms and the emergence of diseases during droughts. Severity could also be measured by a combination of magnitude and persistence (Shiau 2009 and Mirabbasi 2012 in WMO 2018). Extreme weather and climate events in Indonesia could be considered unique. Traditionally, the designation extreme climate refers to the *D* climate type on Koppen's classification (WMO 2018), which describes a condition of significantly different average air temperature, with winter air temperature averaging less than 0 °C and summer air temperature averaging higher than 18 °C. Extreme weather is also related to atmospheric events that are significantly different from normality or its average measurements, and these abnormalities could be observed in a single or multiple weather elements. Extreme weather could also be defined as a rare physical condition or phenomena in a particular location and time, with a relatively short duration. The BMKG categorize weather events as extreme if they fulfill these conditions: (a) Surface air temperature higher than 35 °C or under 15 °C (for lowlands/coastal regions); (b) Wind speed over 25 knots or 45 km/h; (c) Rainfall above or equal to 50 mm in a single day, or; (d) Humidity under 40%.

Figure 4 shows an illustration of a spatial depiction of droughts in Sumatera, Indonesia, which had caused fires in the island using the category developed by McKee 1993, in WMO (2018).

Indonesia experienced an impact from extreme events and disasters including flood, land and forest fires, drought, heavy wind and landslides. The disasters involved the total of 22,886 occurrences in the last 20 years (1998–2018) which are flood as the most frequent (8814) followed by heavy wind/storms (5969), landslides (4946), drought (1872), forest and bush fire (641), tidal wave/Abrasion (298), earthquake (238), Eruption (95) and tsunami (13). Its proportion of the occurrences is shown in Fig. 5 (Fig. 6).

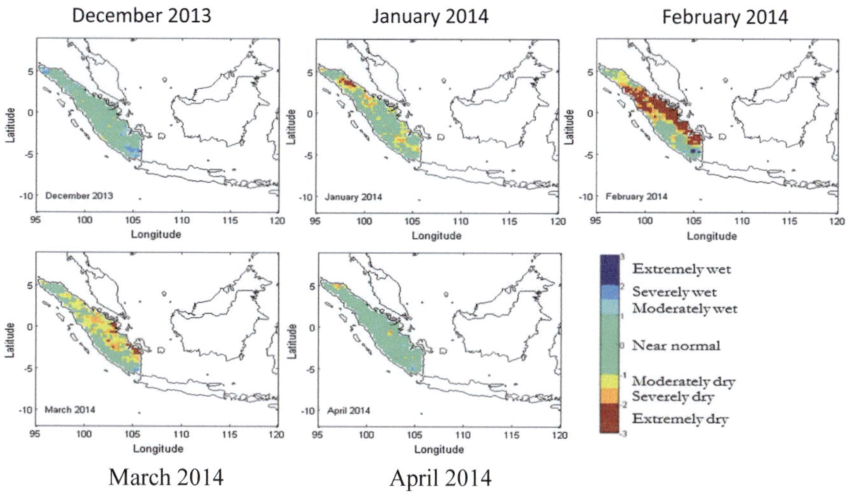

Fig. 4 Distribution of drought in Sumatera, Indonesia, from December 2013 to April 2014. *Source* WMO (2018)

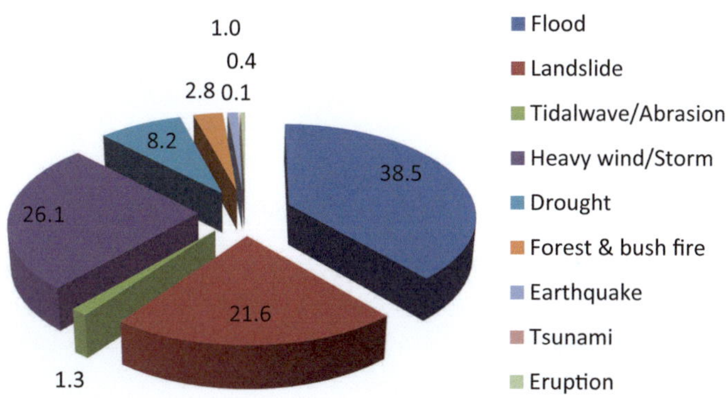

Fig. 5 Proportion of disasters in Indonesia 1998–2018. *Source* Disaster database & information BNPB. http://dibi.bnpb.go.id/dibi/

The 2015 Indonesia's extreme drought shows multitudes of consecutive days without rain, with many regions in Indonesia experiencing over 60 days without rain during that year. These droughts were observed in Bangka-Belitung, South Sumatera, most of Java, Bali, East Nusa Tenggara, South Borneo, South Sulawesi, North Sulawesi, and the northern areas of the Moluccas (Fig. 7).

Another potential risk to the extreme event in Indonesia is the location and dynamics of tropical cyclones in the south-eastern area of the Indian Ocean in January to April, and in the eastern part of the Pacific Ocean in May to December that could create strong winds and massive rainfall. The transition period of the monsoon wind from the north-east to the south-west also associated with remarkably strong wind

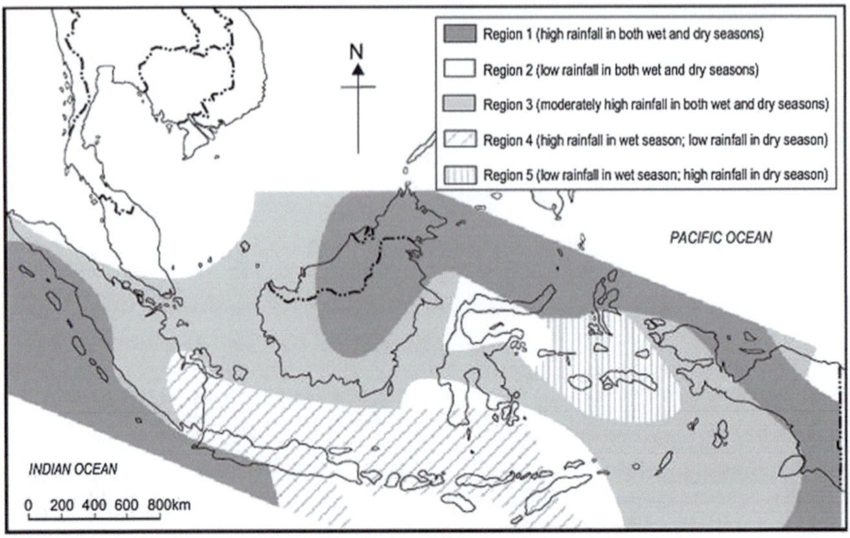

Fig. 6 Indonesia rainfall and dry season patterns by region. *Source* Arcari et al. (2007, Fig. 3, p. 256)

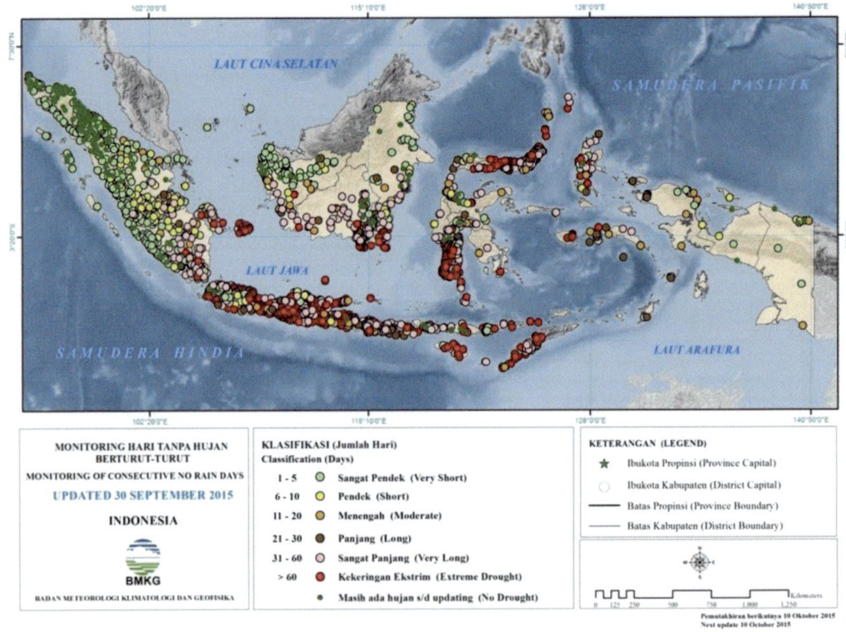

Fig. 7 Continuous dry days map showing the condition for early October 2015. Colors of the circles denote the length of the dry days, ranging from short (green) to long (red). *Source* BMKG (2018)

velocity. As the El Niño event increases in frequency, the ten most devastating El Niño happened after the 1970s and global warming started to gain speed. Despite the current changes Indonesia experiences are due to El Niño or as a result from greenhouse effects, or a mixture of both, it is clear that Indonesia is experiencing a changing climate, and the consequences will be felt generations to come (UNDP 2007).

4 Human Health Impacts and Cost

The links between climate change and disease are very complex. As discussed in the Indonesia Climate Change Roadmap paper on health, they may involve a number of different direct and indirect connections:

- Changes in temperature and rainfall may affect the life cycles of the organisms that cause the disease or the other species that transmit those diseases to humans. Such impacts could be of many types. They could affect their reproductive success, lifespan, the time required for them to reach an age at which they can transmit disease, their spatial dispersion, or other factors that can affect the likelihood of humans receiving the disease.
- Changes in temperature and rainfall will affect agricultural output, causing a risk of harvest failure and malnutrition, with consequent increases in the risk of many other health problems.
- Changes in rainfall and hydrology may affect water supply and sanitation, increasing the risk of water-borne diseases.
- Extreme weather events may destroy settlements or income-generating resources and facilities, leading to direct physical harm, loss of income and consequent well-being, psychological stress, and other direct and indirect human health consequences.

The climate change roadmap focuses on three diseases: dengue fever, malaria, and diarrheal.

Two aspects of disease, direct expenditures for medical care and prevention and foregone income due to illness or death, were analyzed. Estimates of direct costs attributable to climate change were based on data found in the literature on current expenditures related to dengue fever and malaria in Indonesia, and then applied to the number of cases of each disease that could result from climate change. The estimates of forgone income use the so-called disability-adjusted life-year (DALY) to measure the consequence of illness. This standard measure is the sum of the number of years of life lost by those who die of a disease plus a weighted sum of the number of years that people live with that disease. The weights are set to reflect shared views on the burden imposed by living with different medical conditions. Thus, for example, the weight for living with a spinal cord lesion that causes paralysis from the waist down is 0.296. If each year of life is weighted equally, a man with a life expectancy of 60 years who dies in an accident at age 20 will generate 40 DALYs. If his twin sister (with the same life expectancy) in the same accident and lived with paralysis until age 60, she would generate 40 × 0.296 = 11.84 DALYs. If she died at age 40 due to complications of her paralysis, she would generate (40–20) × 0.296 + 20 = 25.92 DALYs. The weights used to calculate DALYs for different diseases were established through extensive global survey research (Salamon et al. 2012). A key issue in that work was whether people at different income levels, or living in different countries and cultures, feel differently about living with disease or disability.

Indonesian Ministry of Health is the leading sector to determine the health crisis, based on the criteria that the district head has declared a certain disaster or the President or if the population impacted above 50 persons and there are a certain number of refugees. Health crisis center is the center under Ministry of Health, which handles health crisis management in Indonesia. All relevant data related to health crisis are collected by health crisis center through information system or called Health Crisis Management Information System. In general, the biggest health impact that should be considered in disaster events is health crisis issues, including fatalities, injuries, nutrition issue, refugees, hygiene and sanitation, the availability of fresh water, health services, infectious diseases, mental health and reproduction health services issues. The health crisis management-related problems that should be resolved as part of health crisis management during disaster events are the information system, coordination amongst stakeholders to solve health-relevant issues, mobilization of donation and assistance to disaster sites, unsupported financial system, early warning system, limited resources to be sent to disaster areas, and the management of local and international aid that poorly supervised. Based on this reason, the integrated health crisis management is needed to solve some health-related issues during the disaster events in Indonesia (Lestari et al. 2017a, b, c).

Human health impact derived from extreme events and disaster in Indonesia is presented in Table 1. During 2017, there were 2263 events and 198 of these events are considered as health crisis. There were 305,837 persons impacted and 198 deaths, major injuries 2314 persons, minor injury 63,578 persons and refugee 243,691 persons.

Extreme events and disasters frequency which has an impact on human health has been recorded as 16–30 events in 5 (five) major provinces in Indonesia: Central Java (29 events), DKI Jakarta (26 events), West Java (19 events), East Java (17 events) and North Sumatra (12 events). These 5 (five) provinces have been considered as Medium to High Disaster Risk Index in Indonesia with score: Central Java (158—

Table 1 Human Health impact from extreme events and disaster in Indonesia (2017)

Description	Number of people
Deaths	198
Major injuries	2314
Minor injuries	63,578
Refugee	243,691

Source Health Crisis Center, Indonesian Ministry of Health, 2017

Table 2 Extreme events and disaster frequency in 5 (five) major provinces in Indonesia with its Disaster Risk Index

Provinces	Extreme Events and disasters frequency	Disaster risk index score	Category of disaster risk index
Central Java	29	158	High
DKI Jakarta	26	103	Medium
West Java	19	166	High
East Java	17	171	High
North Sumatra	12	150	High

Source Health Crisis Center, Indonesian Ministry of Health, 2017

High Disaster Risk Index); DKI Jakarta (103—Medium Disaster Risk Index), West Java (166—High Disaster Risk Index), East Java (171—High Disaster Risk Index) and North Sumatra (150—High Disaster Risk Index). The most frequent locations extreme events and disasters frequency are located in Java Island (Table 2).

The estimated impacts of climate change to the economy cost in 2050 vary substantially both across areas of impact and across provinces. The total cost to the economy is about IDR 132 trillion (approximately US $ 8.8 billion), in which agriculture sector accounts for the majority (53%) followed by health sector and sea level rise (SLR) sector of 34 and 13%, respectively (Fig. 8).

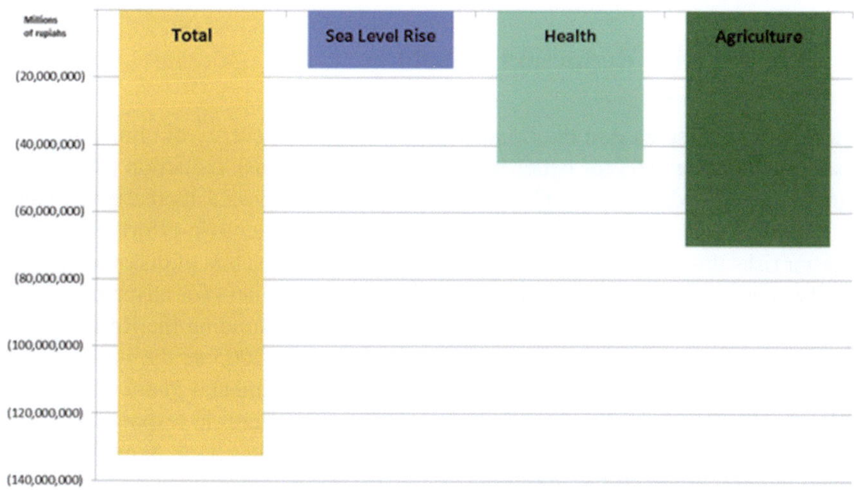

Fig. 8 Cost of climate change in 2050 by area of impact (in millions IDR)

5 The Future Challenges

Natural disasters are becoming more frequent, deadly, and costly. Within the period of 1900–2010, there has been a fivefold increase in the number of natural disasters reported, taking its peak within the period 2000–2010 (EM-DAT 2011a, b). While there is a gradual reduction in the number of deaths, the number of people reported affected increased rapidly in the last 40–50 years and averaged more than 300 million by 2010 (EM-DAT 2011a, b). The costs of natural disasters have increased sharply since the 1980s and the average costs peak at just below US$100 billion by 2010 (EM-DAT 2011a, b). Globally, the frequency and magnitude of weather and climate-related hazards is increasing (IPCC 2007a, b; EM-DAT 2011a, b), with flood as the most frequent disaster and affected most people.

Indonesia's national statistical office, Badan Pusat Statistik (BPS) 2013, projected population by province and by gender to 2035 (BPS 2013). The BPS provincial projections were extended to 2050, ensuring that the total projected population in 2050 would be as close as possible to the United Nation's projections. This was done by projecting the 2020–2035 change in five-year growth rates for each province to the period 2040–2050, using those projections to calculate growth rates for the corresponding periods, and then using those calculations to project population in each province in 2040, 2045, and 2050.[8] BPS also projected the rural urban breakdown out to 2035. This study used similar calculations to extend them to 2050 as well (Table 3).

6 Disaster Risk Reduction Efforts

The anticipated increase in the frequency, intensity, and severity of climate-related disasters therefore calls for better integration of disaster risk reduction (DRR) and climate change adaptation (CCA) to reduce vulnerability and increase resilience to natural disasters. The UNISDR defines DRR as "systematic efforts to reduce disaster risks through analyzing and managing the causal factors of disasters including the reduction of vulnerability, and improved preparedness for adverse events." The "Hyogo Framework for Action (HFA) 2005–2015: Building the Resilience of Nations and Communities to Disasters" was adopted in 2005 as the international framework for DRR. CCA is defined by The Intergovernmental Panel on Climate Change (IPCC) as "an adjustment in natural or human systems in response to actual or expected climatic stimuli or their effects, which moderates harm or exploits beneficial opportunities" (IPCC 2007a, b).

Indonesia's response to climate change remains dominated by mitigation activities, especially through the REDD program (www.redd-indonesia.org/). The Indonesian Climate Change Trust Fund (ICCTF) was created in 2009 to build on the policy framework for climate change mitigation and adaptation and to support its implementation (UNDP Indonesia 2009a, b, c). The formation of the ICCTF addresses three

Table 3 Costs imposed by climate change in 2050 due to impacts of dengue fever/DHF by province (in millions of rupiahs)

Province	Forgone income	Medical expenditures	Total
Aceh	7850	3108	10,958
Sumatera Utara	94,638	29,248	123,887
Sumatera Barat	421,256	154,116	575,372
Riau	1,337,241	135,425	1,472,666
Jambi	444,336	119,836	564,172
Sumatera Selatan	546,086	159,957	706,043
Bengkulu	141,323	66,959	208,282
Lampung	287,807	115,733	403,540
Kepulauan Bangka Belitung	822,888	224,330	1,047,218
Kepulauan Riau	1,650,170	197,813	1,847,984
DKI Jakarta	18,850,295	1,305,419	20,155,714
Jawa Barat	217,785	82,871	300,656
Jawa Tengah	8741	3676	12,417
DI Yogyakarta	11,841	5237	17,078
Jawa Timur	1,889,658	554,485	2,444,143
Banten	80,860	25,749	106,609
Bali	172,332	56,287	228,619
Nusa Tenggara Barat	31,603	20,460	52,063
Nusa Tenggara Timur	13,477	11,504	24,981
Kalimantan Barat	303,687	124,657	428,345
Kalimantan Tengah	518,738	158,330	677,068
Kalimantan Selatan	404,736	138,017	542,753
Kalimantan Timur + Kal. Utara	5,846,528	385,198	6,231,726

(continued)

Table 3 (continued)

Province	Forgone income	Medical expenditures	Total
Sulawesi Utara	291,005	102,222	393,226
Sulawesi Tengah	224,721	85,012	309,733
Sulawesi Selatan	260,453	90,510	350,963
Sulawesi Tenggara	60,472	21,089	81,561
Gorontola	31,389	16,581	47,970
Sulawesi Barat	56,194	28,929	85,123
Maluku	65,986	41,164	107,151
Maluku Utara	2981	1618	4599
Papua Barat	8671	1420	10,091
Papua	73,221	18,566	91,787
Total in Rps 106	35,110,852	4,444,691	39,555,543

major targets: conducting low-carbon economic development, promoting national resilience to climate change, and achieving effective CCA (UNDP Indonesia 2009a, b, c). Indonesia‘s First National Communication submitted to the UNFCCC in 1999 included a national emissions inventory of greenhouse gases (GHG), described sectoral measures to reduce GHG emissions and provided other important information related to climate change. The Second National Communication published in 2009 reported Indonesia‘s progress in adapting and mitigating to climate change and provided a detailed plan for GHG emissions reduction of 26% by 2020 (MoE 2009; UNDP Indonesia 2009a, b, c).

Adaptation activities to-date focus on the planning and formulation of key strategic documents. The 2007 National Action Plan for Climate Change (RAN-PI), aims to create development systems that are resilient to climate change and climate variability, and to implement more sustainable development that decreases the rate of environmental destruction. This action plan outlines Indonesia’s strategies on mitigation and adaptation. The plan specifically states that the country’s current capacity to cope with climate change will strongly affect its capacity in the future, and that it is therefore important to incorporate Indonesia’s current RAN-PRB within the RAN-PI. Bappenas (National Development Planning Agency) outlines in its report “National Development Planning: Indonesia Responses to Climate Change” the possible impacts of climate change on Indonesia, sectoral targets for climate change activities and possible funding mechanisms. It also provides a climate change sectoral roadmap (ICCSR) which identifies nine development priorities that will be strongly linked with climate change mitigation and adaptation activities: energy, forestry, transportation, industry, waste treatment, agriculture, marine and fisheries, water resources, and health (Bappenas 2009).

Most of the detected extreme rainfalls happened on the period of November to January with the following definition of extreme rainfall.

Effort on disaster risk reduction in Indonesia has been aligned with Sendai Framework Disaster Risk Reduction. It is also embedded with Sustainable Development Goals. Indeed, community resilience is the main cause of community sustainability and its assessment is an essential prerequisite for sustainable development. In parallel, several studies on DRR had also being conducted. The assessment of community resilience provides the baseline for recognizing gaps and directing management decisions for the fair, efficient, and effective allocation of the resources needed to improve community resilience. An integrated Framework for Community Resilience Assessment and Measurement has been developed by the FRAME project. Indonesia has been selected as a case studies model amongst other three countries on the effort to implement Disaster Risk Reduction and the development of this framework. This framework provides guidance on the data collection and analysis of three distinct case studies in different countries in Southeast Asia (Malaysia, Indonesia, and Thailand) to establish the current situation with respect to community resilience. Indonesia has been selected as a case study. The three cases have been presented in the workshop that attended by key stakeholders such as ASEAN Humanitarian Agency, Indonesian National Disaster Agency, and AsianAgri (Lestari et al. 2018).

The significant impact of agriculture—negative in most provinces but positive in some—will be important in developing approaches to adapt to climate change, from both an economic and a food security perspective. Where rainfall is expected to rise, farmers may wish to shift to corn, rain-fed rice, or other rain-fed crops whose yields will be higher than in the past. In areas with increased rainfall, the need to make use of existing irrigation systems may decrease unless the increase in volume is also accompanied by an increase in irregularity. The fact that this agricultural analysis (and indeed, all of the studies identified that consider links between agriculture and climate change in Indonesia, including the government's climate change roadmap) relies on the same 2008 study by Handoko and Syaukat means that more research on climate change and agriculture is essential. This issue is also important for policy decisions to be based on a single study whose coefficients are probably applicable only to a marginal change in weather conditions. Additional research into how climate patterns affect agricultural output, covering the whole country rather than just selected provinces, addressing a wider range of changes in temperature and rainfall, and considering more carefully which crops may do better or worse under the new conditions, is essential to design effective adaptation strategies.

The impacts of health problems are second to those of agriculture. The research on health and climate change focuses on how disease may increase with climate change but the analysis finds that in some specific cases, the incidence of disease drops rather than increases with climate change. This is quite clear in the case of dengue fever; where rainfall is expected to drop, the disease may become less prevalent than it is at present. These results suggest that more research on empirical links between rainfall, temperature, and the incidence of malaria across Indonesia, analogous to the Arcari et al. (2007) work on dengue fever, may be useful to better understand the implications of climate change for disease. This is not to suggest that climate change is less of a problem for health than is usually thought. Rather, it is to suggest that to plan for adaptation, it will be important to know more about the causes of lower costs as well as the higher costs, as this is important information for the allocation of resources to address the impacts of climate change.

As expected—and unavoidably—the impacts of SLR are only negative. Indeed, there would be no plausible explanation for these results to be positive. Also not surprisingly, the value of flooded urban land is much greater than that of flooded rural settlements, a natural result of the higher property values in urban areas. Flooded urban land is also much more costly than lost property in agriculture or aquaculture; again, not a surprising result given the property values and income streams from the different activities.

As already discussed, the biggest problem with coastal flooding relates to extreme storms rather than long-term SLR. Gradual SLR is predictable and easier to avoid, but extreme storms may impose much greater costs and destroy properties near the coast before the gradual change ever occurs. From an adaptation perspective, of course, the problem is that the timing, magnitude, and location of extreme events are difficult to predict, and the probability of one occurring in any specific place at any given time is assumed to be low.

If it is not possible to estimate the probability of an extreme storm in any given place, one way to analyze possible impacts is simply to determine what would flood, if there were a storm surge of a given height, without knowing how likely it actually is. If it can be determined what will flood, it is possible to estimate the losses, both direct and, if the models can be built, indirect through economic multiplier effects. To prioritize the choice of areas for investment in adaptation and to assess how much investment is appropriate, it would be valuable to know how likely it is that a storm surge of a given height would actually occur in different parts of the country; knowing what would be lost if it did occur but not how likely it is to occur is not an adequate basis for policy choices.

Initiated by the Ministry of Environment and Forestry (MoEF), on an integrated line ministers union workshop most currently, it was agreed that climate change adaptation measures can be integrated with disaster prevention with development planning then budget implementation become more efficient. The Union will also strengthen cooperation between sectors in implementation of an action plan for adaptation to climate change. "Both can run together, not divided." The MoEF, who met with the BNPB, the National Council on Climate Change, the Ministry of Public Works, and the United Nations Development Program, mentioned there are three main aspects that became the point of unification of adaptation measures of climate change and disaster prevention, identification of opportunities and challenges in unification of adaptation and disaster prevention, the identification of the main focus areas and the formed of roadmap for drafting a framework.

7 Conclusion

Extreme storms and weather pose a great economic threat: Analysis of the probability of extreme storms in different parts of the country would be very valuable in identifying the costs they will impose and setting priorities for investments in adaptation. To the extent that the fields of climatology and oceanography can shed light on this issue, investments in further research will be essential. Policymakers must order a more detailed analysis of the impacts of extreme storms on large urban areas, particularly Jakarta. In particular, work that can consider the macroeconomic or multiplier implications of the loss of key urban infrastructure will be essential to set priorities for investments in adaptation or strategies to minimize the harm caused by such storms. The private sector must anticipate these problems as well; every company whose business relies on transportation through the harbor or airport should create plans for continuity of its operations in the face of such extreme storms.

Work on the probability of extreme storms and the costs they would impose in different parts of the country is also very important, particularly the macroeconomic implications of flooding on nationally important transportation infrastructure in Jakarta. More generally, flooding and drought have direct costs and multiplier effects throughout the economy and these should be studied.

Indonesia cannot wait until 2050 to act: This study projects the situation in 2050 in specific areas of impact under specific parameters. As climate change is gradual, it may be assumed that the conditions described for 2050 will evolve between now and then, and that Indonesia will increasingly and inexorably experience the costs and benefits each year. This means that policy makers should not wait until the future to implement changes that either lessen or take advantage of the impacts.

It is suggested that firstly, there needs to be a re-orientation of the institutional arrangements for DRR and CCA, to increase the effectiveness of planning and implementation; secondly, DRR and CCA activities needed to be stronger supported at the local level, with a specific aim to reduce the underlying causes of vulnerability of communities at risk, and thirdly, non-government organizations play a very important role in integrating DRR and CCA through community-based initiatives. For building resilience in particular, it is recommended to reorient the institutional structure of the national government, strengthen technical and financial support to local governments, and recognize the importance of NGOs and community-based initiatives.

References

Aldrian, E. (2003). *Simulations of Indonesian Rainfall with a Hierarchy of Climate Models*. Examensarbeit Nr. 92. Hamburg: Max Planc Institute of Meteorology.

Arcari, P., Tapper, N., & Pfueller, S. (2007). Regional variability in relationships between climate and dengue/DHF in Indonesia. *Singapore Journal of Tropical Geography, 28*,

251–272. Retrieved from http://journals.lww.com/epidem/Fulltext/2005/09000/the_Importance_of_Climate_for_Dengue_Dhf_in.83.aspx# (no free access).

Bappenas. (2009). *Indonesia climate change sectoral roadmap*. Jakarta: Bappenas.

BNPB. (2016). *Risiko Bencana Indonesia 2016*. Jakarta: BNPB.

BNPB. (2018). *Disaster statistics in Indonesia 1815–2018. Disaster database & information BNPB*. Source: http://dibi.bnpb.go.id/dibi/. Accessed September 30, 2018.

BNPB. (2018). *Bencana menurut jenisnya di Indonesia tahun 1998–2018*. BNPB at http://bnpb.cloud/dibi/tabel1.

BPS. (2013), p. 24 or data available at http://www.bps.go.id/linkTabelStatis/view/id/1274.

Buendnis Entwicklung Hilft, and UNU-EHS. (2011). WorldRiskReport 2011. Buendnis Entwicklung Hilft, Berlin and UNU-EHS, Bonn.

Departemen Geografi Universitas Indonesia dan BNPB. (2013). *Laporan tentang Iklim dan Cuaca Ekstrim di Indonesia*. Jakarta (unpublished).

Djalante, R., & Thomalla, F. (2012). Disaster risk reduction and climate change adaptation in Indonesia: Institutional challenges and opportunities for integration. *International Journal of Disaster Resilience in the Built Environment, 3*(2), 166–180.

EM-DAT. (2011a). *Natural disasters in Indonesia, Table created on: 1 October 2011, Data version: v 12.07*. Brussels: Center for Research on the Epidemiology of Disasters (CRED).

EM-DAT. (2011b). "*The international disaster database". Natural Disasters Trends*, available at: www.emdat.be/natural-disasters-trends. Accessed October 1, 2018.

Germanwatch. (2011). *Global climate risk index 2011*. Bonn: Germanwatch.

IPCC. (2007a). Contribution of working group II to the fourth assessment report of the intergovernmental panel on climate change. In M. L. Parry, O. F. Canziani, J. P. Palutikof, P. J. van der Linden, & C. E. Hanson (Eds.), *Fourth assessment report: Climate change 2007 (AR4)*. Cambridge: Cambridge University Press.

IPCC. (2007b). Summary for policymakers. In O. F. Canziani, M. L. Parry, J. P. Palutikof, P. J. van der Linden, & C. E. Hanson (Eds.), *Climate change 2007: Impacts, adaptation and vulnerability contribution of working group ii to the fourth assessment report of the intergovernmental panel on climate change*. Cambridge: Cambridge University Press.

IPCC. (2012). *Managing the risks of extreme events and disasters to advance climate change adaptation*. In: Field, B. Christopher, V. Barros, T. F. Stocker, Q. Dahe, D. J. Dokken, K. L. Ebi, M. D. Mastrandrea, K. K. Mach, G.-K. Plattner, S. K. Allen, M. Tignor, & P. M. Midgley (Eds.), *A Special report of working groups I & II of the intergovernmental panel on climate change*. Cambridge, UK, and New York, NY, USA: Cambridge University Press.

Lestari, F. (2018). Workshop on framework for community resilience assessment & measurement (FRAME), Sunway, Malaysia, 14–15 August 2018. Collaborative research work between Universitas Indonesia, Lancaster University, UK, Sunway University & Thammasat University. Website: http://www.lancaster.ac.uk/events/framework-for-community-resilience-assessment-and-measurement-frame and http://drrc.ui.ac.id.

Lestari, F., Kurniawan, L., Hidayah, A. F., et al. (2017a). *Bunga Rampai Riset Kebencanaan UI: Kontribusi Pemangku Kepentingan untuk Penurunan Tingkat Risiko Bencana* (p. 2017). Jakarta: UI Press.

Lestari, F., et al. (2017b). *Kontribusi pemangku kepentingan untuk penurunan tingkat risiko bencana*. Jakarta: UIP Penerbit Universitas Indonesia.

Lestari, F., Kusminanti, Y., Firdaus, R., et al. (2017c). *Health Crisis Management in Indonesia*. In Conference on health management in post disaster recovery. Strengthening health care system capacity to disaster: towards a sustainable risk reduction. 22–23 May 2017, Banda Aceh, Indonesia. Website: http://tdmrc.unsyiah.ac.id/wp-content/uploads/2017/10/REVISI_Proceedings-DAAD-Conference.pdf.

Maplecroft. (2010). *Natural disasters index 2010*. Bath: Malpercroft.

Maplecroft. (2011). *Climate change vulnerability index 2011*. Bath: Maplecroft.

MoE. (2009). *Summary for policy makers: Indonesia second national communication under the United Nations framework convention on climate change*, (p. 43) State Ministry of Environment.

Rosenzweig, C., Iglesius, A., Yang, X. B., et al. (2001). *Climate change and extreme weather events—Implications for food production, plant diseases, and pests* (p. 24). Paper: NASA Publications.

Scott, P. A., Allen, M., Christidis, N., Dole, R., et al. (2013). Attribution of weather and climate-related extreme events. In: G. R. Asrar & J. W. Hurrell (Eds.), *Climate science for serving society: Research, modelling and prediction priorities*, (pp. 307–337). Springer.

Salamon, J., et al. (2012). Common values in assessing health outcomes from disease and injury: Disability weights measurement study for the Global Burden of Disease Study 2010. *The Lancet, 380*, 2129–43. Retrieved from http://www.thelancet.com/pdfs/journals/lancet/PIIS0140-6736%2812%2961680-8.pdf.

UNDP. (2007). *Human development report 2007; Fighting climate change: Human solidarity in divided world*. London: Palgrave MacMillan Publisher.

UNDP Indonesia. (2009a). *Gearing up for Copenhagen, Indonesia convenes a high-level dialogue on climate change*, available at: www.undp.or.id/press/view.asp?FileID¼20091123-1&lang¼en. Accessed 1 October 1, 2018.

UNDP Indonesia. (2009b). *The missing link—the province and its role in Indonesia's decentralization*. Jakarta: UNDP Indonesia.

UNDP Indonesia. (2009c). *UNDP appointed as interim trustee for Indonesia Climate Change Trust Fund*. available at: www.undp.or.id/press/view.asp?FileID¼20090914-1&lang¼en. Accessed October 1, 2018.

WMO. (2018, January). *Guideline of the definition and monitoring and extreme weather and climate events*. WMO. (unpublished-final draft).

Yusuf, A. A., & Francisco, H. A. (2009). *Mapping for Southeast Asia, Economy and Environment Program for Southeast Asia (EEPSA)*.

Climate-Related Disasters and Health Impact in Malaysia

Nasrin Aghamohammadi, Logaraj Ramakreshnan, Chng Saun Fong and Nik Meriam Sulaiman

Abstract Urban heat island (UHI) is a typical urban thermal pollution which occurs due to the development of higher ambient air temperatures in densely built cities that deteriorate the air quality of urban area due to anthropogenic activities. The aforementioned urban complexity results in the absorption and retention of heat which is re-emitted after sunset, creating a steep temperature gradient between urban and rural areas. The effect of UHI on public health is reduced outdoor thermal comfort levels of city dwellers in various urban settings. In this study, UHI, air quality and thermal comfort are discussed as climate-related disasters on public health in Malaysia. Understanding of UHI Intensity (UHII) using meteorological network's observations and mobile weather trackers in the urban and suburban areas of Malaysia will explore the hotspot and green lung of each climate zone for implementing a necessary approach to minimize the human exposure to extreme heat and contaminated air. Incorporation of more explanatory variables of UHI from both meteorological and urban factors is a requisite to identify their pivotal association with UHI intensification in the Malaysian context. In spite of the fact that seasonal haze episodes and UHI have become a recurrent seasonal phenomenon in Malaysia, the number of public health studies is still scarce in this region. More intensive studies are still needed to establish veritable evidence for the susceptible populations to devise effective health planning to reduce haze and UHI-associated health impacts in Malaysia. With such acquired knowledge of the UHI and air quality issues and related underlying mechanisms, urban planners, designers and decision-makers can perform more evidence-based decision-making to create, reform or rejuvenate climate-friendly sustainable cities for enhanced liveability in future. Furthermore, such scientific knowledge will become a valuable input and tool for the policy-makers to identify the right policy choices aimed at sustainable healthcare solutions.

N. Aghamohammadi (✉) · L. Ramakreshnan · C. S. Fong
Centre for Occupational and Environmental Health, Department of Social and Preventive Medicine, Faculty of Medicine, University of Malaya, 50603 Kuala Lumpur, Malaysia
e-mail: nasrin@ummc.edu.my; nasrinam01@gmail.com

N. M. Sulaiman
Department of Chemical Engineering, Faculty of Engineering, University of Malaya, 50603 Kuala Lumpur, Malaysia

R. Akhtar (ed.), *Extreme Weather Events and Human Health*,
https://doi.org/10.1007/978-3-030-23773-8_17

Keywords Air pollution · Climate change · Haze · Public health · Sustainability · Urban heat island

1 Introduction

Anthropogenic activities and natural disasters that exude sufficient quantities of greenhouse gases into the atmosphere trap additional heat in the lower atmosphere and affect the global climate. Southeast Asia (SEA) is occasionally experiencing catastrophes such as flash flood, landslides and tsunami that challenged international attempts to reduce health vulnerability related to climate change. Health impacts associated with environmental degradation-induced climate change need to be well studied to strengthen public health interventions and to increase community resilience at local levels. In response to this, this chapter examines the association between climate-related disasters, specifically on air pollution, urban heat island phenomenon and health issues in Malaysia as well as air quality and urban heat island management policies to safeguard the community health.

2 Air Pollution and Public Health Issues

Generally, any substance introduced to the atmosphere including particulate matter, gases, radioactive materials, hazardous air pollutants (HAPs) and many others which cause severe effects to living things and the environment is an air pollution (Arshad et al. 2015). Airborne particulate matter (APM) containing heavy metals is considered as a public health concern as it can enter human lungs through the respiratory system (Arshad et al. 2015). The size of these particles affects the radiation dose through inhalation of airborne dust that determines where radioactivity was deposited, either in the lung or in the tissues. While particles less than 5 μm can reach the pharynx tract, fine particles less than 2.5 μm or ultra-fine particles can travel deep into the lungs with the potential to penetrate tissues and deposit among the interstitial cells (Ny and Lee 2011). In daily activities, humans are exposed to various substances, either toxic or non-toxic through inhaling, digesting and skin contact. Continuous exposure and accumulations of these chemical substances will lead to increased exposure and risks of causing severe diseases and even death (Arshad et al. 2015). There are a limited number of epidemiological studies that make it difficult to estimate the health impact of air pollution in Malaysia. Illnesses related to air pollution were significantly associated with ambient air quality and also proportionate with the duration of time spent in areas with air pollution (Leh et al. 2012). Air pollution in general, whether from traffic-related, industrial or haze-caused pollutants, is considered as a major threat to human health at the population level because of its link to increased mortality (Forouzanfar et al. 2015).

2.1 *Background History of Air Pollution in Malaysia*

The level of environmental concerns in Malaysia has always been low in earlier periods until the haze episodes in 1983, 1984, 1991, 1994 and 1997 in Southeast Asia brought instantaneous increase in awareness among the environmental management groups in Malaysia (Afroz et al. 2003). In response to that, the Malaysian Air Quality Guidelines, the Air Pollution Index and the Haze Action Plan were established to improve the air quality in Malaysia (Afroz et al. 2003). The Air Pollution Index (API) is a system used in Malaysia to disseminate information regarding the level of air pollution in its simplest terms for the general public. This index reflects the effect of air pollution level on human health ranging from good to hazardous and is categorized accordingly as stipulated in the National Haze Action Plan (DOE 2018). The five air pollutants taken into account in the API system is the suspended particulate matter of less than 10 microns in size (PM_{10}), sulphur dioxide (SO_2), carbon monoxide (CO), ozone (O_3) and nitrogen dioxide (NO_2). In Malaysia, Department of Environment (DOE) monitors the API through a network, strategically located in urban, industrial and residential areas in the country.

2.2 *Sources of Air Pollution in Malaysia*

The three major sources of air pollution in Malaysia are mobile sources, stationary sources and open burning sources as illustrated in Fig. 1.

Emissions from mobile sources (motor vehicles) contribute at least 70–75% of the total air pollution. Meanwhile, emissions from stationary sources (industry, power stations, industrial fuel burning processes and domestic fuel burning), open burning and forest fires have contributed 20–25% and approximately 3–5%, respectively, to the total air pollution (Afroz et al. 2003). Besides that, a substantial amount of dust particles (Wei et al. 1999) due to biomass burning activities from nearby Sumatra has led to seasonal transboundary haze episodes in the past few decades (Shokoohi

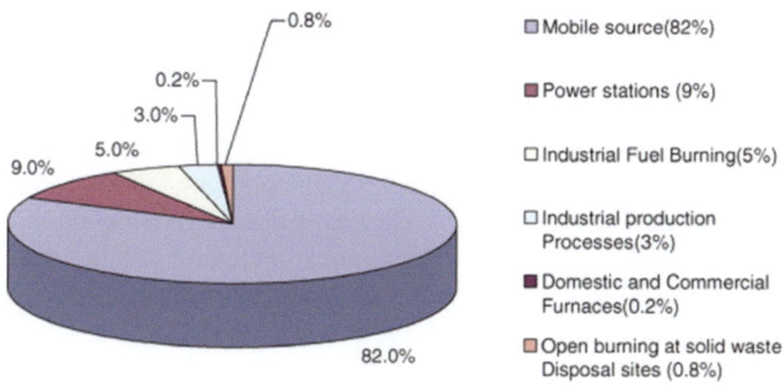

Fig. 1 Sources of air pollution in Malaysia (Afroz et al. 2003)

and Nikitas 2017). The occurrence of haze along with local emissions from motor vehicles, industries and open burning activities further deteriorates the air quality status in Malaysia.

2.3 Health Impacts as Relating to Local Gas Emissions

In terms of local emission as such from landfill, methane (CH_4) and carbon dioxide (CO_2) are the major gases released to the environment. CO_2 concentrations of 10% or more can cause unconsciousness or death, while lower concentrations may cause headache, sweating, rapid breathing and increased heartbeat, shortness of breath, dizziness, mental depression, visual disturbances or shaking (ATSDR 2001). Other gases produced from landfill are carbon monoxide (CO), hydrogen sulphide (H_2S), a trace amount of non-methane organic compounds (NMOC) in varying proportions (Durmusoglu et al. 2010). One of the NMOCs is volatile organic compounds (VOCs) such as benzene, toluene, ethylbenzene and xylenes (m, p, o-xylene) (BTEX). These compounds are classified as hazardous air pollutants (HAPs) (Chiriac et al. 2009; Chiriac et al. 2007; Durmusoglu et al. 2010). Exposure to HAPs can cause a variety of health problems such as cancerous illnesses, respiratory irritation and central nervous system damage (Durmusoglu et al. 2010; Nadal et al. 2009). In addition, sulphur gases which are known to create an unpleasant odour in the atmosphere are sources of undesirable health effects or symptoms such as headaches and nausea (Kamarrudin et al. 2013) and mainly emitted from open burning, biomass burning and treatment plants.

2.4 Health Impact Associated with Traffic-Related Air Pollutions in Malaysia

In Malaysia, there is relatively scarce evidence on the health effects of traffic-related air pollution (Wong et al. 2017). Public transport commuters are exposed to high concentrations of air pollutants, both originated from vehicular traffic and those caused by other urban, industrial or environmental factors. The three highest adverse health experiences during daily normal commute are fatigue and weakness (35.5%), coughing (23.6%) and dizziness (18.2%), while during haze days, the three highest adverse health experiences were headache (85%), coughing (84.8%) and breathing difficulties (84.5%) (Wong et al. 2017). On top of that, psychological effects of air pollution can also lead to psychiatric symptoms, including anxiety disorders and changes in mood, cognition and behaviour, as well as to a reduction in psychological well-being (Genc et al. 2012; Lundberg 1996; Power et al. 2015).

2.5 *Health Impacts Related to Indoor Air Quality in Malaysia*

Recent work has suggested that second-hand smoke (SHS) which contains $PM_{2.5}$ may cause or exacerbate a range of illnesses and disease states (Brook et al. 2010; Eisner et al. 2009; Jaakkola et al. 2003; Kurahashi et al. 2008). Lee et al. (2010) found that the average concentration of $PM_{2.5}$ in 22 indoor environments in Malacca, Malaysia, is 1.8 times higher than the current standard permitted by the Air Quality Guidelines (Lee et al. 2010; WHO 2006). The migration of PM of different sizes from the outdoor to the indoor environment along its impact towards indoor air quality and potential health impacts on indoor occupants remain largely unknown in SEA due to the lack of systematic investigations and observational data.

3 Seasonal Haze and Health Impacts in Malaysia

Haze is an atmospheric phenomenon in which higher concentrations of particulate matter and other aerosols in air obscure the clarity of the sky and results in a measurable reduction of visual range (Latif et al. 2018). In a broader context, haze is a result of interaction between high amounts of secondary gas, such as tropospheric ozone with VOCs to form secondary organic aerosols (Liu et al. 2016; Wu et al. 2016). Even though haze episodes can be associated with various sources of pollutions, forest fires and peat burning is of major concern due to their regular incidences and adverse impacts on regional air quality in SEA (Huang et al. 2013). The resulting thick haze usually engulfs at least six main SEA countries such as Indonesia, Malaysia, Brunei, Thailand, Singapore and the Philippines (Ramakreshnan et al. 2018a; Thomas et al. 2014).

3.1 *Historical Background of Seasonal Haze and Health Studies in Malaysia*

In Malaysia, the first biomass burning induced-transboundary haze erupted in the year 1982 and some severe episodes occurred in 1997, 2005 and 2015 (Latif et al. 2018). During these episodes, the concentrations of particulate matter with an aerodynamic diameter of less than 10 μm (PM_{10}) exceeded the Malaysian Air Quality Guideline for PM_{10} concentration (150 $\mu g/m^3$ for a 24-h average) at many locations throughout Malaysia (Latif et al. 2018). In another case, Pavagadhi et al. (2013) clarified that about 90% of soot exuded from biomass burning is lower than 2.5 μm in size and has a high probability of penetrating deep into the lower respiratory tract (Pavagadhi et al. 2013). In certain studies, prolonged exposure of $PM_{2.5}$, which is defined as minute particles less than or equal to 2.5 μm in aerodynamic diameter, is reckoned to increase the risk of developing lung cancers (Betha et al. 2013; Khan et al.

2016). In this context, identification and apprehension of health impacts caused by air pollutants loaded in transboundary haze episodes in Malaysia are vital to alert public institutions to devise evidence-based prevention strategies and health plannings. The year 1997s transboundary haze was the first event that causes many devastating health impacts in Malaysia which stimulated various haze-induced health impact studies. In general, outpatient visits to hospitals nearly tripled in the peak period of haze in 1997 (WHO 1998). Assembled data indicated an increase in asthma, acute respiratory infection and conjunctivitis cases in major hospitals in Kuala Lumpur (Brauer and Hisham-Hashim 1998). Selangor recorded a significant increase in the total number of respiratory-related cases (from 912 to more than 5000 cases), acute respiratory infection cases (from about 6000 to more than 30,000 cases) and conjunctivitis (from 207 cases to 3496 cases) in between June and October 1997 (Afroz et al. 2003). Meanwhile, Nasir et al. (2000) demonstrated that the total health effects were estimated to include 285,227 asthma attacks, 118,804 cases of bronchitis in children, 3889 cases of chronic bronchitis in adults, 2003 respiratory hospital admission and 26,864 emergency room visits during this period (Nasir et al. 2000).

3.2 Escalating Mortality Incidences During Haze Events in Malaysia

The widely documented physical health effect in Malaysia is escalating mortality rates during severe haze episodes. One of the earliest studies was undertaken by Sastry (2002), who reported a rise in mortality counts when PM_{10} exposure exceeds 210 $\mu g/m^3$ among the elderly class, during the 1997 Southeast Asian wildfire (Sastry 2002). Later on, Marlier et al. (2013) discovered higher frequencies of regional mortality due to cardiovascular diseases during high fire incidences between 1997 and 2006 in Southeast Asia (Marlier et al. 2013). Sahani et al. (2014) used a case-crossover study design to identify the incremental risk of respiratory mortality among elderly males during haze days when the PM_{10} concentration exceeds 100 $\mu g/m^3$. At the same time, a two-fifths increase of mortalities was found among the children below 14 years old after two days of exposure during the haze episodes (Sahani et al. 2014). In another cross-country study across Indonesia, Malaysia and Singapore, Koplitz et al. (2016) determined that the year 2015 recorded an excess of 100,300 mortalities compared to the year 2006 by comparing population-weighted smoke exposure data of $PM_{2.5}$ between both respective year's haze events. Their results revealed that regional smoke-related mortality in 2015 was more than double of the 2006 haze event that enveloped equatorial Asia (Koplitz et al. 2016).

3.3 Haze-Associated Respiratory Cases in Malaysia

Besides this, another widely reported physical health impact of haze exposure is respiratory morbidity and impairment of lung functions. Hashim et al. (1998) compared pulmonary functions of 107 primary school children before, during and after the haze episodes in Malaysia and identified significant reductions in pulmonary functions during and after the haze episodes among the girls. A residual impact of haze exposure on children's pulmonary functions was found even after the haze episode ended (Hashim et al. 1998). In another case, Mott et al. (2005) through the time-series analyses between the pre- and post-haze periods had observed a statistically significant increase in respiratory hospitalizations, specifically those for chronic obstructive pulmonary disease (COPD) and asthma in the 19–39 and 40–64 years' age categories in Kuching, Malaysia. They also identified that elderly people above 65 years old with previous health records were more vulnerable to be re-hospitalized during the follow-up period after exposure to the haze in 1997 (Mott et al. 2005).

3.4 Cardiovascular Morbidity During Haze in Malaysia

A limited number of studies reported the risk of cardiovascular morbidity due to haze exposure. Sastry (2002) witnessed less cardiovascular-related infant deaths compared to the other age categories in his retrospective study in Kuala Lumpur, Malaysia. However, death risk from cardiovascular causes for the elderly above 65 years old is twice as high after the direct exposure to the severe haze episodes (Sastry 2002). In agreement to the previous findings, Mott et al. (2005) observed increasing records of hospitalizations for ischaemic heart disease (IHD) among the individuals aged between 40 and 64 years based on the historical data due to prolonged exposure to the forest fires in Kuching, Malaysia, in 1997 (Mott et al. 2005). In Malaysia, Excessive Lifetime Cancer Risk (ELCR) assessments due to carcinogenic pollutants present in haze are still scanty. In a recent study, Sulong et al. (2017) estimated highest excess lifetime cancer risk among the adult group (above 18 years old) in Kuala Lumpur (KL) during haze period compared to pre- and post-haze periods of 2016 (Sulong et al. 2017). In spite of the fact that seasonal haze episodes have become a recurrent seasonal phenomenon in Malaysia, the number of public health studies is still scarce. More intensive studies are still needed to establish veritable evidence for the susceptible populations to devise effective health plannings to reduce haze-associated health impacts in Malaysia.

4 Urban Heat Island (UHI), Heatwaves and Public Health Issues in Malaysia

Urban Heat Island (UHI) is a typical urban thermal pollution which occurs due to the development of higher ambient air temperatures in densely built cities compared to their rural surroundings (Rajagopalan et al. 2014; Roth and Chow 2012). UHI phenomenon is often triggered by urbanization effects which are closely associated with a larger number of buildings, compact and massive urban structures with narrow street canyons and reduced sky view factor, non-reflective and impermeable surface materials, lack of transpiring vegetation, transport flows, increased energy consumption as well as higher concentrations of urban pollutants (Elsayed 2012; Ramakreshnan et al. 2018b; Shaharuddin et al. 2009). The aforementioned urban complexity results in the absorption and retention of heat which are re-emitted after sunset, creating a steep temperature gradient between urban and rural areas. The effect of UHI on public health is reduced outdoor thermal comfort levels of the city dwellers in various urban settings.

4.1 Historical Background of UHI in Tropics and Malaysia

The concern of UHI came into limelight when a British scientist, Luke Howard, discovered that the city of London is relatively warmer than the surrounding rural areas due to urbanization effects in the 1810s (Roth and Chow 2012). By replicating his urban–rural pair temperature comparisons at two metres height above the ground, many scientists conducted quantitative UHI assessments in various cities of different climates that observed a rapid economic and human capital development due to urbanization. In the tropics, UHI assessment was pioneered by Simon Nieuwolt in Singapore in the late 1960s, which later inspired a serious of UHI work by Sani in 1970s in the neighbouring country, Malaysia (Ramakreshnan et al. 2018b; Roth and Chow 2012). During his two-decadal comprehensive UHI work in and around Kuala Lumpur (KL) using temperature transverse techniques, he identified that the UHI Intensity (UHII) of KL can reach up to 6.7 °C in 1980 (Sani 1984). Later on, Elsayed (2012) recorded an increase of 1.5 °C in UHII for KL City compared to a similar study done by Sani in 1985 (Elsayed 2012; Sani 1986). After some period of time, Ramakreshnan et al. (2019a) reported an intensity of 1.7 °C in one of the congested urban stations (Petaling Jaya) of Greater Kuala Lumpur in 2016.

4.2 State-of-Art Review of UHI Assessments in Malaysia

Concerning about the weaknesses of in situ measurements with poor spatial and temporal resolutions, a growing number of studies successfully employed wide and versatile applications of satellite technology in UHI quantifications in Malaysia. For instance, Yusuf et al. (2014) reported an average gain of 8.4 °C in surface temperature between 1997 and 2013 in Greater Kuala Lumpur (GKL) using Landsat TM/ETM images (Yusuf et al. 2014). By integrating both TERRA/MODIS satellite data and Geographical Information System (GIS), Shaharuddin et al. (2014) found that the average UHII was higher (14.5 °C) during the intermediate monsoon season in GKL (Shaharuddin et al. 2014). Similarly, Amanollahi et al. (2016), Salleh et al. (2013), Shaharuddin et al. (2009) and Hashim et al. (2007) explained the role of land-use changes and conversion of natural, rural and agricultural land to urban surfaces on temperature increment in various city centres of GKL using remote sensing applications (Amanollahi et al. 2016; Hashim et al. 2007; Salleh et al. 2013; Shaharuddin et al. 2009). Contemporary studies also used various modelling and simulation approaches to investigate UHI phenomenon at the local scale. In this context, Tso and Law (1991) successfully performed a simple surface energy balance (SEB) modelling over selected areas of the city of KL represented by heavy concrete-made building structures and observed maximum peak of temperature at 12 p.m. due to heat absorption by concrete mass during intense solar radiation (Tso and Law 1991). Rajagopalan et al. (2014) conducted numerical simulations in Muar and identified that the chaotic development in the city has caused reduced ventilation in urban canyons which subsequently resulted in 3.2 °C of UHII during night time (Rajagopalan et al. 2014). Morris et al. (2015) applied a single-layer urban canopy model coupled with the WRF/NOAH LSM modelling system in Putrajaya and discovered that the UHII varied temporally and spatially with a maximum magnitude of 3.1 °C in 2012 (Morris et al. 2015). In another study, Morris et al. (2016) used the same simulation approach to investigate the local urban climate changes over a decade (1999–2011) in Putrajaya and concluded that the canopy layer temperature of the area was increasing at the rate of 1.66 °C per decade (Morris et al. 2016). Besides, Morris et al. (2017) also used WRF mesoscale model and identified that daily mean UHII of 0.9 °C with a more severe heating (1.9 °C) at night in GKL (Morris et al. 2017).

4.3 Deterioration of Thermal Comfort and UHI-Associated Health Impacts in Malaysia

The most documented direct effect of UHI on public health is reduced outdoor thermal comfort levels of city dwellers in various urban settings (Fong et al. 2019). For instance, Makaremi et al. (2012) quantitatively investigated the outdoor thermal comfort in shaded outdoor spaces within Universiti Putra Malaysia's campus-based upon the measurement of major climatic parameters. The findings revealed that the values of Physiologically Equivalent Temperature (PET) thermal comfort index in the selected shaded outdoor spaces of the campus were higher than the comfort range defined for a tropical climate, especially during morning and late afternoon.

This study also demonstrated that thermal adaptation and psychological parameters play a pivotal role in affecting human thermal comfort level in outdoor spaces (Makaremi et al. 2012). Nasir et al. (2012) conducted microclimatic measurements and systematic interviews in shaded Shah Alam Lake Garden to identify the impacts of weather and personal factors on respondents' perpetual and sensation estimations. The results showed that only one-fifth of the urban park users felt comfortable with the surrounding microclimatic factors of the park, whereas 80% of them felt the warm sensation (Nasir et al. 2012). In another study, Nasir et al. (2013) investigated peoples' perception of their physical activities in Shah Alam Lake Garden during hot hours and identified that a negative correlation ($r = -0.291, p < 0.001$) between people's activities and the time in a day. This indicates that people show less interest in doing outdoor activities during the hot time (Nasir et al. 2013). In addition, Qaid et al. (2016) investigated the outdoor thermal comfort at different areas in a planned city of Putrajaya and identified that the city's human thermal comfort exceeds the natural range of 30 °C. However, they discovered that high-rise residential buildings and the Boulevard Street are thermally comfortable during daytime due to building shading effects (Qaid et al. 2016). Besides thermal comfort assessments, Wong et al. (2017) surveyed about 1050 working communities in the city of Kuala Lumpur from manufacturing and service sectors regarding the health impact of UHI effects. Majority of the respondents (90.2%) reported respiratory problems as the main health impact due to increasing temperature in their surrounding environment. This is followed by heat exhaustion (83.1%) and heat cramps (72.9%). Depression (64.7%) was the experience that was reported the highest among all psychological health experiences queried (Wong et al. 2017). In conclusion, no much research evidence was available to elucidate the association between health impacts and UHI phenomenon effects on the working community in the city of Kuala Lumpur. More scholarly studies are vital to obtaining an in-depth view of UHI-driven physical and psychological health impacts of the urban community in future.

5 Air Quality and Urban Heat Island Management and Policy in Malaysia

5.1 Air Quality Management in Malaysia and Policies

The first Malaysia Ambient Air Quality Guideline has been used since 1989. The New Ambient Air Quality Standard adopts six air pollutants' criteria that include five existing air pollutants which are particulate matter with the size of less than 10 micron (PM_{10}), sulphur dioxide (SO_2), carbon monoxide (CO), nitrogen dioxide (NO_2) and ground-level ozone (O_3) as well as one additional parameter which is particulate matter with the size of less than 2.5 micron ($PM_{2.5}$). Clean Air Action Plan (CAAP) under the DOE, the Clean Air Action Plan (2010–2020) was established in 2011 and it contains five main strategies in order to improve the air quality. The five strategies are described to reduce emissions from motor vehicles, prevent haze pollution from land and forest fires, reduce emissions from industries, build institutional capacity and capabilities and strengthen public awareness and participation.

5.2 *Sustainable Transport in Malaysia*

For achieving significant levels of CO_2 emission reductions, the idea of transit-oriented development started to gain recognition after its inclusion in the Kuala Lumpur Structure Plan and the draft of Kuala Lumpur 2020 City Plan (Kuala Lumpur Structure Plan 2017). The first evidence of any political attention being given to sustainable transportation in KL was as recent as 2010, when the Economic Transformation Program (ETP) was launched on; this programme is an initiative by the Malaysian government to turn the country into a high-income nation and turn KL into a liveable city by 2020 (Shokoohi and Nikitas 2017). KL Car Free Morning started in January 2014 to promote a healthier lifestyle as well as create awareness about the alternative modes of transportation.

5.3 *Second-Hand Smoke (SHS) in Malaysia*

Like in most other countries there are no Malaysian guidance or limit values in relation to SHS exposure either in the workplace or for non-occupational exposure within public spaces. The Malaysian Code of Practice for Indoor Air Quality guidelines for non-industrial settings was introduced by the Malaysian Department of Occupational Safety and Health in 2005 and states that 8-h time-weighted average for PM_{10} was 150 $\mu g/m^3$ (Malaysia Department of Occupational Safety and Health 2005). In 2010 the WHO indicated that their AQG 2005 for $PM_{2.5}$ (25 $\mu g/m^3$) could now be applied to indoor settings.

5.4 *Unleaded Fuel in Malaysia*

Two successful regulations that led to the reduction of negative impact of mobile sources on air quality are the amendments in the Environmental Quality Act (EQA) and the phase-out of leaded gasoline sales. A significant first step towards implementing Malaysia's Clean Air Plan was achieved in 1996 with the approval of two regulations designed to reduce emissions from mobile sources: the Environmental Quality (Control of Emissions from Diesel Engines) Regulations 1996 and the Environmental Quality (Control of Emissions from Petrol Engines) Regulations 1996.

5.5 *Urban Heat Island Management in Malaysia*

There are no policies in Malaysia that specifically address the UHI phenomenon, even in the National Policy on Climate Change which was formulated in 2009 to discuss some serious local issues of climate change. On the other hand, National Urbanization Policies 1 and 2 (2006 and 2016) of Malaysia have provided general statements on the need to address UHI impacts for future sustainable urban development (Ramakreshnan et al. 2019b). However, the implementation of various local guidelines on the development of green and sustainable cities constructed by the Federal Department of Town and Country Planning of Malaysia is able to ameliorate the eventual impacts and severity of UHI. One such guideline is the Green Neighbourhood Guideline as presented in Table 1. Incorporating relevant best practices, standards and guidelines (as shown in Table 1) in policies is one of the best moves

Table 1 General guidelines for Green Township Development in Malaysia

General guidelines	Description
Site selection and site planning	○ Avoid environment sensitive area (ESA) ○ Connect existing street network with adjacent network ○ Introduce integrated infrastructure
Application of green technology in building development	○ Comply with international and national assessment criteria for green building
Walkability and connectivity	○ The pedestrian network is interconnected and accessible to all amenities and services ○ Apply universal design/barrier-free design ○ Destination and amenities are within walking distance
Green network	○ Planting strips and street trees ○ Connecting natural habitat and waterway with landscape or garden ○ Neighbourhood farming plot and home 'green plots'
Short distance to transit	○ Rapid, convenient and efficient transit service
Green infrastructure	○ Provision of green infrastructure through the application of technology or using natural methods
Mixed-used development	○ Diverse land use ○ Mixed use/multiuse building ○ Diversity of activities
High density (compact development)	○ Higher density located in, adjacent to or within close proximity to the core area ○ Crucial in supporting public transportation and as a threshold to support community facilities

Source Hashim and Rosly (2011)

to minimize the effects of UHI while enhancing the urbanite's adaptation to strive well in hotter cities.

6 Conclusion

Haze episodes, extreme heat and weather events have become a recurrent seasonal phenomenon in Malaysia since 1997, and the number of public health and extreme weather studies is still scarce. The association between respiratory morbidity, especially on the aggravation of asthma, and haze exposure was found in limited studies; however, there is still lack of in-depth investigation associated with public health to reach a strong and valid agreement between the patterns of other diseases during and following exposure. Malaysia needs to initiate toxicological studies on the biological mechanism of hazardous pollutants. In fact, long-term exposure assessments need to establish veritable evidence for the susceptible populations for future health planning (Ramakreshnan et al. 2018a). Understanding of UHI Intensity using meteorological network's observations and mobile weather trackers in urban and suburban area of Malaysia will explore the hotspot and green lung of each climate zone for implementing a necessary approach to minimize the human exposure with extreme heat. Incorporation of more explanatory variables of UHI from both meteorological and urban factors is a requisite to identify their pivotal association with UHI intensification in the Malaysian context. With such acquired knowledge of the UHI issue and related underlying mechanisms, urban planners, designers and decision-makers can perform more evidence-based decision-making to create, reform or rejuvenate climate-friendly sustainable cities for enhanced liveability in future. Furthermore, such scientific knowledge will provide valuable input and tools for policy-makers to make the right policy choices aiming at sustainable healthcare solutions.

Acknowledgements The authors are grateful to Sustainable Science Cluster, University of Malaya, since this work is financially supported by University of Malaya Grand Challenges Research Grant (GC002A-15SUS).

References

Afroz, R., Hassan, M., & Ibrahim, N. A. (2003). Review of air pollution and health impacts in Malaysia. *Environmental Research, 92*(2), 71–77.

Amanollahi, J., Tzanis, C., Ramli, M. F., & Abdullah, A. M. (2016). Urban heat evolution in a tropical area utilizing Landsat imagery. *Atmospheric Research, 167,* 175–182.

Arshad, N., Hamzah, Z., Wood, A. K., Saat, A., & Alias, M. (2015). Determination of heavy metals concentrations in airborne particulates matter (APM) from Manjung district, Perak using energy dispersive X-ray fluorescence (EDXRF) spectrometer. In: AIP Conference Proceedings (Vol. 1659, p. 50008).

ATSDR. (2001). Chapter 2: Landfill Gas Basics, C, 1–6. Retrieved from https://www.atsdr.cdc.gov/HAC/landfill/html/ch2.html.

Betha, R., Pradani, M., Lestari, P., Joshi, U. M., Reid, J. S., & Balasubramanian, R. (2013). Chemical speciation of trace metals emitted from Indonesian peat fires for health risk assessment. *Atmospheric Research, 122,* 571–578.

Brauer, M., & Hisham-Hashim, J. (1998). Peer reviewed: Fires in Indonesia: Crisis and reaction. *Environmental Science & Technology, 32*(17), 404A–407A.

Brook, R. D., Rajagopalan, S., Pope, C. A., III, Brook, J. R., Bhatnagar, A., Diez-Roux, A. V., et al. (2010). Particulate matter air pollution and cardiovascular disease: An update to the scientific statement from the American Heart Association. *Circulation, 121,* 2331–2378.

Chiriac, R., Carre, J., Perrodin, Y., Fine, L., & Letoffe, J. M. (2007). Characterisation of VOCs emitted by open cells receiving municipal solid waste. *Journal of Hazardous Materials, 149*(2), 249–263.

Chiriac, R., Carre, J., Perrodin, Y., Vaillant, H., Gasso, S., & Miele, P. (2009). Study of the dispersion of VOCs emitted by a municipal solid waste landfill. *Atmospheric Environment, 43*(11), 1926–1931.

Durmusoglu, E., Taspinar, F., & Karademir, A. (2010). Health risk assessment of BTEX emission in the landfill environment. *Journal of Hazardous Materials, 176,* 870–877.

Eisner, M. D., Iribarren, C., Yelin, E. H., Sidney, S., Katz, P. P., Sanchez, G., & Blanc, P. D. (2009). The impact of SHS exposure on health status and exacerbations among patients with COPD. *International Journal of COPD, 4*(1), 169–176.

Elsayed, I. S. (2012). A study on the urban heat island of the city of Kuala Lumpur, Malaysia. *Journal of King Abdulaziz University, 23*(2), 121.

Fong, C. S., Aghamohammadi, N., Ramakreshnan, L., Sulaiman, N. M., & Mohammadi, P. (2019). Holistic recommendations for future outdoor thermal comfort assessment in tropical Southeast Asia: A critical appraisal. *Sustainable Cities and Society, 46,* 101428.

Forouzanfar, M. H., Alexander, L., Anderson, H. R., Bachman, V. F., Biryukov, S., Brauer, M., et al. (2015). Global, regional, and national comparative risk assessment of 79 behavioural, environmental and occupational, and metabolic risks or clusters of risks in 188 countries, 1990–2013: A systematic analysis for the Global Burden of Disease Study 2013. *The Lancet, 386*(10010), 2287–2323.

Genc, S., Zadeoglulari, Z., Fuss, S., & Genc, K. (2012). The adverse effects of air pollution on the nervous system. *Journal of Toxicology, 2012,* 1–23.

Hashim, J. H., Hashim, Z., Jalaludin, J., Lubis, S. H., & Hashim, R. (1998). Respiratory function of elementary school children exposed to the 1997 Kuala Lumpur Haze. *Epidemiology, 9*(4), 103.

Hashim, N., Ahmad, A., & Abdullah, M. (2007). Mapping urban heat island phenomenon, remote sensing approach. *Journal-The Institution of Engineers, Malaysia, 68*(3), 25–30.

Hashim, N., & Rosly, D. (2011). *Guideline and framework for green township in Malaysia.* Retrieved from http://www.townplan.gov.my/download/guideline_and_framework_for_green_township_in_malaysia_-_paper_3a1861603323.pdf.

Huang, K., Fu, J. S., Hsu, N. C., Gao, Y., Dong, X., Tsay, S. C., & Lam, Y. F. (2013). Impact assessment of biomass burning on air quality in Southeast and East Asia during BASE-ASIA. *Atmospheric Environment, 78,* 291–302.

Jaakkola, M. S., Piipari, R., Jaakkola, N., & Jaakkola, J. J. K. (2003). Environmental tobacco smoke and adult-onset asthma: A population-based incident case-control study. *American Journal of Public Health, 93*(12), 2055–2060.

Kamarrudin, N., Zulkafli, N. H., Sikirman, A., Mahayuddin, N. M., Sigau, B. A., Ku Hamid, K. H., & Akhbar, S. (2013). Concentration and toxicological study on sanitary landfill gases at drilling point closed cell. In *BEIAC 2013—2013 IEEE Business Engineering and Industrial Applications Colloquium* (pp. 333–338).

Khan, M. F., Latif, M. T., Saw, W. H., Amil, N., Nadzir, M. S. M., Sahani, M., et al. (2016). Fine particulate matter in the tropical environment: Monsoonal effects, source apportionment, and health risk assessment. *Atmospheric Chemistry and Physics, 16*(2), 597–617.

Koplitz, S. N., Mickley, L. J., Marlier, M. E., Buonocore, J. J., Kim, P. S., Liu, T., et al. (2016). Public health impacts of the severe haze in Equatorial Asia in September–October 2015: Demonstration of a new framework for informing fire management strategies to reduce downwind smoke exposure. *Environmental Research Letters, 11*(9), 1–10.

Kurahashi, N., Inoue, M., Liu, Y., Iwasaki, M., Sasazuki, S., Sobue, T., et al. (2008). Passive smoking and lung cancer in Japanese non-smoking women: A prospective study. *International Journal of Cancer, 122*(3), 653–657. https://doi.org/10.1002/ijc.23116.

Latif, M. T., Othman, M., Idris, N., Juneng, L., Abdullah, A. M., Hamzah, W. P., et al. (2018). Impact of regional haze towards air quality in Malaysia: A review. *Atmospheric Environment, 177,* 28–44.

Lee, J., Lim, S., Lee, K., Guo, X., Kamath, R., Yamato, H., et al. (2010). Secondhand smoke exposures in indoor public places in seven Asian countries. *International Journal of Hygiene and Environmental Health, 213*(5), 348–351.

Leh, O. L. H., Ahmad, S., & Aiyub, K. (2012). Urban air environmental health indicators for Kuala Lumpur City. *Sains Malaysiana, 41*(2), 179–191.

Liu, Q., Ma, T., Olson, M. R., Liu, Y., Zhang, T., Wu, Y., & Schauer, J. J. (2016). Temporal variations of black carbon during haze and non-haze days in Beijing. *Scientific Reports, 6,* 1–10.

Lundberg, A. (1996). Psychiatric aspects of air pollution. *Otolaryngology-Head and Neck Surgery: Official Journal of American Academy of Otolaryngology-Head and Neck Surgery, 114*(2), 227–231.

Makaremi, N., Salleh, E., Jaafar, M. Z., & GhaffarianHoseini, A. (2012). Thermal comfort conditions of shaded outdoor spaces in hot and humid climate of Malaysia. *Building and Environment, 48,* 7–14.

Marlier, M. E., DeFries, R. S., Voulgarakis, A., Kinney, P. L., Randerson, J. T., Shindell, D. T., et al. (2013). El Niño and health risks from landscape fire emissions in Southeast Asia. *Nature Climate Change, 3,* 131–136.

Morris, K. I., Aekbal Salleh, S., Chan, A., Ooi, M. C. G., Abakr, Y. A., Oozeer, M. Y., & Duda, M. (2015). Computational study of urban heat island of Putrajaya, Malaysia. *Sustainable Cities and Society, 19,* 359–372. https://doi.org/10.1016/j.scs.2015.04.010.

Morris, K. I., Chan, A., Morris, K. J. K., Ooi, M. C., Oozeer, M. Y., Abakr, Y. A., et al. (2017). Impact of urbanization level on the interactions of urban area, the urban climate, and human thermal comfort. *Applied Geography, 79,* 50–72.

Morris, K. I., Chan, A., Salleh, S. A., Ooi, M. C. G., Oozeer, M. Y., & Abakr, Y. A. (2016). Numerical study on the urbanisation of Putrajaya and its interaction with the local climate, over a decade. *Urban Climate, 16,* 1–24.

Mott, J. A., Mannino, D. M., Alverson, C. J., Kiyu, A., Hashim, J., Lee, T., et al. (2005). Cardiorespiratory hospitalizations associated with smoke exposure during the 1997, Southeast Asian forest fires. *International Journal of Hygiene and Environmental Health, 208*(1–2), 75–85.

Nadal, M., Inza, I., Schuhmacher, M., Figueras, M. J., & Domingo, J. L. (2009). Health risks of the occupational exposure to microbiological and chemical pollutants in a municipal waste organic fraction treatment plant. *International Journal of Hygiene and Environmental Health, 212*(6), 661–669.

Nasir, R. A., Ahmad, S. S., & Ahmed, A. Z. (2012). Psychological adaptation of outdoor thermal comfort in shaded green spaces in Malaysia. *Procedia—Social and Behavioral Sciences, 68,* 865–878.

Nasir, R. A., Ahmad, S. S., & Ahmed, A. Z. (2013). Physical activity and human comfort correlation in an urban park in hot and humid conditions. *Procedia - Social and Behavioral Sciences, 105,* 598–609.

Nasir, M. H., Choo, W. Y., Rafia, A. M., Theng, M. R., & Noor, L. C. (2000). Estimation of health damage cost for 1997-haze episode in Malaysia using the Ostro model. In *Malaysian Science and Technology Congress.* Confederation of Scientific and Technological Association, Kuala Lumpur, Malaysia.

Ny, M. T., & Lee, B. K. (2011). Size distribution of airborne particulate matter and associated metallic elements in an urban area of an industrial city in Korea. *Aerosol and Air Quality Research, 11*(6), 643–653.
Pavagadhi, S., Betha, R., Venkatesan, S., Balasubramanian, R., & Hande, M. P. (2013). Physicochemical and toxicological characteristics of urban aerosols during a recent Indonesian biomass burning episode. *Environmental Science and Pollution Research, 20*(4), 2569–2578.
Power, M. C., Kioumourtzoglou, M.-A., Hart, J. E., Okereke, O. I., Laden, F., & Weisskopf, M. G. (2015). The relation between past exposure to fine particulate air pollution and prevalent anxiety: Observational cohort study. *BMJ*, h1111.
Qaid, A., Bin Lamit, H., Ossen, D. R., & Shahminan, R. N. R. (2016). Urban heat island and thermal comfort conditions at micro-climate scale in a tropical planned city. *Energy and Buildings, 133*, 577–595.
Rajagopalan, P., Lim, K. C., & Jamei, E. (2014). Urban heat island and wind flow characteristics of a tropical city. *Solar Energy, 107*, 159–170.
Ramakreshnan, L., Aghamohammadi, N., Fong, C. S., Bulgiba, A., Zaki, R. A., Wong, L. P., & Sulaiman, N. M. (2018a). Haze and health impacts in ASEAN countries: A systematic review. *Environmental Science and Pollution Research, 25*(3), 2096–2111.
Ramakreshnan, L., Aghamohammadi, N., Fong, C. S., Ghaffarianhoseini, A., Ghaffarianhoseini, A., Wong, L. P., Norhaslina, H., & Sulaiman, N. M. (2018b). A critical review of urban heat island phenomenon in the context of greater Kuala Lumpur, Malaysia. *Sustainable Cities and Society, 39*, 99–113.
Ramakreshnan, L., Aghamohammadi, N., Fong, C. S., Ghaffarianhoseini, A., Wong, L. P., & Sulaiman, N. M. (2019a). Empirical study on temporal variations of canopy-level urban heat island effect in the tropical city of greater Kuala Lumpur. *Sustainable Cities and Society, 44*, 748–762.
Ramakreshnan, L., Aghamohammadi, N., Fong, C. S., Ghaffarianhoseini, A., Wong, L. P., Noor, R. M. … Hassan, N. (2019b). A qualitative exploration on the awareness and knowledge of stakeholders towards urban heat island phenomenon in Greater Kuala Lumpur: Critical insights for urban policy implications. *Habitat International, 86*, 28–37.
Roth, M., & Chow, W. T. L. (2012). A historical review and assessment of urban heat island research in Singapore. *Singapore Journal of Tropical Geography, 33*(3), 381–397.
Sahani, M., Zainon, N. A., Wan Mahiyuddin, W. R., Latif, M. T., Hod, R., Khan, M. F., et al. (2014). A case-crossover analysis of forest fire haze events and mortality in Malaysia. *Atmospheric Environment, 96*, 257–265.
Salleh, S. A., Abd. Latif, Z., Mohd, W. M. N. W., & Chan, A. (2013). Factors contributing to the formation of an urban heat island in Putrajaya, Malaysia. *Procedia—Social and Behavioral Sciences, 105*, 840–850.
Sani, S. (1984). Urban development and changing patterns of night-time temperatures in the Kuala Lumpur, Petaling Jaya area, Malaysia. *Jurnal Teknologi, 5*, 27–36.
Sani, S. (1986). Temperatures in Kuala Lumpur and the merging Klang Valley conurbation. Project report prepared for UNESCO. Malaysia: UKM Press.
Sastry, N. (2002). Forest fires, air pollution, and mortality in Southeast Asia. *Demography, 39*(1), 1–23.
Shaharuddin, A., Noorazuan, M. H., Takeuchi, W., & Noraziah, A. (2014). The effects of urban heat islands on human comfort: A case of Klang Valley Malaysia. *Global Journal of Advances in Pure and Applied Science, 2*, 1–8.
Shaharuddin, A., Noorazuan, M. H., & Yaakob, M. J. (2009). Fenomena Pulau Haba Bandar dan isu alam sekitar di Bandaraya Kuala Lumpur. *Geografia: Malaysian Journal of Society and Space, 5*(3), 57–67.
Shokoohi, R., & Nikitas, A. (2017). Urban growth, and transportation in Kuala Lumpur: Can cycling be incorporated into Kuala Lumpur's transportation system? *Case Studies on Transport Policy, 5*(4), 615–626.

Sulong, N. A., Latif, M. T., Khan, M. F., Amil, N., Ashfold, M. J., Wahab, M. I. A., et al. (2017). Source apportionment and health risk assessment among specific age groups during haze and non-haze episodes in Kuala Lumpur, Malaysia. *Science of the Total Environment, 601,* 556–570.

Thomas, G., Sherin, A. P., Ansar, S., & Zachariah, E. J. (2014). Analysis of urban heat island in Kochi, India, using a modified local climate zone classification. *Procedia Environmental Sciences, 21,* 3–13.

Tso, C. P., & Law, K. Y. (1991). Evaluation of the thermal environment in the building areas of Kuala Lumpur city by a new twin energy balance method. *Energy and Buildings, 15*(1), 271–277.

Wei, F., Teng, E., Wu, G., Hu, W., Wilson, W. E., Chapman, R. S. … Zhang, J. (1999). Ambient concentrations and elemental compositions of PM_{10} and $PM_{2.5}$ in four Chinese cities. *Environmental Science and Technology*, *33*(23), 4188–4193.

WHO. (1998). *Report of the Biregional Workshop on Health Impacts of Haze-Related Air Pollution*. Manila, Philippines: WHO.

WHO. (2006). *Air Quality Guidelines: Global Update 2005*. World Health Organization.

Wong, L. P., Alias, H., Aghamohammadi, N., Aghazadeh, S., & Sulaiman, N. M. N. (2017). Urban heat island experience, control measures and health impact: A survey among working community in the city of Kuala Lumpur. *Sustainable Cities and Society, 35,* 660–668.

Wu, R., Li, J., Hao, Y., Li, Y., Zeng, L., & Xie, S. (2016). Evolution process and sources of ambient volatile organic compounds during a severe haze event in Beijing, China. *Science of the Total Environment, 560,* 62–72.

Yusuf, Y. A., Pradhan, B., & Idrees, M. O. (2014). Spatio-temporal assessment of urban heat island effects in Kuala Lumpur metropolitan city using landsat images. *Journal of the Indian Society of Remote Sensing, 42*(4), 829–837.

The Impact of Extreme Weather Events on Health and Development in South Africa

Christie Nicole Godsmark and James Irlam

Abstract High climate variability and deep societal inequalities render South Africa a climate-vulnerable nation. Extreme weather events are increasing and impacting health and development in multiple ways. This chapter describes the impacts of extreme heat; floods and storms; drought; and wildfires on health and development in South Africa. It identifies socio-economic and occupational groups that are most vulnerable. It describes important recent policy and legislative responses that have the potential to create a more climate-resilient nation. It concludes with recommendations for action on climate and health by health professionals in South Africa.

Keywords Climate change · Public health · Extreme weather events · Environmental health

1 Introduction

South Africa is a climate-vulnerable nation, due largely to high levels of existing climate variability and to deep and persistent health and development inequalities (Department of Environment Affairs 2017). Global climate change is predicted to further existing trends of rising temperatures, increased rainfall variability, sea-level rise, and more frequent and intense extreme weather events in South Africa (Davis-Reddy 2017).

Under a high emissions' scenario (RCP8.5),[1] mean annual temperature may rise by about 5.1 °C on average from 1990 to 2100 (and by 1.4 °C under the low emissions scenario of RCP2.6). The annual average of heatwaves (defined as 7 or more consecutive days with maximum temperatures above the 90th percentile threshold for that time of the year) may increase from fewer than 5 days to as much as 145 days

[1]Model projections from CMIP5 for RCP8.5 and RCP2.6.

C. N. Godsmark · J. Irlam (✉)
UCT Faculty of Health Sciences, E47 OMB Groote Schuur Hospital, Cape Town, South Africa
e-mail: James.Irlam@uct.ac.za

C. N. Godsmark
e-mail: Christie.Godsmark@gmail.com

R. Akhtar (ed.), *Extreme Weather Events and Human Health*,
https://doi.org/10.1007/978-3-030-23773-8_18

during the same period (or 25 days under RCP2.6). The longest dry spell is projected to increase by about 30 days to approximately 110 days in the year 2100 (or by fewer than 10 days under RCP2.6), with large year-to-year variability (World Health Organisation 2015). The Mediterranean-type climatic region in the south-western Cape is at particular risk of drying (Department of Environment Affairs 2017). Without significant adaptation under a high emissions' scenario, an average of 13,900 South Africans per year between 2070 and 2100 may be affected by flooding due to rising sea levels. An additional 8500 people annually above the estimated baseline of 45,900 in 2010 may be at risk of inland river flooding by 2030 because of climate change.[2] No change in the number of days with very heavy precipitation of 20 mm or more is projected, however, remaining around 6–7 days on average (World Health Organisation 2015). Although annual rainfall trends are weak overall and non-significant, there is a tendency towards fewer rain days in almost all hydrological zones. This implies a tendency for more intense rainfall events and longer dry spells (Department of Environment Affairs 2013).

South Africa carries a quadruple burden of disease from four broad categories of causes: injuries; non-communicable diseases; HIV/AIDS and tuberculosis (TB); and communicable diseases, perinatal conditions, maternal causes, and nutritional deficiencies (Gray and Vawda 2017). Climate change will most likely exacerbate this existing burden rather than introduce many new diseases (Myers and Rother 2012), but the following groups will be most affected: those living in the hottest parts of the country; coastal populations prone to flooding from rising sea levels; those of poor socio-economic status; those living in informal housing; migrants and displaced groups; rural women; infants and children, especially those in child-headed households; chronically and mentally ill and physically challenged people; and outdoor workers (Department of Environment Affairs 2017; Department of Environmental Affairs 2011; Department of Health 2014).

In addition to South Africa's variable climate, deteriorating ecosystems, and its unequal and severe burden of disease, there are many development-related factors that contribute to national climate vulnerability. Significant socio-economic inequality, among the worst in the world, increases household vulnerability and unequal access to resources, which is exacerbated by rapid rural to urban migration, inadequate infrastructure, poor delivery of basic services, limited information, and inflexible institutions (Department of Environment Affairs 2017).

Global economic stagnation compounded by national leadership failures, corruption, and the undermining of state-owned entities all challenge South Africa's growth prospects, which undermines the ability to finance development of a more resilient economy. Recent political developments in South Africa and new leadership to address these challenges provide some hope of a recommitment to national development as outlined in the National Development Plan (National Planning Commission 2012), which is also cognisant of the national climate policy framework (Department of Environmental Affairs 2011). Furthermore, the South African con-

[2]World Resources Institute Aqueduct Global Flood Analyzer, which assumes continued current socio-economic trends and a 25-year flood protection. http://www.wri.org.

stitution recognises the right to an environment that is not harmful to health and well-being and that is protected for future generations (RSA 1996).

2 Health and Development Impacts of Extreme Weather Events in South Africa

Due to South Africa's vulnerability to climate change, there is a strong imperative for urgent adaptation towards greater climate resilience. The multi-sectoral Long-Term Adaptation Scenarios (LTAS) Research Programme (Department of Environment Affairs 2013) was mandated by the National Climate Change Response White Paper (Department of Environmental Affairs 2011) to develop scenarios for long-term adaptation planning.

LTAS classified the main impacts of extreme weather events in South Africa as extreme heat, floods and storms, drought, and wildfires. These are summarised in Table 1 and discussed below.

2.1 *Extreme Heat*

Heat-related deaths and illnesses are likely to increase in South Africa as the climate warms and the number of hot days increases (Scovronick et al. 2018), which is consistent with global evidence. Heatwaves are known to increase food insecurity and malnutrition (Bloem et al. 2010), food-borne diseases such as salmonellosis (Kovats et al. 2004), water-borne diseases such as cholera (Baker-Austin et al. 2016), air pollution (Dawson et al. 2007), non-communicable diseases (Friel et al. 2011), and mental ill-health (Doherty and Clayton 2011; Hansen et al. 2008). Civil violence, interpersonal violence, and other criminal behaviours have also been associated with raised temperatures (Breetzke and Cohn 2012; Ranson 2014). The influence of temperature on the spread of vector-borne diseases, such as malaria, is equivocal, and attribution is often weakened by control programs. Temperature, along with other environmental variables such as rainfall and humidity, is a key driver of vector-borne disease transmission, affecting the vector life cycle, proliferation of pathogens, and geographic range of vector habitats (Beck-Johnson et al. 2013; Siraj et al. 2014).

2.2 *Floods and Storms*

Severe summer thunderstorms in many parts of South Africa, especially when accompanied by strong winds, flooding, hail, lightning, and occasionally tornadoes, pose a considerable social and economic risk to agriculture, property, infrastructure, and

Table 1 Main impacts of extreme weather events in South Africa

Exposure	Main health, social and development impacts	References
Extreme heat	• Heat-related deaths and illnesses for humans and livestock • Reduced work capacity • Cardiovascular disease • Respiratory disease • Renal disease • Food-borne and vector-borne diseases • Pest-related diseases and increased chemical exposures • Violence and conflict • Risk of wildfires	• Kovats and Hajat (2008), Thornton et al. (2009) • Kjellstrom et al. (2009) • Scovronick et al. (2018) • Glaser et al. (2016) • Gage et al. (2008), Myers and Rother (2012) • Boxall et al. (2009), Costello et al. (2009) • Hsiang et al. (2013) • Davis-Reddy (2017)
Floods and Storms	• Water-borne and diarrhoeal diseases • Food insecurity, malnutrition • Damage to crops and farmland • Damage to critical infrastructure and contamination of water supply • Infectious diseases • Environmental refugees (population displacement) • Mental ill-health, e.g. PTSD, depression, suicide • Hazard-specific injuries and deaths, e.g. drowning • Respiratory diseases	• Cann et al. (2013), Levy et al. (2016) • Phalkey, et al. (2015) • Rosenzweig et al. (2001) • Mirza (2003), Davis-Reddy (2017) • Patz et al. (2003), Shuman, (2010) • Myers (2002) • Berry et al. (2010) • Davis-Reddy (2017) • Ayres et al. (2009)
Drought	• Food insecurity, malnutrition • Reduced agricultural yield • Loss of livelihoods and jobs particularly in agriculture and tourism • Environmental refugees (population displacement) • Mental ill-health, e.g. depression, suicide • Conflicts for resources • Risk of wildfires	• Phalkey et al. (2015) • Davis-Reddy (2017) • Myers (2002) • Berry et al. (2010) • Davis-Reddy (2017)
Wildfires	• Hazard-specific injuries and deaths, e.g. burns and smoke inhalation • Exacerbation of respiratory conditions, e.g. asthma • Production of carcinogenic agents • Loss of property and livelihoods • Mental ill-health, e.g. PTSD	• Stefanidou et al. (2008) • Naeher et al. (2007) • Pharoah (2012) • Berry et al. (2010)

transport systems and may lead to loss of life (Blamey et al. 2017). There is an average of between 1.5 and 8.8 lightning-related deaths per million per year in South Africa, which is considerably higher than the global average (Blamey et al. 2017). Yet despite projected increases due to global climate change, the frequency and distribution of severe storms is poorly understood (Blamey et al. 2017).

Frequent floods and storms may increase or decrease vector abundance by impacting their habitats and may increase the risk of respiratory and diarrhoeal diseases, drownings, and injuries (Hoshen and Morse 2004; Thomson et al. 2006). Damage to sewerage systems from heavy rainfall and storms is a risk factor for diarrhoeal disease through the contamination of water sources. Water sources might also be contaminated by increased run-off of agricultural pesticides and other industrial chemicals (Lipp et al. 2002; Patz et al. 2008).

Floods and storms may also displace people in large numbers (Myers 2002), disrupt the food supply (Lobell et al. 2008), and increase malnutrition (Bloem et al. 2010). Landslides following fires and floods can cause huge loss of life and property, leading to the spread of infectious diseases and more mental ill-health. Post-traumatic stress disorder (PTSD), grief over loss of loved ones, and depression and anxiety from the disruption of livelihoods (Doherty and Clayton 2011; Friel et al. 2011) often have long-term impacts (Hales et al. 2003).

2.3 Drought

Decreases in rainfall and the number of rainfall days over parts of South Africa (Davis-Reddy 2017), and a long-term (1921–2015) drying trend in the north and north-east regions (Davis-Reddy 2017), have increased droughts and impacted agriculture, ecosystems, and the economy. Severe impacts have been observed on vector abundance (Thomson et al. 2006), food security and malnutrition (Bloem 2010; Lobell et al. 2008), water supplies, and infectious diseases, all of which affect health and well-being. Mental health is particularly impacted by loss of livelihoods and displacement of people (Berry et al. 2010; Doherty and Clayton 2011; Morrissey and Reser 2007; Myers 2002). The current severe drought in the Western Cape province (see case study below) may be a foretaste of more frequent droughts in future as a result of projected declines in regional winter rainfall (Davis-Reddy 2017).

2.4 Wildfires

Although trauma and burns are common effects of wildfires, smoke inhalation is often more common and severe (Stefanidou et al. 2008). In urbanised and industrial areas, the combustion of plastics and chemicals may also produce very harmful secondary toxins (Stefanidou et al. 2008). More wildfires as a result of drought and heat extremes in South Africa may therefore increase all-cause and cardio-respiratory mortality, hazard-specific injuries, emergency hospital admissions, mental ill-health, and even adverse birth outcomes (Jayachandran 2009; Papanikolaou et al. 2011; Stefanidou et al. 2008; Youssouf et al. 2014). There are also indirect yet devastating health impacts of wildfires due to secondary events such as landslides, which occur more readily after heavy rainfall where the vegetation cover has been burnt away (Santi et al. 2013).

3 Groups Vulnerable to Extreme Weather Events

Those most vulnerable to the direct and indirect impacts of extreme weather events are usually also most vulnerable to the multiple impacts of climate change. Adults and children with weakened immune systems are particularly vulnerable (Doherty and Clayton 2011; Gardner 1980; Rocklöv et al. 2010; Katsumata et al. 1998). People with pre-existing chronic diseases and respiratory conditions, such as asthma and chronic obstructive pulmonary disease (COPD); smokers; children; the elderly; and those of low socio-economic status are also vulnerable (Delfino et al. 2009; Rappold et al. 2012; Youssouf et al. 2014).

Poor communities are vulnerable to extreme weather events for a host of reasons. These include poor housing, poor environmental health and basic services, and occupancy of marginal land that may be at high risk from flooding, storms, and other extreme weather events (Shirley et al. 2012). Extreme weather events may cause structural damage to homes, transport, and health infrastructure, thus exacerbating negative health outcomes (Friel et al. 2011). Those migrating as environmental or socio-economic refugees to rapidly urbanising areas where the demand is high for inadequate social services, may become victims of conflict over resources and basic services such as land, housing, water, sanitation, and employment (Hales et al. 2003).

In South Africa, many workers in heat-exposed sectors like agriculture, construction, and mining are vulnerable to heat stress (Mathee et al. 2010). Rescue workers deployed during weather disasters and firefighters are at increased risk of death, injury, and mental ill-health (Schulte and Chun 2009).

4 National Policy and Legislative Responses to the Climate-Development Challenge

South Africa has adopted several significant national climate policies and adaptation plans over the past 7 years to guide adaptation towards a more climate-resilient future. Reference has been made above to the 2011 *National Climate Change Response White Paper (2011)* (Department of Environmental Affairs 2011), which mandated the multi-sectoral *Long-Term Adaptation Scenarios (LTAS) Research Programme* (Department of Environment Affairs 2013) to develop scenarios for long-term adaptation planning. The multi-sectoral *South African National Climate Change Adaptation Strategy (October 2017 draft)* (Department of Environment Affairs 2017) and the health sector-specific *National Climate Change Health Adaptation Plan 2014–2019* also provide important guidance for climate-resilient development (Department of Health 2014).

South Africa's National Climate Change Bill was gazetted in June 2018 (Department of Environment Affairs 2018). The purpose of the Bill is to "*build an effective climate change response and ensure the long-term, just transition to a climate-resilient and lower carbon economy and society*",[3] within the context of sustainable development for South Africa. The Bill acknowledges that climate change is an urgent threat to society and the environment and that anticipated domestic impacts have the potential to undermine the country's development goals. An effective, progressive, and well-coordinated response is therefore required, supported by a robust legislative framework. The Bill provides for national, provincial, and local government coordination, led by the Minister of Environmental Affairs and a ministerial committee on climate change, including Health, Rural Development, Energy, and Water and Sanitation, among others.

The Bill addresses national climate change adaptation, mitigation of greenhouse gas emissions, and alignment of laws and policy. It calls for climate change adaptation to be guided by a National Adaptation Strategy that must include a national assessment of "*vulnerability to climate change and related risks at a sectoral, cross-sectoral, and geographic level, including a consideration of relevant national disaster risk assessments in terms of the Disaster Management Act 57 of 2002*". This Act includes the formation of disaster management advisory forums at national, provincial, and municipal levels to prepare context-specific disaster management plans. At the sub-national level, provincial and municipal obligations include a climate change needs and response assessment to inform a climate change response implementation plan, both of which are to be reviewed every five years (Department of Environment Affairs 2018).

5 Recommendations for Health Professionals in South Africa

5.1 Education for Advocacy by Public Health Professionals

Health is the bottom line of climate change. It is therefore incumbent upon health professionals to inform themselves, their colleagues, and their patients about climate-health risks and the opportunities to act decisively and urgently to protect the public from the worst impacts. Health professionals have a significant voice as agents of

---

[3]Statement from the Department of Environmental Affairs 11 June 2018.

change in society and should be equipped to exercise this agency effectively and strategically.

Health science curricula should therefore include climate and health issues at undergraduate and postgraduate levels. Public health professional associations also have an important role in mobilising and developing capacity among their members. An example is the Climate Energy and Health Special Interest Group (CEH SIG) of the Public Health Association of South Africa (PHASA), which recognises the health and development challenges of the carbon-intensive energy sector in South Africa.

5.2 Climate-Health Research

Climate-resilient development requires greater capacity for making decisions that take climate change into account, despite the challenges of uncertainty. Reliable climate change information and effective communication to key decision-makers are thus required.

The prediction of extreme weather events in South Africa is an evolving science, but more data and research on the health impacts are needed, especially among vulnerable populations. Attributing specific health outcomes to extreme weather events is challenging due to incomplete reporting. The Long-Term Adaptation Scenarios (LTAS) for South Africa therefore urge greater collaboration between stakeholders, comprehensive risk and vulnerability assessments, improved reporting and communication, and stronger climate and health surveillance systems. Specific metrics for immediate impacts could include population displacement, hospital admissions, and deaths (Department of Environment Affairs 2013).

6 Conclusion

This chapter has described the health and development impacts in South Africa of extreme heat; floods and storms; drought; and wildfires and has identified vulnerable groups. It has also described some of the important policy and legislative responses that can, if properly implemented and enforced at all levels of society, significantly diminish this risk and create a more climate-resilient nation. A greater role for public health professionals in raising public awareness of climate-health risks has also been recommended.

The 40th anniversary of the International Declaration on Primary Health Care (PHC) in 2018 (World Health Organization 2018) reminds us that the achievement of "Health for All" requires all sectors of society to collaborate in addressing the environmental and social determinants of good health, with particular attention to the most vulnerable, and a deep commitment to social justice.

Effective action on climate change requires a similar approach to transform it from a significant threat to a great opportunity to protect and improve public health and development now and in future generations. South Africa has the will and capacity to achieve this, but it will require strong leadership, good intersectoral collaboration, evidence-based policies, and constant vigilance by the public health community and civil society.

7 Case Study: Western Cape Water Crisis 2017–2018

The Western Cape Province of South Africa is experiencing a severe and prolonged drought at the time of writing (July 2018): 22 of its 25 local and metropolitan municipalities are under water restrictions, with five municipalities facing very severe restrictions. The City of Cape Town is currently under the most stringent Level 6B water restrictions in the province, restricting residents and businesses to 50 l of water per person per day. Climate change, poor rainfall in the catchment areas over the last three years, the growing population and economy, and a shortfall in new dam storage capacity are considered primary reasons for the water crisis (Western Cape Government 2018).

The City of Cape Town has implemented a number of water augmentation strategies, namely the building of desalination plants, harvesting groundwater through aquifers, and water recycling/reclamation from treated effluent. As part of its public awareness campaign, the City has created an online water dashboard, which displays the projected date of Day Zero (when municipal water supply would be switched off and people would need to queue for their daily water rations), as well as current combined dam levels supplying the City, and links to extensive water-saving resources (City of Cape Town 2018).

According to local news agency reports, the drought has already had direct and indirect health impacts. As summer temperatures rise in Cape Town, diarrhoea cases usually surge among children under five years old; a noticeable increase in cases this past summer may have been due to worsening sanitation and personal hygiene associated with water-saving (Evans 2018a, b). A national outbreak of listeriosis, a food-borne disease, killed almost 200 people and resulted in the treatment of more than 900 confirmed cases; 12% of these were from the Western Cape, highlighting the risks of poor water and food hygiene. Decreased hand-washing and poor sanitation may also stimulate antibiotic resistance through the improper use of medications and may increase the transmission of bacterial and viral infections (Green 2018).

Farmers in the Western Cape have suffered large financial loses as a result of the drought and severe water restrictions. Globally, farmer suicides linked to droughts have been reported in Australia (Hanigan et al. 2012) and locally, according to a news agency, there was an increase in farmer suicides in the Northern Cape Province in 2016, likely due to drought-related mental and financial stress (Child 2017). Many seasonal agricultural workers and employees in the tourism sector lost their jobs during the drought (Phakathi 2018), which coupled with increasing food prices might worsen their vulnerability to ill-health and malnutrition.

Drought conditions often leave the soil compacted and less porous, so there is a higher risk of flash flooding during heavy rainfall. In mid-February 2018 for example, nine people drowned in a flash flood near Theewaterskloof Dam, one of Cape Town's largest water sources (Evans 2018a, b). A combination of high temperatures, low relative humidity and precipitation, strong winds, and desiccated vegetation provides fuel for runaway wildfires in the area (Climate Systems Analysis Group 2016; Flannigan and Harrington 1988). Climate change is anticipated to exacerbate these conditions, and therefore, an increase in the frequency and intensity of wildfires is predicted (Flannigan et al. 2013; Flannigan et al. 2009). This in turn increases water demand in an already water-stressed region and places firefighters and residents at greater risk of burns, trauma, and cardio-respiratory diseases (Stefanidou et al. 2008).

References

Ayres, J. G., Forsberg, B., Annesi-Maesano, I., Dey, R., Ebi, K. L., Helms, P. J., et al. (2009). Climate change and respiratory disease: European Respiratory Society position statement. *European Respiratory Journal, 34*(2), 295.

Baker-Austin, C., Trinanes, J. A., Salmenlinna, S., Löfdahl, M., Siitonen, A., Taylor, N. G. H., et al. (2016). Heat wave-associated vibriosis, Sweden and Finland, 2014. *Emerging Infectious Diseases, 22*(7), 1216–1220. https://doi.org/10.3201/eid2207.151996.

Beck-Johnson, L. M., Nelson, W. A., Paaijmans, K. P., Read, A. F., Thomas, M. B., & Bjørnstad, O. N. (2013). The effect of temperature on Anopheles mosquito population dynamics and the potential for malaria transmission. *PLoS ONE, 8*(11), e79276. https://doi.org/10.1371/journal.pone.0079276.

Berry, H. L., Bowen K., Fau-Kjellstrom, T., & Kjellstrom, T. (2010). Climate change and mental health: A causal pathways framework (1661-8564 (Electronic).

Blamey, R., Middleton, C., Lennard, C., & Reason, C. (2017). A climatology of potential severe convective environments across South Africa. *Climate Dynamics, 49*(5–6), 2161–2178.

Bloem, M. W., Semba, R. D., & Kraemer, K. (2010). Castel gandolfo workshop: An introduction to the impact of climate change, the economic crisis, and the increase in the food prices on malnutrition. *Journal of Nutrition, 140*(1), 132S–135S.

Boxall, A. B. A., Hardy, A., Beulke, S., Boucard, T., Burgin, L., Falloon, P. D., et al. (2009). Impacts of climate change on indirect human exposure to pathogens and chemicals from agriculture. *Environmental Health Perspectives, 117*(4), 508–514. https://doi.org/10.1289/ehp.0800084.

Breetzke, G. D., & Cohn, E. G. (2012). Seasonal assault and neighborhood deprivation in South Africa: Some preliminary findings. *Environment and Behavior, 44*(5), 641–667.

Cann, K. F., Thomas, D. R., Salmon, R. L., Wyn-Jones, A. P., & Kay, D. (2013). Extreme water-related weather events and waterborne disease. *Epidemiology and Infection, 141*(4), 671–686. https://doi.org/10.1017/S0950268812001653.

Child, K. (2017). Farmer suicides soar as worst drought in decades drives them to ruin. . *Rand Daily Mail.* Retrieved from https://www.businesslive.co.za/rdm/news.

City of Cape Town. (2018). Water dashboard. Retrieved from http://coct.co/water-dashboard.

Climate Systems Analysis Group. (2016). *Climate change projections for the City of Cape Town, an update based on the most recent science.* Retrieved from http://www.csag.uct.ac.za/~cjack/CSAG_CCT%20report.pdf.

Costello, A., Abbas, M., Allen, A., Ball, S., Bell, S., Bellamy, R., et al. (2009). Managing the health effects of climate change. *The Lancet, 373*(9676), 1693–1733. https://doi.org/10.1016/S0140-6736(09)60935-1.

Davis-Reddy, C. L. Vincent, K. (2017). *Climate Risk and Vulnerability: A Handbook for Southern Africa*. Retrieved from https://www.csir.co.za/sites/default/files/Documents/SADC%20Handbook_Second%20Edition_full%20report.pdf.

Dawson, J. P., Adams, P. J., & Pandis, S. N. (2007). Sensitivity of ozone to summertime climate in the eastern USA: A modeling case study. *Atmospheric Environment, 41*(7), 1494–1511.

Delfino, R. J., Brummel, S., Wu, J., Stern, H., Ostro, B., Lipsett, M. … Gillen, D. L. (2009). The relationship of respiratory and cardiovascular hospital admissions to the southern California wildfires of 2003. *Occupational and environmental medicine, 66*(3), 189–197. https://doi.org/10.1136/oem.2008.041376.

Department of Environment Affairs. (2013). *Climate change implications for human health in South Africa*. South Africa: Pretoria.

Department of Environment Affairs. (2017). *National climate change adaptation strategy*. Pretoria, South Africa: South African Government Printer. Retrieved from https://www.environment.gov.za/sites/default/files/reports/nationalclimate_changeadaptation_strategyforcomment_nccas.pdf.

Department of Environment Affairs. (2018). *National climate change bill 2018*. Pretoria: South African Government Printer. Retrieved from http://pmg-assets.s3-website-eu-west-1.amazonaws.com/Climate_Chane_Bill.pdf.

Department of Environmental Affairs. (2011). *National climate change response white paper*. Pretoria: Republic of South Africa.

Department of Environmental Affairs (DEA). (2011). *National climate change response white paper*. Pretoria: Republic of South Africa.

Department of Health. (2014). *National climate change health adaptation plan 2014–2019*. Pretoria, South Africa. Retrieved from http://nhrec.org.za/index.php/shortcodes/2015-03-29-10-42-47/2015-04-30-08-29-27/2015-04-30-08-32-49/category/435-climate-change?download=2609:national-climate-change-and-health-adaptation-plan.

Doherty, T. J., & Clayton, S. (2011). The psychological impacts of global climate change. *American Psychologist, 66*(4), 265.

Evans, J. (2018a). Diarrhoea numbers up in drought-stricken Cape Town. *Health24*. Retrieved from https://www.health24.com.

Evans, J. (2018b). Joy over rain marred as 9 drown in Western Cape floods. *News24*. Retrieved from https://www.news24.com.

Flannigan, M., Cantin, A. S., de Groot, W. J., Wotton, M., Newbery, A., & Gowman, L. M. (2013). Global wildland fire season severity in the 21st century. *Forest Ecology and Management, 294,* 54–61. https://doi.org/10.1016/j.foreco.2012.10.022.

Flannigan, M. D., & Harrington, J. B. (1988). A study of the relation of meteorological variables to monthly provincial area burned by wildfire in Canada (1953–80). *Journal of Applied Meteorology, 27*(4), 441–452.

Flannigan, M. D., Krawchuk, M. A., de Groot, W. J., Wotton, B. M., & Gowman, L. M. (2009). Implications of changing climate for global wildland fire. *International Journal of Wildland Fire, 18*(5), 483–507. https://doi.org/10.1071/WF08187.

Friel, S., Bowen, K., Campbell-Lendrum, D., Frumkin, H., McMichael, A. J., & Rasanathan, K. (2011). Climate change, noncommunicable diseases, and development: The relationships and common policy opportunities. *Annual Review of Public Health, 32*(1), 133–147. https://doi.org/10.1146/annurev-publhealth-071910-140612.

Gage, K. L., Burkot, T. R., Eisen, R. J., & Hayes, E. B. (2008). Climate and vectorborne diseases. *American Journal of Preventive Medicine, 35*(5), 436–450. https://doi.org/10.1016/j.amepre.2008.08.030.

Gardner, I. D. (1980). The effect of aging on susceptibility to infection. *Reviews of Infectious Diseases, 2*(5), 801–810.

Glaser, J., Lemery, J., Rajagopalan, B., Diaz, H. F., García-Trabanino, R., Taduri, G., et al. (2016). Climate change and the emergent epidemic of CKD from heat stress in rural communities: The case for heat stress nephropathy. *Clinical Journal of the American Society of Nephrology, 11*(8), 1472–1483. https://doi.org/10.2215/cjn.13841215.

Gray, A., & Vawda, Y. (2017). Health policy and legislation. *South African Health Review, 2017*(1), 13–24.

Green, A. (2018). Cape Town's drought to fuel antibiotic resistance. *IOL News*. Retrieved from https://www.iol.co.za.

Hales, S., Edwards, S., & Kovats, R. (2003). Impacts on health of climate extremes. *Climate Change and Human Health: Risks and Responses*, 79–96.

Hanigan, I. C., Butler, C. D., Kokic, P. N., & Hutchinson, M. F. (2012). Suicide and drought in new South Wales, Australia, 1970–2007. *Proceedings of the National Academy of Sciences, 109*(35), 13950–13955.

Hansen, A., Bi, P., Nitschke, M., Ryan, P., Pisaniello, D., & Tucker, G. (2008). The effect of heat waves on mental health in a temperate Australian city. *Environmental Health Perspectives, 116*(10), 1369.

Hoshen, M. B., & Morse, A. P. (2004). A weather-driven model of malaria transmission. *Malaria Journal, 3*(1), 32.

Hsiang, S. M., Burke, M., & Miguel, E. (2013). Quantifying the influence of climate on human conflict. *Science, 341*(6151).

Jayachandran, S. (2009). Air quality and early-life mortality evidence from Indonesia's wildfires. *Journal of Human resources, 44*(4), 916–954.

Katsumata, T., Hosea, D., Wasito, E. B., Kohno, S., Hara, K., Soeparto, P., et al. (1998). Cryptosporidiosis in Indonesia: A hospital-based study and a community-based survey. *The American Journal of Tropical Medicine and Hygiene, 59*(4), 628–632.

Kjellstrom, T., Holmer, I., & Lemke, B. (2009). Workplace heat stress, health and productivity—An increasing challenge for low and middle-income countries during climate change. *Global Health Action, 2*(1), 2047. https://doi.org/10.3402/gha.v2i0.2047.

Kovats, R., Edwards, S., Hajat, S., Armstrong, B., Ebi, K., & Menne, B. (2004). The effect of temperature on food poisoning: A time-series analysis of salmonellosis in ten European countries. *Epidemiology and Infection, 132*(3), 443–453.

Kovats, R. S., & Hajat, S. (2008). Heat stress and public health: A critical review. *Annual Review of Public Health, 29*(1), 41–55. https://doi.org/10.1146/annurev.publhealth.29.020907.090843.

Levy, K., Woster, A. P., Goldstein, R. S., & Carlton, E. J. (2016). Untangling the impacts of climate change on waterborne diseases: A systematic review of relationships between diarrheal diseases and temperature, rainfall, flooding, and drought. *Environmental Science and Technology, 50*(10), 4905–4922. https://doi.org/10.1021/acs.est.5b06186.

Lipp, E. K., Huq, A., & Colwell, R. R. (2002). Effects of global climate on infectious disease: The Cholera model. *Clinical Microbiology Reviews, 15*(4), 757.

Lobell, D. B., Burke, M. B., Tebaldi, C., Mastrandrea, M. D., Falcon, W. P., & Naylor, R. L. (2008). Prioritizing climate change adaptation needs for food security in 2030. *Science, 319*(5863), 607–610.

Mathee, A., Oba, J., & Rose, A. (2010). Climate change impacts on working people (the HOTHAPS initiative): findings of the South African pilot study. *Global Health Action, 3*(1), 5612.

Mirza, M. M. Q. (2003). Climate change and extreme weather events: Can developing countries adapt? *Climate Policy, 3*(3), 233–248. https://doi.org/10.3763/cpol.2003.0330.

Morrissey, S. A., & Reser, J. P. (2007). Natural disasters, climate change and mental health considerations for rural Australia. *Australian Journal of Rural Health, 15*(2), 120–125.

Myers, N. (2002). Environmental refugees: A growing phenomenon of the 21st century. *Philosophical Transactions of the Royal Society of London B: Biological Sciences, 357*(1420), 609–613.

Myers, J. E., & Rother, H. (2012). Public health impact of and response to climate change in South Africa. *South African Health Review, 13,* 127–138.

Naeher, L. P., Brauer, M., Lipsett, M., Zelikoff, J. T., Simpson, C. D., Koenig, J. Q., et al. (2007). Woodsmoke health effects: A review. *Inhalation Toxicology, 19*(1), 67–106.

National Planning Commission. (2012). South Africa: National development plan. *Africa Research Bulletin: Economic, Financial and Technical Series, 48,* 19338B–19338C.

Papanikolaou, V., Adamis, D., Mellon, R. C., & Prodromitis, G. (2011). Psychological distress following wildfires disaster in a rural part of Greece: A case-control population-based study. *International journal of emergency mental health, 13*(1), 11–26.

Patz, J., Githeko, A., McCarty, J., Hussein, S., Confalonieri, U., & De Wet, N. (2003). Climate change and infectious diseases. *Climate change and human health: Risks and responses, 6,* 103–137.

Patz, J. A., Vavrus, S. J., Uejio, C. K., & McLellan, S. L. (2008). Climate change and waterborne disease risk in the Great Lakes Region of the U.S. *American Journal of Preventive Medicine, 35*(5), 451–458. https://doi.org/10.1016/j.amepre.2008.08.026.

Phakathi, B. (2018). Farmers lose R14bn as Cape drought bites. *Business Day.* Retrieved from https://www.businesslive.co.za.

Phalkey, R. K., Aranda-Jan, C., Marx, S., Höfle, B., & Sauerborn, R. (2015). Systematic review of current efforts to quantify the impacts of climate change on undernutrition. *Proceedings of the National Academy of Sciences, 112*(33), E4522.

Pharoah, R. (2012). Fire Risk in Informal Settlments in Cape Town, South Africa. In M. Pelling (Ed.), *Disaster risk reduction: Cases from Urban Africa.* Routledge. Retrieved from https://books.google.co.za/books?id=zrgeBAAAQBAJ&lpg=PA105&dq=loss%20of%20property%20and%20livelihood%20fire&lr&pg=PA105#v=onepage&q&f=false.

Ranson, M. (2014). Crime, weather, and climate change. *Journal of Environmental Economics and Management, 67*(3), 274–302. https://doi.org/10.1016/j.jeem.2013.11.008.

Rappold, A. G., Cascio, W. E., Kilaru, V. J., Stone, S. L., Neas, L. M., Devlin, R. B., et al. (2012). Cardio-respiratory outcomes associated with exposure to wildfire smoke are modified by measures of community health. *Environmental Health, 11*(1), 71.

Rocklöv, J., Ebi, K., & Forsberg, B. (2011). Mortality related to temperature and persistent extreme temperatures: A study of cause-specific and age-stratified mortality. *Occupational and environmental medicine, 68*(7), 531–536.

Rosenzweig, C., Iglesias, A., Yang, X. B., Epstein, P. R., & Chivian, E. (2001). Climate change and extreme weather events; Implications for food production, plant diseases, and pests. *Global Change and Human Health, 2*(2), 90–104. https://doi.org/10.1023/A:1015086831467.

RSA. (1996). Constitution of the Republic of South Africa. In: Government Printer Pretoria.

Santi, P., Cannon, S., & DeGraff, J. (2013). Wildfire and landscape change.

Schulte, P. A., & Chun, H. (2009). Climate change and occupational safety and health: Establishing a preliminary framework. *Journal of occupational and environmental hygiene, 6*(9), 542–554.

Scovronick, N., Sera, F., Acquaotta, F., Garzena, D., Fratianni, S., Wright, C. Y., et al. (2018). The association between ambient temperature and mortality in South Africa: A time-series analysis. *Environmental Research, 161,* 229–235.

Shirley, W. L., Boruff, B. J., & Cutter, S. L. (2012). Social vulnerability to environmental hazards. In *Hazards Vulnerability and Environmental Justice* (pp. 143–160). Routledge.

Shuman, E. K. (2010). Global climate change and infectious diseases. *New England Journal of Medicine, 362*(12), 1061–1063. https://doi.org/10.1056/NEJMp0912931.

Siraj, A. S., Santos-Vega, M., Bouma, M. J., Yadeta, D., Carrascal, D. R., & Pascual, M. (2014). Altitudinal changes in malaria incidence in Highlands of Ethiopia and Colombia. *Science, 343*(6175), 1154.

Stefanidou, M., Athanaselis, S., & Spiliopoulou, C. (2008). Health impacts of fire smoke inhalation. *Inhalation Toxicology, 20*(8), 761–766.

Thomson, M., Doblas-Reyes, F., Mason, S., Hagedorn, R., Connor, S., Phindela, T., et al. (2006). Malaria early warnings based on seasonal climate forecasts from multi-model ensembles. *Nature, 439*(7076), 576.

Thornton, P. K., van de Steeg, J., Notenbaert, A., & Herrero, M. (2009). The impacts of climate change on livestock and livestock systems in developing countries: A review of what we know and what we need to know. *Agricultural Systems, 101*(3), 113–127. https://doi.org/10.1016/j.agsy.2009.05.002.

Western Cape Government. (2018). Latest Western Cape dam levels. Retrieved from https://www.westerncape.gov.za/general-publication/latest-western-cape-dam-levels.

World Health Organisation. (2015). *Climate and health country profile 2015: South Africa* Retrieved from http://www.who.int/globalchange/resources/country-profiles/en/#S.

World Health Organization. (2018). Global conference on primary health care. Retrieved from http://www.who.int/primary-health/conference-phc.

Youssouf, H., Liousse, C., Roblou, L., Assamoi, E.-M., Salonen, R. O., Maesano, C. … Annesi-Maesano, I. (2014). Non-accidental health impacts of wildfire smoke. *International Journal of Environmental Research and Public Health, 11*(11), 11772–11804.

The Complex Effects of Extreme Weather Events in the Greater Horn of Africa

Manpreet Singh, Mala Rao and Andy Bechingham

Abstract This chapter uses a primary, secondary, and tertiary effects model to describe the effects of extreme weather events in the Greater Horn of Africa. Primary effects are the direct effects of climate change on health; secondary effects are indirect; and tertiary effects operate at the intersection of climate, policy, and ecology. We consider the tertiary effects of extreme weather on health through the lens of a complex adaptive system. We draw on academic literature, and more contemporaneous sources, including newspapers and humanitarian situation reports, to detail the impact of recent extreme weather events, including 2016–2017 droughts and subsequent famine.

Keywords Horn of Africa · Climate change · Complex adaptive systems · Heatwaves · Famine · Migration

1 Introduction

This chapter explores the complex effects of extreme weather events in the Greater Horn of Africa.

The Greater Horn of Africa, as defined by the Intergovernmental Authority on Development (IGAD), consists of nine countries: Djibouti, Eritrea, Ethiopia, Kenya, Somalia, South Sudan, Sudan, Uganda, and Somaliland which withdrew from union with Somalia but remains unrecognised in the international arena. This geographical grouping had a population of 257 million people in 2015 (over 20% of the total population of Africa) (United Nations Department of Economic and Social Affairs, Population Division 2017). Using UNDP's medium variant estimates of population growth, the Greater Horn of Africa will have a population of 545 million people in 2050 (again, over 20% of the projected total population of Africa) (United Nations Department of Economic and Social Affairs, Population Division 2017). Countries in the Greater Horn of Africa represent a wide range of political and economic contexts. The region includes the two most fragile states as defined by the 2018 Fragile State

M. Singh (✉) · M. Rao · A. Bechingham
Seattle, USA
e-mail: manpreet1@gmail.com

R. Akhtar (ed.), *Extreme Weather Events and Human Health*,
https://doi.org/10.1007/978-3-030-23773-8_19

Index (South Sudan and Somalia) (The Fund for Peace 2018); it also includes some of the fastest growing economies in the world—Djibouti, Ethiopia, and Kenya all experienced greater than five per cent GDP growth in 2017 (African Development Bank 2018). Notably when considering the effects of climate change on health, the region currently has four of the ten biggest source countries for refugees (in order: South Sudan, Somalia, Sudan, and Eritrea), and three of the 10 biggest refugee hosting countries (in order: Uganda, Ethiopia, and Kenya) (UNHCR 2017). Since large proportions of the region's population depend on rain-fed agriculture, and their livelihoods become more precarious due to climate change, the rate of rural-urban migration may be expected to grow still further, with many ending up in camps and informal settlements. This may expose large numbers of people to a variety of new risks, including outbreaks of infectious disease, food price increases and violence (Serdeczny et al. 2017).

Given the political and economic situations outlined above, it is no surprise that countries in the Greater Horn of Africa represent a wide range of health profiles. Somalia and South Sudan both have a maternal mortality ratio above 700 deaths per 100,000 live births (among the highest in the world) (Countdown to 2030 2017). Although some countries in the Greater Horn have demonstrated rapid improvements in health outcomes (e.g. Ethiopia's annual rate of reduction in child mortality from 2006 to 2016 was over 5%), none of the eight countries is on track to hit the 2030 SDG targets for maternal and child health (McArthur et al. 2018). Importantly when considering the effects of climate change on health, malnutrition is the biggest risk factor for death and disability in all eight countries (IHME 2018). The effects of climate change on nutrition and food security will be explored in more detail below.

We define extreme weather events using the definition of 'climatic extremes' from the IPCC SREX report of 2012 (IPCC 2012). This includes both short-term events (on a timescale of days to weeks) and longer-term events (on a timescale of weeks to years). The definition includes (but is not limited to) temperature extremes, precipitation extremes, droughts and floods.

We use two interconnected frameworks to assess the effects of extreme weather events on human health. The first is a primary, secondary, and tertiary effects model (Butler and Harley 2010). In this framework, primary effects are the direct impacts of the physical system on human health (e.g. heatwaves, trauma); secondary effects are the indirect impacts (e.g. changes in the transmission cycles of vector-borne disease, causing changes in infectious disease epidemiology); tertiary effects operate at the intersection of climate, politics, and ecology (Butler and Harley 2010). A similar framework was used by the authors of the 2017 Lancet Countdown on Health and Climate Change (Watts et al. 2018). When considering tertiary effects, we bring in a second framework, drawn from the science of complex adaptive systems—this moves beyond traditional linear models of cause and effect and considers poor health outcomes to be an emergent property of the complex interaction of multiple interconnected elements (Rutter et al. 2017). Given this complexity, it is important to consider how the impact of extreme weather events interacts with other pre-existing biological and social vulnerabilities.

2 Extreme Weather Events in the Greater Horn of Africa

Throughout this chapter, we will use recent extreme weather events to illustrate the predicted impacts of future extreme weather events. Specifically, we will use examples drawn from the 2011 and 2016–2017 droughts in the Horn of Africa. From October to December 2016, rainfall across Somalia and South-eastern parts of Ethiopia was less than 30% of the 1981–2010 average (UNOCHA 2017a, b). This had dramatic effects on the availability of water for human and livestock consumption—in January 2017, most monitored water points in Kenya, Ethiopia, and Somalia were at ‘alert’ or ‘near-dry’ levels (UNOCHA 2017a, b). Recovery from the 2016–2017 drought has been varied across the Greater Horn of Africa. Although Ethiopia and Somalia had a favourable rainy season in October 2017, Ethiopia has continued to experience below average rainfall in 2018 (Famine Early Warning Systems Network 2018b). Parts of Kenya have experienced three consecutive poor rainy reasons, with impacts on maize production and livestock productivity (Famine Early Warning Systems Network 2018a, b).

These extreme events are already occurring (and will continue to occur) in the context of long-term climate trends. Historically, there has been a decrease in rainfall in East Africa over the last three decades, partly linked to warming in the Indian Ocean (IPCC 2014). Long-term effects of climate change on precipitation in the Greater Horn of Africa are highly debated—many regional climate models predict increases in precipitation (Souverijns et al. 2016; Tierney et al. 2015). This disconnect between historically observed droughts and models predicting increases in rainfall has been dubbed the ‘East African climate paradox’. Given this uncertainty, for the purposes of this chapter, we explore the effects of both droughts and flooding. This is driven by recent analyses, which suggest that there will be an increase in precipitation in South-eastern Ethiopia and Somalia, and a decrease in precipitation over the Ethiopian highlands, Great Lakes region, and Sudan (Souverijns et al. 2016).

There is clear evidence from the climate literature of an increase in heatwaves. Regional climate models predict that even a scenario that limits global mean warning to 2 °C (the outcome of the Paris agreement) will result in an increase in the number and intensity of heatwaves in the Greater Horn of Africa (Weber et al. 2018; Dosio 2017). The impact of a high emissions scenario is even greater—projecting that days of ‘heatwaves’ in Uganda will increase from 10 days per year in 1990 to 225 days per year by 2100 (World Health Organization 2015c); in Ethiopia heatwaves will increase from 10 days per year in 1990 to 250 days per year by 2100 (World Health Organization 2015a).

Populations in the region have adapted over millennia to harsh climate, but in recent times, catastrophes have grown in frequency. In Somaliland in the twentieth century, droughts were seen about once every ten years, but in the twenty-first these now occur every other year. Increases in temperature and precipitation changes have resulted in droughts that have killed huge numbers of animals and created destitution for many households. But more permanent impacts are also seen—wind speed increases have eroded topsoil, increased dune formation, and increased desertification. Increased run-off has depleted soils and drowned animals (Hartmann et al. 2009). The situation is likely to deteriorate without better drought risk management (Abdulkadir 2017).

3 Health Effects of Extreme Weather Events

Primary

Primary impacts refer to the direct effects of climate change on health, typically heatwaves and trauma (Butler and Harley 2010). As detailed above, regional climate models predict an increase in the number and intensity of heatwaves in the Greater Horn of Africa. The health effects of heatwaves are well established at a global level and include both direct heat stress and an increase in cardiovascular mortality (Watts et al. 2018). Elderly populations (those over 65) are particularly vulnerable from the direct impacts of heatwaves. Under a high emissions' scenario in Ethiopia, heat-related deaths in the elderly are projected to increase from 3 per 100,000 people per year between 1961 and 1990 to 65 per 100,000 people per year by 2080 (World Health Organization 2015a). In Kenya, the projected increase is from 2 deaths per 100,000 people per year in 1961–1990 to 45 deaths per 100,000 people per year in 2080 (World Health Organization 2015b); in Uganda, from 2 deaths per 100,000 people per year in 1961–1990 to 81 deaths per 100,000 people per year in 2080 (World Health Organization 2015c).

Likewise, extreme weather events (e.g. floods) can directly cause deaths through trauma. Between 2000 and 2018, 2300 deaths from flooding were reported in the Greater Horn of Africa (Université Catholique de Louvain 2018), and the number of people at risk is increasing through climate change. WHO estimates that in Ethiopia, 154,400 people are currently at risk of inland river floods; by 2030, an additional 248,200 will be at risk due to climate change (World Health Organization 2015a). In Kenya, an additional 75,100 people will be at risk of annual river floods due to climate change by 2030 (from a 2010 baseline of 29,600 people at risk) (World Health Organization 2015b); in Uganda, an additional 34,600 will be at risk of annual river floods due to climate change by 2030, from a 2010 baseline of 15,500 people (World Health Organization 2015c).

Secondary

The secondary impacts of extreme weather events on health are indirect, operating through a number of complex and interconnected pathways. In this chapter, we explore the effects of extreme weather events on water, agriculture and food security, income, and infectious disease transmission.

Water

Extreme weather events affect both water quantity and quality—therefore causing secondary impacts on health (e.g. through diarrhoeal disease), as well as impacting agriculture and food security (detailed below). This is particularly true for rural poor populations where the water and sanitation infrastructure (e.g. pit latrines and rainwater harvesting) is both more vulnerable to climate impacts and less adaptable to respond to climate threats (Howard et al. 2016).

Considering water quality first—there is good evidence that as temperature rises, common bacterial pathogens replicate more rapidly, and the incidence of diarrhoea caused by bacterial infection rises (Howard et al. 2016). As a result of these temperature impacts, WHO models suggest that, in eastern Africa, there will be 11,000 more deaths of children below the age of 15 from diarrhoea illness by 2030, in a base-case climate change scenario (World Health Organization 2014).

There is also a hypothesized relationship between droughts (and water availability) and diarrhoeal disease, although this has not been included in WHO's most recent models, due to lack of quantitative evidence and statistical power (World Health Organization 2014). The causal link seems apparent—reduced water availability affects ability to hand-wash, and availability of drinking water for local populations (Wall and Michael 2016).

Agriculture and food security

Droughts and other extreme weather events cause crop failure, livestock deaths, and food insecurity. This has direct effects on mortality (e.g. through severe acute malnutrition) and long-term effects on child development (e.g. through stunting). Food insecurity also has short-term effects on household income.

The most recent USAID figures (from August 2017) report that 18 million people were facing food insecurity in Ethiopia, Kenya, and Somalia, as a result of the 2016–2017 drought (USAID 2017). Somalia was the worst affected—in Somalia, the countrywide global malnutrition level was 17%, and 3.3 million people were facing crisis levels of acute food insecurity (USAID 2017). South Sudan is even worse off—at the time of writing this chapter, 5.3 million people (48% of the population) were facing crisis or worse levels of acute insecurity, with a real risk of famine over a large area (Famine Early Warning Systems Network 2018a, b).

Chronic food insecurity and childhood malnutrition cause stunting, with associated impacts on long-term neuro-developmental, educational, and economic outcomes. Stunting has been declining throughout the Greater Horn of Africa, partly driven by improvements in economic growth. Models that account for both future socio-economic change and the effects of climate change on food security, show projected changes in stunting that are close to current prevalence rates. As the IPCC puts it, 'climate change would counteract the beneficial consequences of socio-economic development [on stunting]' (IPCC 2014, 201). Across East Africa (including Kenya and Uganda from the Greater Horn of Africa), every 1 °C increase in temperature is associated with a 12% increase in moderate stunting (Singer et al. 2016). This has significant secondary and tertiary effects on long-term economic growth—for example, there is a strong association between stunting at age 2 and impaired cognitive outcomes in childhood (ages 5–11) (Sudfeld et al. 2015).

Droughts also affect short-term household income—shortages in water and food availability cause herders to sell their livestock; and some livestock die directly

(Maystadt and Ecker 2014). As all herders in a geographical area face the same conditions (and many will be forced to sell their herd), cattle prices can plummet during droughts. In addition, lean or undernourished animals sell for a lower price. The end result is that pastoral farmers are forced to sell their animals, at below the usual market rate, at the time of their greatest need. The 2010–2011 drought in Somalia reduced cattle prices by 30–50% (Maystadt and Ecker 2014). Pastoralism has mostly adapted to erratic climates, conserving an array of ecosystems in semi-arid areas. But this century it has proved vulnerable to global warming, and extreme weather has caused food insecurity among pastoral communities in Somaliland. Livestock numbers were tremendously reduced in some areas, and many people became impoverished (Hartmann et al. 2009).

Infectious disease

A common secondary effect of climate change is the result of infectious disease outbreaks. For example, climate change will likely influence malaria transmission in East Africa—causing longer transmission seasons and expansion of the disease into previously non-endemic highlands (Onyango et al. 2016). This is not related to extreme weather events, but rather related to the long-term impacts of climate change.

Extreme climatic events also have effects on infectious disease outbreaks. For example, outbreaks of Rift Valley fever (a viral disease transmitted by mosquito vectors) are closely associated with climatic conductions—high rainfall causes outbreaks (Anyamba et al. 2009).

Tertiary

The tertiary effects of extreme climatic events operate at the intersection of climate, policy, and ecology. We assess these impacts at a sub-national, national, and then international level.

Sub-national conflict

There is strong global evidence of a causal relationship between climatic events and conflict—this includes interpersonal violence, as well as intergroup conflict (Hsiang et al. 2013). Specific examples of this have been seen in the Greater Horn of Africa. One study showed that higher temperatures increase the risk of conflict across East Africa, although reflecting the complexity discussion below, other political and geographical conditions are more predictive than climate events alone (O'Loughlin et al. 2012). In Somalia, there is similar causal evidence showing a relationship between droughts and conflict—droughts cause downturns in the price of livestock, and losses to the income of pastoralist farmers, making them more likely to engage in local conflict (Maystadt and Ecker 2014).

National political response

Extreme weather already influences national politics in the region. In 2017, Kenya held national elections against a backdrop of drought, food insecurity, and widespread rises in food prices. The price of maize flour repeatedly featured in the Kenyan press, and became a major political point of attack by the opposition party (Andae 2017). In February 2017, the drought was declared a national emergency, and government food aid was allotted to drought-affected areas (Okiror 2017).

Government action on adaptation needs to be multi-sectoral, since climate change affects such a wide variety of factors, groups, and sectors. Effective adaptation requires policy and practical collaboration between many parties, in the private sector as well as in civil society. It also needs the clout to enforce implementation of measures—often difficult in fragmented nations coping with internal conflict—plus democratic legitimation (Winsvold et al. 2009). While only a few low-income countries have introduced comprehensive climate policies, Ethiopia's 2011 Climate Resilient Green Economy policy integrated mitigation and adaptation programmes for economic growth across government ministries. This was possible because of high levels of institutional autonomy within government departments. Ethiopia's centralised technocratic leadership prioritized climate policy and fostered co-ordination across agencies and regions via centralised training, championing the idea of 'green growth development'. The strategy sets an example for policy-makers in the region: an alternative model for low-income countries seeking sustainable development despite climate change (Paul and Weinthal 2018).

International migration

Climate change causes population movement through three main channels—forced displacement as a result of extreme conditions, resettlement, and migration as an adaptive response (McMichael et al. 2012). Globally, movement of people following an extreme weather event is usually temporary, as people choose to return to their home countries and cities once conditions allow (McMichael et al. 2012). This dynamic is likely to be altered in the Greater Horn of Africa—countries in this region are already a major source for international refugees and also major recipients of international refugees. As a result of the complex protracted emergency, UNOCHA reports that there were 4.4 million refugees in the Greater Horn and Great Lakes region in 2017 (UNOCHA 2017a, b). However, repeated extreme weather catastrophes deepen poverty and create socio-economic 'traps' from which escape by migration may become impossible for the most vulnerable groups (Bowles et al. 2014).

4 Complexity in the Context of Extreme Weather Events and Health Impacts

There are three main defining characteristics of complex adaptive systems: emergence, feedback, and adaptation. Emergence is the property whereby outcomes are more than the sum of their parts and emerge from the complex interactions of multiple interrelated elements. Feedback loops can be both positive and negative (e.g.

a positive feedback loop represents the case where a change in a system reinforces further change). Adaptation refers to changes in individual and societal behaviour in response to changes in conditions (Rutter et al. 2017).

All three of these are features of complex systems and responses caused by humanitarian emergencies. This section does not attempt to exhaustively detail or model the complex effects of extreme weather events on climate change, but it does attempt to outline a few examples of why complex adaptive systems matter for climate change.

Emergence

Some of the secondary and tertiary health impacts detailed above are emergent properties of the complex systems in which extreme weather events occur. A classic example is famine. As Amartya Sen wrote in 'Development as Freedom', no famine has taken place in a functioning democracy (Sen 2000). Famines emerge through the complex interaction of drought and food insecurity, poverty and poorly functioning markets, and social and political isolation. Another example is related to infectious disease outbreaks. Complex emergencies can reduce coverage of routine immunization. If and when outbreaks emerge, they can spread more quickly in crowded humanitarian settings and have worse effects as people suffering from acute malnutrition are more vulnerable to mortality. In 2017, over 7000 children contracted measles in drought-affected parts of Ethiopia, Kenya, and Somalia (UNICEF 2017).

Feedback loops

Feedback loops are a common aspect of the health impact of extreme weather events. Feedback loops have been extensively studied at a climatic level—for example, ice albedo is an example of a positive feedback loop. Ice reflects heat; as ice melts, more surface area of water is visible; water is darker than ice and absorbs more heat, leading to even more ice melting (Curry et al. 1995). Using our primary, secondary, and tertiary framework, similar feedback loops can be observed at a population level, particularly when human behaviour is taken into account. For example, as discussed above, drought reduces crop yields, can force pastoral farmers to sell livestock at lower-than-usual prices, and can drive up violent conflict. Conflict makes humanitarian response harder: preventing food imports to drought-afflicted regions, increasing the price of staple food crops, worsening food security, and exacerbating the negative impact on human health. These types of complex human feedback loops cannot easily be modelled, but demonstrate an example of how the long-term effects of extreme weather events on human health may be even greater than current projections.

Adaptation

Another feature of complex adaptive systems is adaptation—i.e. how behaviour adapts to the system change as it is implemented. Many of the tertiary effects described above are the result of human adaptation to extreme weather events (e.g. forced migration). Adaptation is one aspect of complex adaptive systems that the climate literature has studied extensively; population-level adaptation to climatic changes is now a major focus of many agriculture and/or climate change development programmes. For example, in 2018, the African Development Bank launched a programme titled Rural Livelihood's Adaptation to Climate Change in the Horn of Africa (RLACC II), a programme focused on improving resilience of pastoral households to respond to climate shocks.

In Sudan, a 'sustainable livelihoods' programme to address poverty reduction incorporated many elements that addressed climate change. This focused on the means by which people derive and sustain their livelihoods and the assets they need to move out of poverty. Since persistent drought contributes to poverty, water scarcity, and famine, the programme supported the ways that rural communities used to cope with drought. They included measures to increase water harvesting, introduce new types of crops and livestock, improve productivity of land, and increase communities' skills via training. They increased participation in decision-making, engaging village committees and women's groups, empowered marginalised people via literacy programmes, assisted irrigated gardens for women, and improved credit funds. Taken together, these increased the resilience of communities to recurrent drought, in a holistic way (Elhassan et al. 2012).

Vulnerability and inequality

A fourth important element of complex adaptive system responses is vulnerability. Limited capacity to adapt financially, reliance on natural resources (which can be devastated by extreme weather), and a lack of governmental 'safety nets' make it extremely difficult for many populations in Africa to adapt (Thomas et al. 2007). It is well established that the poorest people in developing countries will bear the brunt of the impact of climate change, exacerbating existing inequities between rich and poor (Costello et al. 2009). This social vulnerability manifests in three ways: Poorer people are more at risk of climate-mediated shocks (e.g. they are more likely to live in vulnerable areas); shocks have a greater impact on poor people (e.g. they lose a larger portion of their income through shocks); and poor people have more difficulty in coping with shocks (as they have fewer savings and less access to affordable credit) (Winsemius et al. 2018; Daoudet al. 2016). Climate change affects poverty by influencing assets, prices, opportunities, and productivity and generally makes poverty reduction more difficult (Neher and Miola 2015). For example, while floods and hailstorms in Ethiopia reduce crop yields during excessive rainy seasons, droughts can reduce household farm production by up to 90% of a normal year output. Ethiopian farmers have developed different coping strategies, such as the sale of animals or loans from relatives, while pastoralists in semi-arid areas temporarily migrated, changed seasonal grazing patterns, or diversified their herds during excessively harsh conditions. The government also provided safety-net finance packages. But the households that tended to survive drought-related famines were the ones with higher than average incomes, with more diverse income bases and with more valuable livestock (Deressa et al. 2014).

The social vulnerability of poverty is magnified by other vulnerabilities—e.g. gender (Costello et al. 2009). In Africa, extreme weather events impact disproportionately more on women's health than on men's (Roy 2018). Countries with greater gender equality face on average lower fatalities when climate change-related extreme weather events strike (Neher and Miola 2015). In Ethiopia, gender is a significant component of food insecurity—adolescent girls receive less food than boys in households that are severely food insecure (Hadley et al. 2008). Adolescence is a critical time for reproductive health, and under-nutrition of adolescent girls can have intergenerational consequences.

Climate- and non-climate-related stresses contribute significantly to Eritrea's food scarcity, and the country's lack of adaptive capacity exacerbates the challenges of drought for the country's farmers of either gender. But the social and cultural view of women in some ethnic groups has prevented them from being regarded as equal farmers. Limited flexible response by the government to agriculture problems combined with a historical patriarchal culture has restricted the resources available to women farmers and increased their vulnerability to rainfall variability and drought (Tesfamariam and Hurlbert 2017).

5 Conclusions and Policy Responses

A complex adaptive system approach to climate change and health provides a way of understanding the problems, not the potential solutions. The Horn of Africa has been failed by the global multilateral system in two separate ways. Firstly, there is the well-understood failure to address global carbon emissions—disproportionately affecting some of the world's poorest and most vulnerable people. Secondly, there is a failure to respond to crises with appropriate alacrity.

Many of the climatic and extreme weather events detailed in this chapter have an insidious onset, with predictable impacts on human health. An appropriate response needs to take place at global, national, and local levels. At local levels, multi-sectoral programmes to reduce extreme weather impacts may need to be integrated with wider poverty reduction strategies, combining agriculture development with economic support to increase the resilience of the most disadvantaged communities. At national levels, strong multi-ministry action by governments could usefully combine economic, agricultural, and other programmes to drive sustainable national programmes to improve countries' resilience to climate change. And at a global level, donors and multilateral agencies need to prioritize the most vulnerable, even when political/policy environments may not be the easiest to work in. For example, WHO produces health and climate country profiles—there is some focus on the Greater Horn (e.g. profiles exist for Ethiopia, Kenya, and Uganda), but no profiles have been produced for the most vulnerable countries (e.g. Somalia and South Sudan).

Addressing the complex effects of extreme weather events on health will require understanding the distribution and needs of the most vulnerable populations and making sure that global policy responses are appropriately targeted and tailored.

References

Abdulkadir, G. (2017). Assessment of drought recurrence in Somaliland: Causes, impacts and mitigations. *Journal of Climatology and Weather Forecasting, 5*(204), 2.

African Development Bank. (2018). *African Economic Outlook 2018*. https://www.afdb.org/fileadmin/uploads/afdb/Documents/Publications/African_Economic_Outlook_2018_-_EN.pdf.

Andae, G. (2017). *Uhuru Embraces Kibaki Law to Rein in Rising Unga Prices*. Business Daily. 2017. https://www.businessdailyafrica.com/economy/Uhuru-embraces-Kibaki-law-rising-unga-prices-/3946234-4241990-146cugcz/index.html.

Anyamba, A., Chretien, J.-P., Small, J., Tucker, C. J., Formenty, P. B., Richardson, J. H., et al. (2009). Prediction of a rift valley fever outbreak. *Proceedings of the National Academy of Sciences of the United States of America, 106*(3), 955–959. https://doi.org/10.1073/pnas.0806490106.

Bowles, D. C., Reuveny, R., & Butler, C. D. (2014). 14 moving to a better life? Climate, migration and population health. *Climate Change and Global Health 135*.

Butler, C. D., & Harley, D. (2010). Primary, secondary and tertiary effects of eco-climatic change: The medical response. *Postgraduate Medical Journal, 86*(1014), 230–234. https://doi.org/10.1136/pgmj.2009.082727.

Costello, A., Abbas, M., Allen, A., Ball, S., Bell, S., Bellamy, R., et al. (2009). Managing the health effects of climate change: Lancet and university college London institute for global health commission. *The Lancet, 373*(9676), 1693–1733. https://doi.org/10.1016/S0140-6736(09)60935-1.

Countdown to 2030. (2017). *Tracking progress towards universal coverage for women's children's and adolescents' health: The 2017 Report*. Washington, D.C.: UNICEF and WHO. http://countdown2030.org/pdf/Countdown-2030-complete-with-profiles.pdf.

Curry, J. A., Schramm, J. L., & Ebert, E. E. (1995). Sea ice-albedo climate feedback mechanism. *Journal of Climate, 8*(2), 240–247. https://doi.org/10.1175/1520-0442(1995)008%3c0240:SIACFM%3e2.0.CO;2.

Daoud, A., Halleröd, B., & Guha-Sapir, D. (2016). What is the association between absolute child poverty, poor governance, and natural disasters? A global comparison of some of the realities of climate change". *PLOS ONE, 11*(4), e0153296. https://doi.org/10.1371/journal.pone.0153296.

Deressa, T. T., Fuhrer, J., Gregory, P. (2014). What are the factors that dictate the choice of coping strategies for extreme climate events? The case of farmers in the Nile Basin of Ethiopia. In *Climate change impact and adaptation in agricultural systems: Soil ecosystem management in sustainable agriculture* (pp. 271–80). CAB International, UK.

Dosio, A. (2017). Projection of temperature and heat waves for Africa with an ensemble of CORDEX regional climate models. *Clim Dyn, 49*(1–2), 493–519. https://doi.org/10.1007/s00382-016-3355-5.

Famine Early Warning Systems Network. (2018a). *Kenya food security outlook: Despite mixed performance, March–May rains expected to mitigate worse outcomes*. USAID and WFP. http://fews.net/sites/default/files/documents/reports/KENYA%20Food%20Security%20Outlook_February%202018_Final.pdf.

Famine Early Warning Systems Network. (2018b). *Ethiopia food security alert: Poor spring (Gu) rains likely to prolong food security emergency in Southeastern Ethiopia*. USAID. http://fews.net/sites/default/files/documents/reports/FEWS%20NET%20Ethiopia_Alert_20180301.pdf.

Hadley, C., Lindstrom, D., Tessema, F., & Belachew, T. (2008). Gender bias in the food insecurity experience of Ethiopian adolescents. *Social Science Medicine, 66*(2), 427–438. https://doi.org/10.1016/j.socscimed.2007.08.025.

Hartmann, I., Sugulle, A. J., Awale, A. I., Mohamed, A. S. A., Awale, M. H., Samale, A. M., et al. (2009). The Impact of Climate Change on Pastoral Societies of Somaliland. In *Candlelight for health, education & environment,* Hargeysa, Somalia.

Howard, G., Calow, R., Macdonald, A., & Bartram, J. (2016). Climate change and water and sanitation: Likely impacts and emerging trends for action. *Annual Review of Environment and Resources, 41*, 253–276.

Hsiang, S. M., Burke, M., & Miguel, E. (2013). Quantifying the influence of climate on human conflict. *Science, 341*(6151), 1235367.

IHME. (2018). *Global burden of disease. Country profiles*. 2018. http://www.healthdata.org/results/country-profiles.

IPCC. (2012). *Special report on managing the risks of extreme events and disasters to advance climate change adaptation*. Cambridge, UK and New York, USA: Cambridge University Press.

IPCC. (2014). *Climate change 2014: impacts, adaptation, and vulnerability. Part B: regional aspects. contribution of working group II to the fifth assessment report of the intergovernmental panel on climate change*. Cambridge, UK and New York, USA: Cambridge University Press.

Maystadt, J.-F., & Ecker, O. (2014). Extreme weather and civil war: Does drought fuel conflict in Somalia through livestock price shocks? *American Journal of Agricultural Economics, 96*(4), 1157–1182.

McArthur, J. W., Rasmussen, K., & Yamey, G. (2018). How many lives are at stake? Assessing 2030 sustainable development goal trajectories for maternal and child health. *BMJ, 360*(February), k373. https://doi.org/10.1136/bmj.k373.

McMichael, C., Barnett, J., & McMichael, A. J. (2012). An ill wind? Climate change, migration, and health. *Environmental Health Perspectives, 120*(5), 646–654. https://doi.org/10.1289/ehp.1104375.

Neher, F., & Miola, A. (2015). *The role of social inequalities for the vulnerability to climate related extreme weather events*. Joint Research Centre: Publications Office of the European Union, Luxembourg.

O'Loughlin, J., Witmer, F. D. W., Linke, A. M., Laing, A., Gettelman, A., & Dudhia, J. (2012). Climate variability and conflict risk in East Africa, 1990–2009. *Proceedings of the National Academy of Sciences, 109*(45), 18344–18349. https://doi.org/10.1073/pnas.1205130109.

Okiror, S. (2017). *Drought takes centre stage in Kenya's election campaign as food prices rise. The Guardian*, June 2, 2017, sec. Global development. http://www.theguardian.com/global-development/2017/jun/02/drought-centre-stage-kenya-election-campaign-food-prices-rise.

Onyango, E. A., Sahin, Oz, Awiti, A., Chu, C., & Mackey, B. (2016). An integrated risk and vulnerability assessment framework for climate change and malaria transmission in East Africa. *Malaria Journal, 15*(November), 551. https://doi.org/10.1186/s12936-016-1600-3.

Osman-Elasha, B., Goutbi, N., Spanger-Siegfried, E., Dougherty, B., Hanafi, A., Zakieldeen, S., Sanjak, E. A., Atti, H. A., & Elhassan, H. M. (2012). Community development and coping with drought in rural Sudan. In *Climate Change and Adaptation* (pp. 104–22). Routledge.

Paul, C. J., & Weinthal, E. (2018). The development of Ethiopia's climate resilient green economy 2011–2014: Implications for rural adaptation. *Climate and Development, 11*, 1–10.

Roy, S. (2018). Linking gender to climate change impacts in the global south. In *Linking Gender to Climate Change Impacts in the Global South.*

Rutter, H., Savona, N., Glonti, K., Bibby, J., Cummins, S., Finegood, D. T., et al. (2017). The need for a complex systems model of evidence for public health. *The Lancet, 390*(10112), 2602–2604. https://doi.org/10.1016/S0140-6736(17)31267-9.

Sen, A. (2000). *Development as Freedom. 1. Anchor* (Books ed.). New York: Anchor Books.

Serdeczny, O., Adams, S., Baarsch, F., Coumou, D., Robinson, A., Hare, W., et al. (2017). Climate change impacts in sub-saharan Africa: From physical changes to their social repercussions. *Regional Environmental Change, 17*(6), 1585–1600.

Singer, T., Moore, F., & Luby, S. (2016). Potential effects of increased global temperatures on neurological development factors in children under 5 years in East Africa: A modelling study. *The Lancet Global Health, 4*(April), S21. https://doi.org/10.1016/S2214-109X(16)30026-2.

Souverijns, N., Thiery, W., Demuzere, M., & Van Lipzig, N. P. M. (2016). Drivers of future changes in East African precipitation. *Environmental Research Letters, 11*(11), 114011. https://doi.org/10.1088/1748-9326/11/11/114011.

Sudfeld, C. R., McCoy, D. C., Danaei, G., Fink, G., Ezzati, M., Andrews, K. G., et al. (2015). Linear growth and child development in low- and middle-income countries: A meta-analysis. *Pediatrics, 135*(5), e1266–e1275. https://doi.org/10.1542/peds.2014-3111.

Tesfamariam, Y., & Hurlbert, M. (2017). Gendered adaptation of Eritrean dryland farmers. *International Journal of Climate Change Strategies and Management, 9*(2), 207–224.
The Fund for Peace. (2018). *Fragile states index 2018*. http://fundforpeace.org/fsi/data/.
Thomas, D. S., Twyman, C., Osbahr, H., & Hewitson, B. (2007). Adaptation to climate change and variability: Farmer responses to intra-seasonal precipitation trends in South Africa. In *African Climate and Climate Change* (pp. 155–78). Springer, New York.
Tierney, J. E., Ummenhofer, C. C., & deMenocal, P. B. (2015). Past and future rainfall in the horn of Africa. *Science Advances, 1*(9), e1500682. https://doi.org/10.1126/sciadv.1500682.
UNHCR. (2017). *Global trends: Forced displacement in 2016*. Geneva, Switzerland: UNHCR. http://www.unhcr.org/5943e8a34.pdf.
UNICEF. (2017). *Horn of Africa measles outbreak humanitarian situation report*. UNICEF. https://reliefweb.int/report/ethiopia/unicef-horn-africa-measles-outbreak-humanitarian-situation-report-17-may-2017.
United Nations Department of Economic and Social Affairs, Population Division. (2017). *World population prospects: The 2017 revision, volume 1: comprehensive tables*. https://esa.un.org/unpd/wpp/publications/Files/WPP2017_Volume-I_Comprehensive-Tables.pdf.
Université Catholique de Louvain. (2018). *EM-DAT: The emergency events database*. 2018. http://emdat.be/emdat_db/.
UNOCHA. (2017a). *Horn of Africa: Humanitarian impacts of drought—Issue 11. UNOCHA*. https://reliefweb.int/sites/reliefweb.int/files/resources/HOA_drought_updates_snapshot_Nov032017.pdf.
UNOCHA. (2017b). *Regional outlook for the Horn of Africa and great lakes: April–June 2017. UNOCHA*.
USAID. (2017). *Horn of Africa—Complex emergency. 1-Fiscal Year 2018. USAID*.
Wall, N., & Michael, H. (2016). Drought and health in the context of public engagement. In *Extreme Weather, Health, and Communities* (pp. 219–44). Springer.
Watts, N., Amann, M., Ayeb-Karlsson, S., Belesova, K., Bouley, T., Boykoff, M., et al. (2018). The lancet countdown on health and climate change: From 25 years of inaction to a global transformation for public health. *The Lancet, 391*(10120), 581–630. https://doi.org/10.1016/S0140-6736(17)32464-9.
Weber, T., Haensler, A., Rechid, D., Pfeifer, S., Eggert, B., & Jacob, D. (2018). Analyzing regional climate change in Africa in a 1.5, 2, and 3 °C global warming world. *Earth's Future, 6*(4), 643. https://doi.org/10.1002/2017EF000714.
Winsemius, H. C., Jongman, B., Veldkamp, T. I., Hallegatte, S., Bangalore, M., & Ward, P. J. (2018). Disaster risk, climate change, and poverty: Assessing the global exposure of poor people to floods and droughts. In *Environment and Development Economics* (pp. 1–21).
Winsvold, M., Stokke, K. B., Klausen, J. E., & Saglie, I. L. (2009). Organizational learning and governance in adaptation in urban development. In *Adapting to climate change: Thresholds, values, governance*, (pp. 476–90).
World Health Organization. (2014). *Quantitative risk assessment of the effects of climate change on selected causes of death, 2030s and 2050s*. World Health Organization.
World Health Organization. (2015a). *Climate and health country profile 2015: Ethiopia*.
World Health Organization. (2015b). *Climate and health country profile 2015: Kenya*.
World Health Organization. (2015c). *Climate and health country profile 2015: Uganda*. World Health Organization.

Extreme Weather Events Under a Changing Climate: A Brief Context for Brazil and the Role of the Health Sector

Júlia Alves Menezes, Rhavena Barbosa dos Santos, Felipe Vommaro, Ulisses Confalonieri, Martha Macêdo de Lima Barata and Carina Margonari

Abstract Disasters have economic and human impacts that can become more intense with climate change. In Brazil, disasters have increased, both in number and in cost, in recent decades. This situation is favored, in part, by urbanization, continued vulnerable social conditions, and the prioritization of crisis management at the expense of risk management. The health sector has an important role in assisting and improving the management of disasters for the prevention of morbidity, mortality, and health risks. This is especially true regarding actions that mitigate the direct (e.g., deaths, disabilities) and indirect (e.g., infrastructure, conditions) impacts of extreme events. This chapter presents a brief reflection on the current state of disasters in Brazil, how they impact population health, and how the issue has been addressed by public policy.

Keywords Disaster · Health · Policies · Disaster risk reduction · Vulnerability · Climate change

Acronyms

COE-Health	Operative Health Committee
CRED	Center for Research on the Epidemiology of Disasters
IPCC	Intergovernmental Panel on Climate Change
PAHO	Pan American Health Organization
PNPDEC	National Policy of Protection and Civil Defense
S2ID	Integrated Disaster Information System
SEDEC	National Secretariat of Civil Defense
SINPDEC	National System of Civil Protection

J. A. Menezes (✉) · F. Vommaro · M. M. L. Barata
Escola Nacional de Saúde Pública Sergio Arouca (ENSP/FIOCRUZ), Rio de Janeiro, Brazil
e-mail: julimenezes@yahoo.com.br

R. B. dos Santos · U. Confalonieri · C. Margonari
Instituto René Rachou (IRR/Fiocruz), Belo Horizonte, Brazil

R. Akhtar (ed.), *Extreme Weather Events and Human Health*,
https://doi.org/10.1007/978-3-030-23773-8_21

SUDENE	Superintendence for the Development of the Northeast
SUS	Unified Health System
VIGIDESASTRES	Environmental Health Surveillance Related to Risks Arising from Disasters

1 Contextualizing Disasters in Brazil

Natural disasters are considered manifestations of nature that modify the earth's surface, such as cyclones, hurricanes, earthquakes, landslides, floods, droughts, and storms, among others. The Special Report on Managing the Risks of Extreme Events and Disasters to Advance Climate Change Adaptation considers a disaster to be an impact arising from the interaction between a hazard—a natural phenomenon that adversely affects the human system—and vulnerable social conditions that alter the normal functioning of a community (IPCC 2012). Storms, hurricanes, and droughts are expected to become more frequent than in previous years (IPCC 2012, 2013, 2014). Evidence compiled by the Fifth Report of the Intergovernmental Panel on Climate Change (IPCC) shows that between 1951 and 2010 there was a reduction, at the global level, in the number of cold nights/days and an increase in the number hot nights/days. Climate change projections for the end of the twenty-first century point to an increase in the planet's average temperature, accompanied by intensified occurrence of heat waves and natural phenomena such as El Niño, and an uneven distribution of rainfall (IPCC 2014).

Data from the Center for Research on the Epidemiology of Disasters (CRED) show that the highest incidence of disasters in the Americas has been in recent decades, having risen from 216 records affecting 43 million people between 1970 and 1979 to 922 disasters affecting 71 million people between 2000 and 2009 (OPAS 2014). This trend was also observed in Brazil, although the Brazilian Atlas of Natural Disasters itself underscores the historical fragility of the Civil Defense System database, which likely represents an underestimate. Of the 39,996 records between 1991 and 2005, 8515 (22%) occurred in the 1990s; 21,741 (56%) occurred in the 2000s; and 8740 (22%) occurred just between 2010 and 2012 (Universidade Federal de Santa Catarina – UFSC/ Centro Universitário de Estudos e Pesquisas sobre Desastres - CEPED 2013). Box 1 shows the Brazilian government's attempt to improve the quality of disaster records in the country.

Box 1. The Brazilian Integrated Disaster Information System

The Integrated Disaster Information System (S2ID) was recently created by the National Secretariat of Civil Defense (SEDEC) to establish quality and transparency to information regarding disasters in Brazil. The system was derived from a project that surveyed official disaster documents recorded during 1991–2010, and which gave rise to "Database and Disaster Records" and "Brazilian Atlas of Natural Disasters." The S2ID aims to continue this database and provide more reliability to the information based on a geo-tool that will enable policy makers and researchers to develop knowledge in risk management and disaster. Thus, the history of natural and technological disasters in the country has gradually been built from data inserted in S2ID, which is open to civil society.

Among the factors that help to explain the increasing frequency of disasters in developing countries are, in addition to the threat of global environmental change itself, increasing unplanned urbanization of cities, environmental degradation, weak responsiveness of institutions and the population, the lack of infrastructure, and poverty (Thouret 2007). These conditions, combined with the occupation of locations with high exposure to natural hazards, have generated places with intense vulnerability and low response capacity in Brazil. Thus, increases in the frequency and magnitude of disasters in the country have been observed since 1950, mainly associated with morphodynamic, hydrological, and climatic phenomena, such as landslides, floods, and droughts (Almeida and Pascoalino 2009).

The Brazilian Atlas of Natural Disasters compiled and systematized national occurrence records for the period between 1991 and 2012, which has permitted an initial assessment of the situation in Brazil over the last 20 years. As a result, drought was found to be the disaster with the greatest effect on the Brazilian population (51%), followed by flash floods and floods (33%), with the latter being responsible for the greatest number of deaths (71.55%) (Universidade Federal de Santa Catarina – UFSC/ Centro Universitário de Estudos e Pesquisas sobre Desastres - CEPED 2013). Figure 1 shows the distribution of disaster events for the main geographical regions of Brazil and the contribution of each disaster type to all the disasters recorded for the analyzed period.

Data show that droughts are the most frequently recorded disaster in Brazil; between 1991 and 2012 there were 19,517 records, which represents about 48% of the total events reported (UFSC-CEPED 2013). The Northeast Region is among the regions most impacted by drought due to the length of such events in the region and its marked climatic variability, especially in the semi-arid region (Camarinha et al. 2017). The high susceptibility of the Northeast to drought is due to sparse rains, reduced percolation, and runoff of surface water, with severe episodes being recorded as early as the seventeenth century (Brant 2007; Marengo 2009). These environmental factors, in combination with others of geographical and political nature, have culminated, historically, in the occurrence of low-income urban populations with serious social problems. The impacts of such conjuncture are apparent in household subsistence, quality of life, health condition, and economic development. In

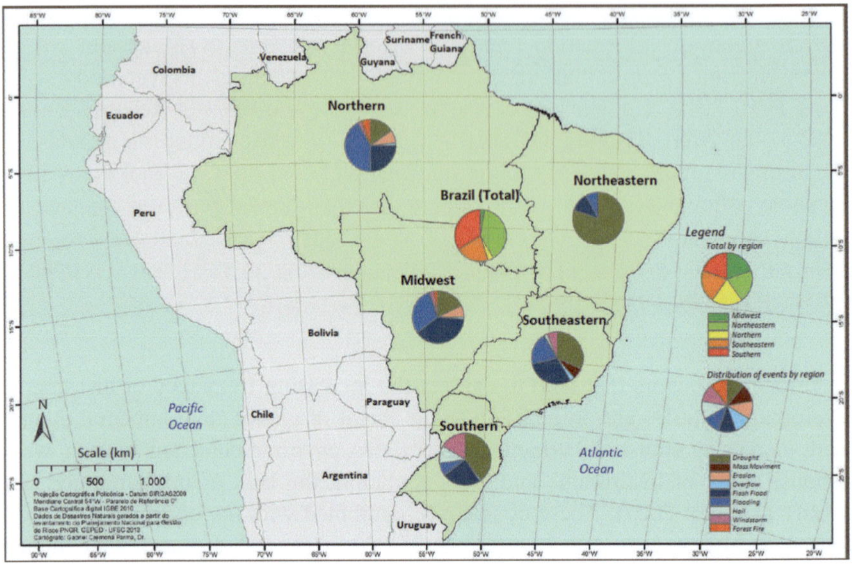

Fig. 1 Distribution of the number and types of natural disasters among geographical regions of Brazil during 1991–2012. The central circle "Brazil (Total)" gives the total distribution of disasters by region and shows that the Southern and Northeast regions experienced the most disasters during the period. The other circles show the distribution of each type of disaster by region. *Source* Universidade Federal de Santa Catarina – UFSC/ Centro Universitário de Estudos e Pesquisas sobre Desastres - CEPED (2013)

other words, social indicators of the Northeast Region are generally lower than those observed in other regions of the country (Sena et al. 2014).

This situation is even more critical given the context of climate change and the predicted increase in water scarcity for the coming decades. Projections by IPCC show, with a moderate level of confidence, that droughts will become more intense by the end of the century for various regions of the globe, including the Northeast Region of Brazil, for which increases in average temperature of up to 4 °C are expected (IPCC 2013). Other national studies have converged with these projections, predicting temperature increases of between 1 and 4 °C, and a reduction in precipitation of 10–15%, accompanied by significant increases in the number of consecutive dry days and drought risk (Marengo 2009; Marengo et al. 2010).

Floods, including riverside and urban events, are the second most common type of disaster in Brazil (Universidade Federal de Santa Catarina – UFSC/ Centro Universitário de Estudos e Pesquisas sobre Desastres - CEPED 2013). Riverside flooding is a natural process of a river flooding its banks, the severity of which is determined by the intensity and duration of precipitation and the initial conditions of humidity. The main impacts to a population result from: (1) lack of knowledge about the occurrence of flood levels and the occupation of the riverbed; and (2) failure to plan the occupation of areas according to flood risk. Urban floods result from the replacement of soil with an impermeable landscape and the channeling of streams, which increases the volume and peaks of floods (Tucci 2007). However, the transformation of floods into disasters is mainly a result of anthropic factors, including the lack of urban infrastructure and social development, which favor the irregular occupation of areas normally subject to natural flooding.

The Amazon Region (largely coincident with the North Region of Brazil) is recognized as one of the regions most affected by floods in Brazil, which is peculiar given the relationship between the population and the overflow of its rivers. In recent years, "floods of the century" were recorded in the region. In 2009, the annual flood reached a peak of 29.77 meters (1.97 meters above average), which far exceeded the expectations of the resident population whose reference for home stilt construction was the quota reached in 1953 (OPAS 2014). In 2012 the flood peak was even higher, reaching 29.97 meters. Although flooding is a seasonal and habitual phenomenon for riverside populations of the North Region (Amazon), these "floods of the century" demonstrate that: (1) river levels during these floods have surpassed historical averages; and (2) these floods affect an increasing number of people due to population density (Freitas et al. 2014).

1.1 Socio-Environmental Vulnerability and the Perpetuation of Disasters in Brazil

According to the CRED database, disasters, such as epidemics, windstorms, forest fires, and droughts, among others, have increased in both frequency and intensity since 1930. It has been estimated that between 1986 and 2006 more than 2.5 billion people were affected by floods, 1.5 million by drought, and around 650,000 by windstorms throughout the world (Adikari and Yoshitani 2009). Explanations for this increase include increased exposure of people and economic and natural assets to the risk of disasters, mainly due to rapid economic growth and urbanization in susceptible areas (e.g., coastal areas, floodplains), a condition that tends to worsen with climate change (Hay and Mimura 2010).

According to Freitas et al. (2012), the genesis of a disaster requires a detonating element that exposes human populations to potential damage or injuries, which vary in seriousness depending on the vulnerability of the affected population. These conditions may be related to factors such as employment, housing, health, environmental degradation, pollution, among others, which can help reveal critical territories according to their physical and socio-environmental particularities (Freitas et al. 2014). In this context, exposure becomes a key concept that will determine, along with the underlying conditions of socio-environmental vulnerability, the heterogeneous distribution of areas most susceptible to the occurrence of hazards and disasters (Vázquez 2001). Not surprisingly, the most common case is the association of low-economic and social development with areas of high exposure to, and risk of, disasters (poor countries with poor populations), which reduces the capacity for risk reduction (Marandola and Hogan 2009; Narváez et al. 2009; OPAS 2014).

In the context of Brazil, the semi-arid region (Northeast) illustrates this kind of relationship where climatic adversities go hand-in-hand with historical-political factors (i.e., water resources management and inefficient disaster prevention and preparedness policies) that give rise to serious social problems in what is one of the most populated semi-arid regions of the world. In fact, the social, economic, and cultural characteristics of the semi-arid region combine to influence the occurrence of problems related to health and governance and that perpetuate situations of public calamity for decades (Sena et al. 2015). Currently, the semi-arid region encompasses 1262 municipalities located in nine states of the Northeast Region and part of the Southeast Region; however, with each new evaluation the scope of this area increases due to persistent public calamities induced by long droughts (Box 2).

Box 2. The Territorial Delimitation of the Brazilian Semi-arid Region

The first criterion for delimiting the semi-arid region was proposed in 1989 based on an annual average rainfall of 800 mm or less in the municipalities within the SUDENE jurisdiction (Superintendence for the Development of the Northeast), which covered an area of 892,309.4 km^2 (Brazil, Law 7827/and 1989). In 2005, the following new criteria were suggested: (1) average annual rainfall of 800 mm or less; (2) aridity index of up to 0.5 (period 1961–1990); and (3) drought risk greater than 60%. While preserving the continuity of the region, this new delimitation included 1135 municipalities (969,589.4 km^2). More recently, a 10-year revaluation of the boundaries of the semi-arid region that was recommended because of climate change resulted in the inclusion of a total of 1189 municipalities. However, after requests from some states to include new territories, the Brazilian semi-arid region was re-defined in November 2017 to include 1262 municipalities and encompass an area of more than 1.03 million km^2 (Brazil, Resolution nº 115/2017).

From the perspective of public policy, understanding the possible interactions among disaster, health and society is essential for focusing actions and resources on reducing disaster risk and health risks in the context of extreme weather events. Surveys that estimate the values of loss and damage for major disasters in Brazil must include the topic of risk management in the agendas of governments and society, especially considering the greater exposure of infrastructure and economic assets brought on by climate change and population growth (Universidade Federal de Santa Catarina – UFSC/ Centro Universitário de Estudos e Pesquisas sobre Desastres - CEPED 2016).

In this sense, the "Material Damage and Losses Report due to Natural Disasters in Brazil—1995–2014" has become a useful tool for accounting the transfer of funds held by the Union to states and municipalities for coping with, and recovering from, situations of disaster (Universidade Federal de Santa Catarina – UFSC/ Centro Universitário de Estudos e Pesquisas sobre Desastres - CEPED 2016). The report estimates that between 2008 and 2013, about R$4 billion was transferred to federal agencies for actions of assistance, rehabilitation, and reconstruction in affected areas (Universidade Federal de Santa Catarina – UFSC/ Centro Universitário de Estudos e Pesquisas sobre Desastres - CEPED 2016). Report data show that total damage and losses accumulated in Brazil during 1995–2014 were on the order of R$182.7 billion. Material damage (i.e., infrastructure, housing, and other facilities) totaled R$45.4 billion, whereas losses (i.e., in both public and private sectors—agriculture, livestock, industry, and services) totaled R$ 137.3 billion. Figure 2 shows the incidence of the total material damage (R$) (A) and to each major disaster type recorded in Brazil (B). Overall, the "Material Damage and Losses Report due to Natural Disasters in Brazil—1995–2014" states that losses of more than R$9 billion are reported annually in Brazil—about R$ 800 million are spent monthly on natural disasters—demonstrating the relevance of the economic impacts of these events in a country still mystified by the absence of significant natural disasters.

1.2 A Brief Description of the National Disaster Policy

The second decade of the twenty-first century in Brazil has been characterized by the acknowledgment of the principles of disaster reduction, highlighting, among other actions of risk and disaster management, the following (Brasil 2016a):

- Implementation of the National Plan for Civil Protection and Defense;
- Construction of the Database of Disaster Records;
- Mapping of Disaster Risks;
- Preparation of the Brazilian Atlas of Natural Disasters;
- Approval of the National Policy of Protection and Civil Defense—PNPDEC;
- Implementation of an Information System and the Monitoring of Disasters;
- Organization of the National System of Civil Protection—SINPDEC.

This new approach of the public sector to disaster risk management in Brazil is recent and still needs improvement. For example, the river basin needs to be adopted as the unit of analysis and planning since the main unit currently in use, the municipality, leads to isolated actions (Freire 2014). Efforts have been made to map priority areas and improve coordination among the various organizations and/or sectors that perform actions that are in some way related to disaster risk management, such as the national health system (Freire 2014).

Currently, civil defense and protection activities in Brazil are carried out under SINPDEC, which consists of organizations and entities of the federal government, the states, the Federal District and the municipalities, as well as public and private entities that play a significant role in the area. Law No. 12,608 establishes that it is the duty of all administrative levels to adopt the necessary measures to reduce disaster risk by incorporating actions of protection and civil defense in their respective plans (Brasil 2016b).

Although all states have an agency responsible for civil defense actions within their territory, the National Secretariat for Civil Protection and Defense is also encour-

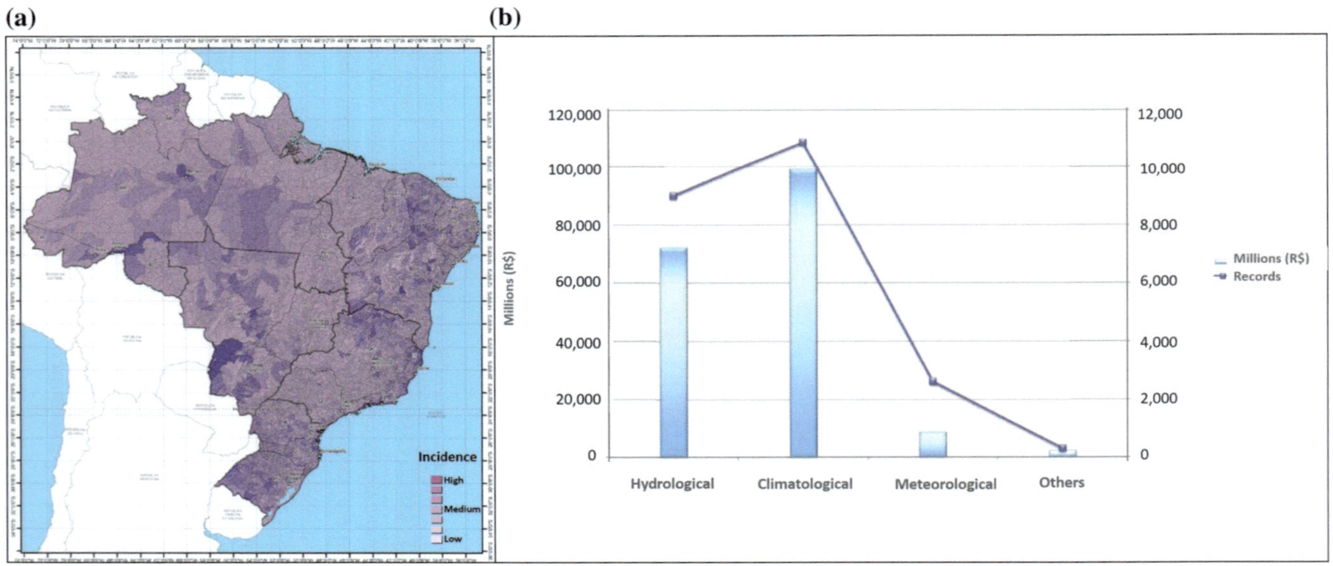

Fig. 2 Total material damages and losses in Brazil between 1995 and 2014, according to estimates made and reported by municipalities in accordance with the Report on Material Damage and Losses Due to Disasters in Brazil. **a** Incidence of the total material damage and losses from disasters in the country (R$); and **b** total damage and losses (R$) per disaster group according to municipal records. Adapted from Universidade Federal de Santa Catarina – UFSC/ Centro Universitário de Estudos e Pesquisas sobre Desastres - CEPED (2016)

aging the implementation of such an organization in each municipality in order to execute, coordinate, and mobilize necessary actions. The main role of a municipal civil protection and defense agency would be to know and identify the risks posed by disasters in the municipality when the participation of the population is essential for mitigating adverse effects. Populations must be organized, prepared, and oriented to know what to do, when to do it, and how to do it. Thus, a municipal civil defense protection agency would serve as a joint channel between the municipal authority and the population that facilitates community participation in the planning and execution of civil defense actions. This local approach would make it possible to prepare institutions and populations to face the situations brought by disaster by developing specific plans to prevent and mitigate the risks of disaster events.

2 The Impacts of Disasters on Health

The impacts disasters have on health depend on the intensity and type of event, the vulnerability of the exposed population (i.e., housing conditions, income, access to services, among others), and the preparedness of local institutions to deal with disaster. Thus, the most common types of disasters in Brazil—droughts and floods—require different actions and preparations depending on their time of occurrence, location, and extent. In general, deaths, disabilities, social and psychological disorders, loss of property, and damage to infrastructure are only some of the consequences that can occur in the short or medium/long term, with important implications for public health. Direct effects can include fractures, drownings, deaths, asphyxia, hypothermia, burns, as well as damage to the physical structures of hospitals, ambulances, and first aid stations, which can increase mortality (Freitas et al. 2014; Pourhosseini et al. 2015).

However, especially in the medium/long term and for the disasters considered as extensive[1], such as drought, many health consequences are related to the deterioration of the social determinants of health[2]—the socio-environmental and economic conditions of the affected territories—culminating, indirectly, in increased morbidity and mortality from chronic diseases (i.e., hypertension, diabetes), infectious diseases, and other health-related problems (e.g., mental disorders). Table 1 shows the consequences of drought and flood in natural and economic systems and how this affects

[1]Extensive disasters are characterized by the severity of the losses and damages as well as their high frequency. They do not necessarily cause significant numbers of deaths, although may account for a large proportion of damage to local infrastructure, housing, and living conditions (e.g., droughts). Intensive disasters are characterized by the low frequency of events, which are geographically concentrated and have great potential for losses, damage, and mortality (e.g., Haiti earthquake; landslides in the mountain region of Rio de Janeiro). Source: Freitas et al. (2012).

[2]The social determinants of health are linked to economic and social conditions (money, power, and resources) and are mostly responsible for health inequities—the unfair and avoidable differences in health status seen within and between countries. More information is available at http://www.who.int/social_determinant.

Table 1 Some consequences of flood and drought disasters for environment, society, and health

Disaster	Environmental impacts	Socioeconomic impacts	Health impacts
Floods	• Biological and chemical contamination of soil and water for human consumption, food production • Deterioration of water supply network and other alternative sources of water • Deterioration of basic sanitation services (garbage, sewage, and drinking water) • Alterations to cycles of vectors, hosts, and reservoirs of diseases	• Full or partial interruption of bridges, streets and roads • Total or partial interruption of the provision of water, electricity, gas, transportation, and communication services • Total or partial interruption of the operation of schools, commerce, funeral services, health services, and others • Total or partial deterioration of agricultural activities • Total or partial destruction of property, houses and buildings; and sources of income and labor • Loss of personal property and sentimental value	• Drowning • Electric shock • Leptospirosis • Hypothermia • Water and food-borne diseases (diarrhea, gastroenteritis, shigellosis, etc.) • Vector-borne diseases (Dengue, Malaria) • Mental disorders (posttraumatic stress, amnesia, insomnia) • Chronic diseases (hypertension, diabetes)
Droughts	• Contamination of water for human consumption and food production • Deterioration of water reservoirs; • Salinization of groundwater • Air contamination by forest fire particles, harmful cyanobacterial/algal blooms, and toxins present in the soil • Alteration in the cycles of vectors, hosts, and reservoirs of diseases	• Total or partial interruption of water supply • Total or partial deterioration of agricultural and forestry activities • Total or partial destruction of income and labor • Loss of personal property and sentimental value due to human displacements	• Malnutrition • Respiratory diseases (asthma, allergy) • Intoxication • Water-borne diseases (hepatitis A, amebiasis, salmonellosis, cholera) • Diarrhea and gastroenteritis • Skin and eye infections (scabies, conjunctivitis) • Vector-borne diseases (Dengue, Malaria, Encephalitis, Chikungunya, Schistosomiasis) • Mental disorders (depression, anxiety, insomnia) • Chronic diseases (hypertension, diabetes)

Adapted from Freitas and Ximenes (2012), Brasil (2018), Stanke et al. (2013)

the health of the population and essential services. Regardless of the impacts, the health and well-being of the affected population will be either directly or indirectly disrupted by successive events during or following the disaster. For droughts, the most common impacts are increased acute respiratory infections, poor access to drinking water, food insecurity, and losses to subsistence fishing, agriculture, and livestock (Confalonieri and Fonseca 2013; Franca 2015). In the case of floods, impacts on health include higher incidences of water-borne diseases, the interruption of public services, damage to households, and the loss of agricultural and extractive activities over the short term (Marengo et al. 2013; Franca 2015).

The effects of droughts and floods may overlap, but are quite different with regard to the ability to cause deaths and injuries, damage health facilities and water supply, and displace populations. Since floods are an intensive disaster, they are often associated with high incidence of deaths and damage to infrastructure in the short term. On the other hand, droughts are an extensive phenomenon, and severely impact water supply and food security, and can lead to large human displacements (Brasil 2018).

3 Organization of the Health Sector to Respond to Disasters in Brazil

The health issue in Brazil is treated in an aggregate manner, covering both individual and collective medical care, prevention, and health promotion. Within the scope of the Unified Health System (SUS), the service network is organized into three levels of care (primary, secondary, and tertiary) to guarantee the integrality of medical assistance, including for disasters, since demand for services may increase, compromising the functioning of health facilities. Thus, the Ministry of Health and its state and municipal secretariats prepare and execute national and municipal disaster preparedness plans to reduce: (i) the vulnerability of health services, (ii) risk factors, and (iii) mortality and impacts to the health of the population. These entities also work with other institutions, such as civil defense, police forces, community organizations, and nongovernmental organizations (NGOs) (Shoaf 2014).

Emergency and disaster preparedness plans comprise a permanent multisectoral activity whose essential components include, among several processes and mechanisms, the establishment of the Operative Health Committee (COE-Health) in the stage of risk reduction and which persists throughout the management of the disaster. As a vital operational unit during disaster response, COE-Health represents the focal point for sectoral coordination and decision-making. It encompasses all the areas of the health sector that are involved in preparation, warning, response, and rehabilitation, thus organizing SUS actions in disasters at the local level (Knowledge Center on Public Health and Disasters—Fiocruz). The insertion of COE-Health within SINPDEC is shown in Fig. 3, which also illustrates how actions and institutions articulate in dealing with disasters.

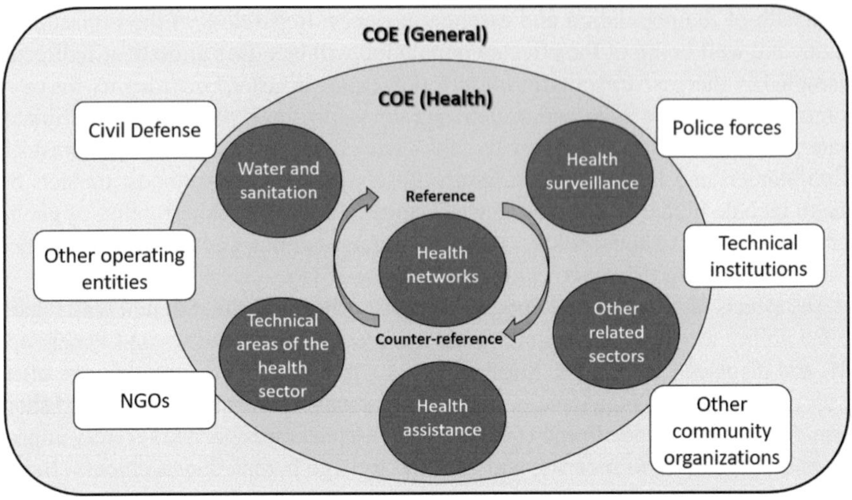

Fig. 3 Organization of COE-Health inside the national disaster management system (COE-General). The COE-General unites all the institutions involved in response to a disaster, and whose structure is replicated in states and municipalities. The COE-Health defines the actions of the health sector, in which health networks guide users among the different levels of care (reference and counter-reference system). *Source* Adapted from Knowledge Center on Public Health and Disasters—Fiocruz; Freitas & Rocha (2014)

Vigidesastres[3] (environmental health surveillance related to risks arising from disasters) seeks to develop a set of actions for public health (SUS) in the context of disasters. It plays a central role in disaster risk management since it is a continuous program focused on prevention, mitigation, preparedness, response, and rehabilitation to reduce the impact of disasters on public health (Freitas et al. 2014; Brazil 2014, 2017). Figure 4 shows the operational model of Vigidesastres with the responsibilities of the health sector in the different stages of risk management. In general, actions are linked to hospital and outpatient care, pharmaceutical assistance, epidemiological and environmental surveillance, and health surveillance, thus allowing the linkage of specific actions in the reduction stage (educational actions and risk monitoring); the management stage (emergency care of those affected); and recovery stage (monitoring and rehabilitation of the health condition of the population). This articulation of actions takes place from the federal level, passing through the states and to the municipalities, where strategies designed for different levels of assistance are implemented and may overlap during decision-making processes.

Among the three levels of assistance, primary care plays a central role in conducting integrated disaster risk management since it represents the level closest to the

[3]The Vigidesastres is a program under the responsibility of the General Coordination of Surveillance in Environmental Health/Ministry of Health. Its aim is to develop a set of actions to be continuously adopted by public health authorities to the reduce the risk of exposure of both the population and health professionals, reduce disease and injury arising out of them as well as damage to the health infrastructure.

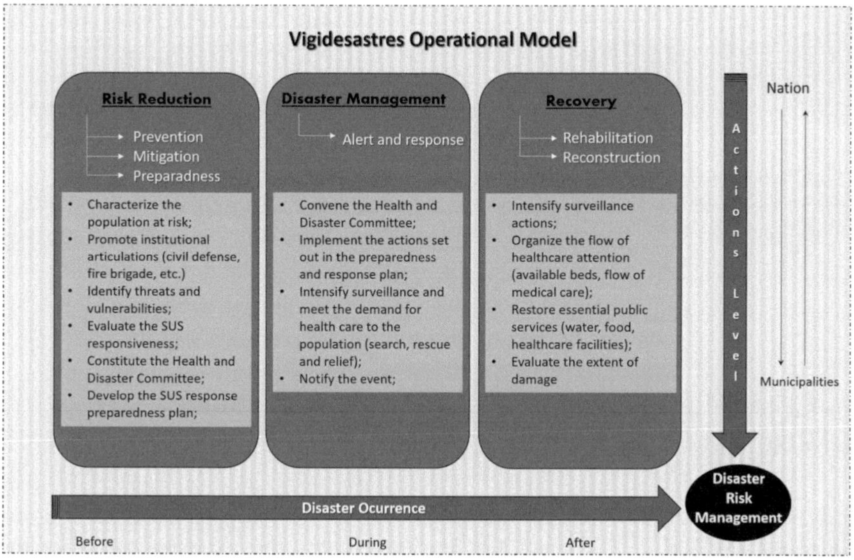

Fig. 4 The operational model of Vigidesastres showing the health sector actions implemented in the different stages of risk management in Brazil. *Source* Adapted from Ministry of Health (http://portalms.saude.gov.br/vigilancia-em-saude/vigilancia-ambiental/vigidesastres/modelo-de-atuacao)

population. It is made up of Basic Health Units whose main function is to promote and protect health through multidisciplinary teams (nurses, physicians, community health agents). These professionals are able to assist in the monitoring and investigation of the health condition of the affected population as well as in the identification of its main vulnerabilities. In this way, Basic Health Units and its multidisciplinary teams play an essential role for other institutions, such as Vigidesastres and Civil Defense (Brasil 2012, 2017). The secondary and tertiary levels guarantee health care in disaster situations that imply more specialized procedures or therapies (i.e., surgeries, traumatology, neurology, rehabilitation) and hence contain more complex health units (i.e., emergency care, hospitals). Regardless of the level of assistance, all health facilities must be safe from disasters, which encompasses not only their physical structure but also their accessibility and functioning at full capacity following a disaster or an emergency (Box 3).

Box 3. Safe Hospitals—Mitigating Disaster Risks in the Health Sector

For all levels of assistance, ensured safety of health facilities is central to the continuity of medical care for the affected population, and for which Brazil has adopted the "safe hospitals in the face of disasters" campaign. According to the Pan American Health Organization (PAHO), more than 67% of the nearly 18,000 hospitals in Latin America and the Caribbean are located in areas at high risk of disasters (Organización Panamericana de la Salud–OPS 2011). An example of an action aimed at hospital safety during disasters is the Hospital Safety Index developed by the PAHO. This index allows the evaluation of the probability that a hospital can continue to function after a disaster and has been applied successfully in Brazil, Bolivia, Ecuador, Cuba, and Peru (Organización Panamericana de la Salud–OPS 2004; Salles and Cavalin 2012; Pereira and Barata 2014).

4 Crisis Management or Risk Management? Some Relevant Cases from the Last Decade

Historically, Brazil has been the scene of extreme weather events and natural disasters, the development of which has been reflected in the exposure of vulnerable populations and in economic and human losses. Although the country has been investing in a more comprehensive disaster risk reduction policy in recent years, the "learning from experiences" profile still prevails at the expense of articulating and/or planning actions based on known vulnerabilities and risks. This can be seen as a prioritization of "crisis management" over risk management and reveals the lack of resiliency, adaptive capacity, and efficiency of the public policies of the country. Major events of the past decade will be discussed here in an effort to illustrate this condition in some local contexts.

4.1 The Itajaí Valley, Santa Catarina State

In the South Region of Brazil, the Itajaí Valley in the state of Santa Catarina recorded an excess of rainfall during the second half of November 2008, which resulted in floods and landslides. The cumulative rainfall per hour was classified as moderate for most of the month, with a few periods without rain. Thus, a historical record of over 600 mm was registered only in four days (November 20 and 24) for a region where the climatological monthly average is about 150 mm (Dias 2009). According to the December 31, 2008, report of the Civil Defense of Santa Catarina, 32,853 people were displaced and 5617 left homeless; by then, landslides and floods had resulted in 135 fatalities. The report also listed 63 out of a total of 293 municipalities as in an emergency situation and 14 in a state of public calamity. The total number of people affected was on the order of 1.5 million (CEPED 2015).

This case demonstrates the vulnerability of relatively small urban centers to extreme precipitation events—of the 54 municipalities comprising Vale do Itajaí, only four had a population of over 100,000 inhabitants. It also revealed the urgent need to improve infrastructure for disasters, such as investments in systems for observing, processing, and disseminating weather-related information in Brazil (Dias 2009). It should be noted that the main causative element of the landslides and floods in this case was concentrated rainfall in an area of rugged terrain, which, in association with the dynamics of land use, revealed historical processes of high social-environmental vulnerability supported by land occupation and environmental degradation (FAPESC 2009).

The health impacts during and after the event were devastating, with an increase in demand for hospital visits for reasons directly or indirectly related to the floods of November 2008. Total admissions for infectious diseases underwent a significant change (Xavier et al. 2014b). The number of hospitalizations for leptospirosis, for example, an infectious disease associated with urine and water contaminated by bacteria of the genus *Leptospira*, was considered low prior to the disaster, but then increased in the subsequent months (Ko et al. 1999; Freitas and Ximenes 2012). During such a critical event, the breakdown of normal living conditions of the population and the impairment of sanitary infrastructure facilitated (i) the exposure of people and food to harmful infectious agents present in water and soil, (ii) contact with animal hosts (i.e., rats), and (iii) the agglomeration of affected people in precarious shelters, conditions that frequently worsen during the slow process of reconstruction and reestablishment of daily life. Data show that the effects of the disaster seemed to persist as the processes of rehabilitation of services, recovery of infrastructure, and reconstruction of the life of the affected communities took place slowly and precariously (Xavier et al. 2014a, b).

4.2 Mountainous Region of Rio de Janeiro State

A similar scenario—catastrophe and slow recovery—was observed in the mountainous region of the state of Rio de Janeiro in 2011, a place covered by the Atlantic Forest. Upon receiving 70% of the precipitation expected for the entire month of January in just two days, the State's Secretariat of Civil Defense recorded floods and landslides that culminated in 918 deaths, 8795 homeless, and 22,604 displaced. More than 300,000 people were affected, which ranked the event among the ten largest landslides in the world since 1900 (CEPED 2015). Regarding health, the post-disaster period between January and March 2011 produced 525 suspected cases of leptospirosis (177 with diagnostic confirmation) and 23 deaths (Pereira et al. 2014b). However, dengue fever was more notorious due to the large number of confirmed cases—between January and December 2011, 937 cases were reported and attributed to the disaster by municipal health surveillance (Pereira et al. 2014a). The landslides caused major changes in the environment of the affected areas such that along with the pre-existing issues of poor sanitation and lack of urban cleanliness, there was an increase in the availability of breeding sites for the vector mosquito.

The magnitude of the event in Rio de Janeiro aroused a discussion about the Brazilian Forest Code and the consequent retraction of forests, mainly the Atlantic

Forest, as a potentiating agent of events such as the landslides that occurred in this mountainous region. The state of Rio de Janeiro had experienced a similar tragedy in February 1996, when a total of 380 mm of rain was recorded in less than 24 h at the Mayrink Chapel Station, located in the Floresta da Tijuca in the city of Rio de Janeiro. Hundreds of landslides occurred, of which 14% were on slopes covered by conserved forest; 42% on slopes under degraded forest; and 43% on slopes with grasses (Oliveira et al. 1996; Coelho Netto 2011). This mountainous region combined environmental (e.g., rugged terrain, abundant water, subsoil composed of rock and a small layer of soil) and social (disordered urban occupation) characteristics with high environmental degradation (i.e., native wood extraction, deforestation, and the use of exotic plant species for commercial purposes), thus making the region vulnerable to the occurrence of landslides and floods (Alves et al. 2010; CEPED 2011; Freitas et al. 2012). For this reason, five of the 11 municipalities affected by the 2011 disaster are still among the 251 municipalities most vulnerable to landslides in the country—Petrópolis, Sumidouro, Nova Friburgo, Teresópolis, and Bom Jardim (Brasil 2011).

This critical situation mobilized federal, state, and municipal governments, as well as organized civil society, to respond immediately to the disaster, such that the broad media coverage of the tragedy favored building a network of solidarity—government, NGOs, churches, and businesses. According to information from the state government, more than 500 actors were involved in the disaster response. However, despite its size, the aid did not achieve the necessary effectiveness. On the one hand, there was no coordination of actions and responses of the various stakeholders to enable effective governance of crisis management, and second, corruption and misappropriation of resources threatened the implementation of reconstruction works, causing serious impacts to the quality of life of the population and to the credibility of government institutions (Busch and Amorim 2011; Silva et al. 2012).

4.3 The Amazon Region

The Northern Region of Brazil has also suffered extremes of drought and flood. Considered to be events of the century for the Brazilian Amazon Region, the years 2005 and 2010 presented severe droughts that affected mainly the state of Amazonas, which possesses abundant water resources. Later, in the years 2009 and 2012, gradual floods occurred that surpassed the historical levels of the overflow of the main rivers (OPAS 2014). According to Marengo et al. (2008), the causes of the drought in the Amazon in 2005 were not related to El Niño, but to some other potential factors such as an anomalously warmer than normal Tropical North Atlantic, and a reduction in the intensity of humidity transported by the northeast trade winds toward the south of the Amazon Region during the peak of the summer season. When the forest is subjected to anomalously dry periods, the probability of the occurrence of forest fires increases, which release smoke and aerosols into the atmosphere, polluting the air over large areas and affecting the health of the population and the beginning of the rainy season (Andreae et al. 2004). The impacts of forest fires on human health are well established, especially with regard to increased hospitalizations (e.g., respiratory infections, asthma, pneumonia), and the particular effects to children and the elderly (Gonçalves et al. 2012; Menezes et al. 2018).

With regard to flooding, the Amazon was the scene of a historic flood in 2012, where several rivers had simultaneous peak flows due to the anticipated and intensive occurrence of the La Niña phenomenon. Pacific sea surface temperatures reached about −2 °C, thus favoring an increase in convergence and transport of moisture, which resulted in prolonged rainfall and huge floods (Espinoza et al. 2013; Pereira and Szlafsztein 2015). In addition to structural and material damage, this flooding modified the natural environment due to increased erosion, thus causing large social impacts. Riverine populations, who are directly dependent on ecosystem services, experience the greatest transformation of daily life due to floods. In general, during such flood events the frequencies of cases of water-borne diseases (leptospirosis, hepatitis A, acute diarrheal diseases), accidents with poisonous animals, and infections with vector-borne diseases (malaria) increase. Such flooding also negatively impacts productive areas, thus interfering with the availability of staple foods and generating economic losses (Pereira and Szlafsztein 2015; Souza and Nascimento 2017). Floods in the Amazon are a peculiar case; on the one hand, a large part of the population has developed its own way of dealing with river flooding (stilts, floating houses, planting of permanently dry areas), and on the other hand, the altimetric quotas have been exceeded year after year, undermining adaptive skills that have been shaped by previous experience of past events.

4.4 The Municipality of Mariana, Minas Gerais State

In addition to impacts associated with natural disasters, there was a national-scale incident that threw light on a subject little explored in Brazil—technological disasters that unfold into major social and environmental disasters. This was the case of an event in the municipality of Mariana in the state of Minas Gerais in 2015. The event not only revealed the institutional weakness of Brazil to legislate issues related to industrial activities and their social and environmental impacts, but also highlighted the complicity of the state in the company's inability to estimate, contain, and mitigate the damage from the rupture of the ore dam of the company Samarco—a joint venture between Brazil's Vale Group and the Anglo-Australian BHP Billiton. The Environmental Impact Study/Environmental Impact Report (EIA/Rima) required by Brazilian environmental legislation, for example, did not project scenarios and possible effects of an event such as the dam rupture, thus leaving fundamental and dangerous gaps, that became apparent in the absence of effective emergency actions (Wanderley et al. 2016).

More than 70 million cubic meters of mud and iron waste escaped, producing a massive flow that buried the municipality of Barra Longa immediately below the tailings dam and caused countless victims—17 bodies were found, at least two were never recovered, and more than 1200 people were forced to leave their homes (Porto

2016; Wanderley et al. 2016). The mud traveled 663 km down the Doce River basin through two Brazilian states (35 cities in Minas Gerais and four in Espírito Santo) until it reached its mouth and the marine ecosystem. About 1.2 million people were left without water and 11 tons of fish died, five species of which were likely driven to extinction. Since a considerable part of the Doce River has been silted up, an increased risk of flooding is expected, as well as changes to flood dynamics in affected urban areas—locations that were not previously occupied by water during floods—may now be affected (Wanderley et al. 2016).

5 Final Remarks

Records of natural disasters, and their social and economic losses, show the need for the Brazilian federal government to promote disaster risk management instead of focusing only on response and reconstruction actions for affected areas. In April 2012, the PNPDEC was established in Brazil, which changed the status of the theme "disaster" by establishing as one of its guidelines "priority for preventive actions related to disaster minimization" (article 4º, III). Programs and plans designed by the federal government have since tried to better organize the sector. Among these efforts, the National Plan for Risk and Disaster Response Management, launched in August 2012, stands out with an emphasis on risk reduction actions focused on four main areas: prevention, mapping, monitoring, and warning and disaster response.

Disasters require inter-disciplinary and inter-sectorial action for their proper management and prevention. Thus, the health sector must operate directly on the socio-environmental determinants of disasters to concretely assist in mitigating damage, a responsibility for which environmental health surveillance is fundamental because it consolidates this multi-sectorial aspect of SUS. Although Brazil has become more aware of the importance of disaster risk management due to the establishment of PNPDEC in 2012, the perspective is still focused on "crisis management" with the main purpose of attenuating the harmful consequences of these events. It is necessary to advance knowledge regarding the populations and places in the country that are most vulnerable to disasters. Furthermore, policies need to work both horizontally and vertically with regard to protection and civil defense in the coming decades, with a focus on reducing the high social and economic costs associated with disasters in Brazil.

References

Adikari, Y., & Yoshitani, J. (2009). *Global trends in water-related disasters: An insight for policy-makers*. Paris: UNESCO. ISBN 978-92-3-104109-9.

Almeida, L. D., & Pascoalino, A. (2009). *Gestão de risco, Desenvolvimento e (meio) Ambiente no Brasil-um estudo de caso sobre os desastres naturais de Santa Catarina, XIII Simpósio Brasileiro de Geografia Física Aplicada* (1 Ed., Universidade Federal de Viçosa, 6 a 10 de julho de 2009), Viçosa.

Alves, H. P. F., Alves, C. D., Pereira, M. N., & Monteiro, A. M. V. (2010). Dinâmicas de urbanização na hiperperiferia da metrópole de São Paulo: análise dos processos de expansão urbana e das situações de vulnerabilidade socioambiental em escala intraurbana. *Revista Brasileira de Estudos Populacionais, 27*(1), 141–159.

Andreae, M. O., Rosenfeld, D., Artaxo, P., Costa, A. A., Frank, G. P., Longo, K. M., et al. (2004). Smoking rain clouds over the Amazon. *Science, 303*(5662), 1337–1342.

Brant, S. (2007). *Assessing vulnerability to drought in Ceará, Northeast Brazil* (pp. 50). School of Natural Resources and Environment, Vol. Master's Thesis University of Michigan, Michigan.

Brasil, Casa Civil. (1989). Lei nº 7.827, de 27 de setembro de 1989. Disponível em http://www2.camara.leg.br/legin/fed/lei/1989/lei-7827-27-setembro-1989-365476-norma-pl.html. Acessado em March 12, 2018.

Brasil, Ministério da Integração Nacional. (2016a). *Histórico da Defesa Civil.* Availabe at http://www.mi.gov.br/historico-sedec.

Brasil, Ministério da Integração Nacional. (2016b). *Política Nacional de Proteção e Defesa Civil–P-NPDEC.* Available at http://www.mi.gov.br/web/guest/defesa-civil/pnpdec.

Brasil, Ministério da Integração Nacional. (2017). Resolução nº 115, de 23 de novembro de 2017. Available at http://sudene.gov.br/images/arquivos/semiarido/arquivos/resolucao115-23112017-delimitacaodosemiarido-DOU.pdf. Acessado em March 12, 2018.

Brasil, Ministério da Saúde. (2012). *Política Nacional de Atenção Básica* (pp. 110). Brasília: Ministério da Saúde.

Brasil, Ministério da Saúde. (2014). *Plano de Resposta às Emergências em Saúde Pública* (pp. 44). Brasília: Ministério da Saúde.

Brasil, Ministério da Saúde, Secretaria de Vigilância em Saúde. (2017). *Guia de preparação e resposta à emergência em saúde pública por inundação* (p. 5). Brasília: Ministério da Saúde.

Brasil, Ministério da Saúde, Secretaria de Vigilância Epidemiológica. (2018, March). Desastres naturais e saúde: análise do cenário de eventos hidrológicos no Brasil e seus potenciais impactos sobre o Sistema Único de Saúde. *Boletim epidemiológico, 49*, 13.

Brasil, Ministério das Minas e Energia–MME. (2011). *Seleção dos municípios críticos a deslizamentos – nota explicativa* (pp. 5). Brasília: Serviço Geológico do Brasil/Secretaria de Geologia, Mineração e Transformação Mineral.

Busch, A., & Amorim, S. N. D. (2011). *A tragédia da região serrana do Rio de Janeiro em 2011: procurando respostas* (pp. 20). Casoteca: Gestão de Crise e Gestão de Riscos, ENAP.

Camarinha, P. I., Debortoli, N. S., & Hirota, M. (2017). *Índice de vulnerabilidade aos desastres naturais relacionados às secas no contexto da mudança do clima: sumário executivo* (p. 36). Brasília, DF: Ministério do Meio Ambiente, Ministério da Integração Nacional, WWF-Brasil, MMA.

Centro de Estudos e Pesquisas em Engenharia Civil e Desastres–CEPED. (2015). *Arquivos da categoria de desastres.* Disponível em: http://www.ceped.ufsc.br/category/grandes-desastres/. Acessado em 3 de março de 2018

CEPED. (2011). *Diretrizes em Redução de Riscos de Desastres Região Serrana do Rio de Janeiro.* Florianópolis: CEPED-UFSC.

Coelho Netto, A. L. C. (2011). A interface florestal-urbana e os desastres naturais relacionados à água no maciço da tijuca: desafios ao planejamento urbano numa perspectiva sócio-ambiental. *Revista do Departamento de Geografia, 16*, 46–60

Confalonieri, U. E., & Fonseca, A. F. Q. (2013). Health security in Amazonia (pp. 63). *Report for Global Canopy Programme and International Center for Tropical Agriculture as part of the Amazonia Security Agenda project.*

Dias, M. A. F. S. (2009). As chuvas de novembro de 2008 em Santa Catarina: um estudo de caso visando a melhoria do monitoramento e da previsão de eventos extremos. In *Nota técnica* (pp. 67p). São José dos Campos: INPE.

Espinoza, J. C., Ronchail, J., Frappart, F., Lavado, W., Santini, W., & Guyot, J. L. (2013). The major floods in the Amazonas River and tributaries (Western Amazon basin) during the 1970–2012 period: A focus on the 2012 flood. *Journal of Hydrometeorology, 14*(3), 1000–1008.

Franca, R. R. D. (2015). *Eventos pluviais extremos na Amazônia meridional: riscos e impactos em Rondônia*. Programa de Pós-Graduação em Geografia, Vol. Doctoral's dissertation Universidade Federal do Paraná, Curitiba, pp. 190.

Freire, A. F. R. (2014). *A Política Nacional de Proteção e Defesa Civil e as Ações do Governo Federal na Gestão de Riscos de Desastres*. Master's Thesis Escola Nacional de Saúde Pública Sergio Arouca, Rio de Janeiro, pp. 137.

Freitas, C. M., Carvalho, M. L. D., Ximenes, E. F., Arraes, E. F., & Gomes, J. O. (2012). Vulnerabilidade socioambiental, redução de riscos de desastres e construção da resiliência: lições do terremoto no Haiti e das chuvas fortes na Região Serrana, Brasil. *Ciência & Saúde Coletiva, 17*, 1577–1586.

Freitas, C. M., & Rocha, V. (2014). Agentes locais em desastres naturais: defesa civil e saúde na redução de riscos – livro do aluno (169p). Rio de Janeiro: Fiocruz.

Freitas, C. M., & Ximenes, E. F. (2012). Enchentes e saúde pública: uma questão na literatura científica recente das causas, consequências e respostas para prevenção e mitigação. *Ciência & Saúde Coletiva, 17*, 1601–1616.

Freitas, C. M. D., Silva, D. R. X., Sena, A. R. M. D., Silva, E. L., Sales, L. B. F., Carvalho, M. L. D., et al. (2014). Desastres naturais e saúde: uma análise da situação do Brasil. *Ciência & Saúde Coletiva, 19*, 3645–3656

Fundação de Apoio à Pesquisa Científica e Tecnológica do Estado de Santa Catarina – FAPESC, Secretaria de Estado de Desenvolvimento Econômico e Sustentável. (2009). *Plano Integrado de Prevenção e Mitigação de Riscos e Desastres Naturais na Bacia Hidrográfica do Rio Itajaí, Florianópolis* (pp. 52).

Gonçalves, K. D. S., Castro, H. A. D., & Hacon, S. D. S. (2012). Ground-clearing fires in the amazon and respiratory disease. *Ciencia & Saude Coletiva, 17*(6), 1523–1532.

Hay, J., & Mimura, N. (2010). The changing nature of extreme weather and climate events: Risks to sustainable development. *Geomatics, Natural Hazards and Risk, 1*(1), 3–18.

Intergovernmental Panel on Climate Change (IPCC). (2012). *Managing the risks of extreme events and disasters to advance climate change adaptation. A Special Report of Working Groups I and II of the Intergovernmental Panel on Climate Change* (pp. 582). Cambridge University Press, Cambridge, UK, and New York, NY, USA.

IPCC. (2013). *Climate Change 2013: The Physical Science Basis, Contribution of Working Group I to the Fifth Assessment Report of the Intergovernmental Panel on Climate Change* (pp. 1535). Cambridge University Press, Cambridge, United Kingdom and New York, NY, USA.

IPCC. (2014). *Climate Change 2014: Synthesis Report. Contribution of Working Groups I, II and III to the Fifth Assessment Report of the Intergovernmental Panel on Climate Change* (pp. 151). Geneva, Switzerland.

Knowledge Center on Public Health and Disasters-Fiocruz. Organização do setor saúde para resposta. Available at: http://andromeda.ensp.fiocruz.br/desastres/content/organizacao-do-setor-saude-para-resposta. Accessed on July 24, 2018.

Ko, A. I., Reis, M. G., Dourado, C. M. R., Johnson, W. D., Jr., & Riley, L. W. (1999). Urban epidemic of severe leptospirosis in Brazil. *The Lancet, 354*(9181), 820–825.

Marandola, Jr., E., Hogan, D. J. (2009). Vulnerabilidade do lugar vs. vulnerabilidade sociodemográfica: implicações metodológicas de uma velha questão. *Revista Brasileira de Estudos de População, 26*(2), 161–181

Marengo, J. A. (2009). Vulnerability, impacts and adaptation (VIA) to climate change in the Semi-Arid region of Brazil. *Brazil and climate change: Vulnerability, impacts and adaptation* (pp. 137–164). Brasilia: Center for Strategic Studies and Management.

Marengo, J. A., Ambrizzi, T., Da Rocha, R. P., Alves, L. M., Cuadra, S. V., Valverde, M. C., et al. (2010). Future change of climate in South America in the late twenty-first century: Intercomparison of scenarios from three regional climate models. *Climate Dynamics, 35*(6), 1073–1097.

Marengo, J. A., Borma, L., Rodriguez, D. A., Pinho, P., Soares, W. R., & Alves, L. M. (2013). Recent extremes of drought and flooding in Amazonia: Vulnerabilities and human adaptation. *American Journal of Climate Change, 2*(02), 87.

Marengo, J. A., Nobre, C. A., Tomasella, J., Oyama, M. D., Sampaio de Oliveira, G., Oliveira, R., et al. (2008). The drought of Amazonia in 2005. *Journal of Climate, 21*(3), 495–516.

Menezes, J. A., Margonari, C., Santos, R. B., & Confalonieri, U. (2018). *Air pollution, climate change, and human health in Brazil, in climate change and air pollution* (pp. 375–403). Cham: Springer.

Narváez, L., Pérez Ortega, G., & Lavell, A. (2009). *La gestión del riesgo de desastres. Un enfoque basado en processos* (1 Ed., pp. 106). Limas: Secretaría General de la Comunidad Andina.

Oliveira, R. R., Avelar, A. S., Oliveira, C. A., Rocha Leão, O. M., Freitas, M. M., Coelho Netto, A. L. (1996). *Degradação florestal e deslizamentos ocorridos em Fevereiro 1996 no maciço da Tijuca, RJ* (pp. 353). Nova Friburgo, Anais do XLVII Congresso Nacional de Botânica.

Organización Panamericana de la Salud–OPS. (2011). *Hospitales seguros y riesgo urbano en Centroamérica: ciudades más seguras con un sistema de salud mejor preparado para responder a desastres* (p. 24). Panamá: OPS/OMS

Organização Pan-Americana da Saúde–OPAS. (2014). *Desastres Naturais e Saúde no Brasil* (pp. 49). Série Desenvolvimento Sustentável e Saúde. Brasília, DF: OPAS, Ministério da Saúde.

Organización Panamericana de la Salud–OPS. (2004). *Manual de evaluación de daños y necesidades en salud para situaciones de desastre* (pp. 193). Ecuador: OPS.

Pereira, C. A. R., & Barata, M. M. D. L. (2014). Organização dos serviços urbanos de saúde frente à mudança do clima e ao risco de desastres na América Latina. *Saúde em Debate, 38,* 624–634.

Pereira, C. A. R., Barata, M. M. L., Hoelz, M. P. C., Medeiros, V. N. L. O., Vommaro, F. C. M., Costa Neto, C., et al. (2014a). Avaliação Econômica dos Casos de Dengue Atribuídos do Desastre de 2011 em Nova Friburgo (RJ/Brasil). *Ciência e Saúde Coletiva, 19*(9), 3693–3704.

Pereira, C., Barata, M., & Trigo, A. (2014b). Social cost of leptospirosis cases attributed to the 2011 disaster striking Nova Friburgo. *Brazil, in International journal of environmental research and public health, 11*(4), 4140–4157.

Pereira, D. M., & Szlafsztein, C. F. (2015). Natural hazards and disasters in South Western Amazon Region: Analysis of Purus River Basin. *Raega Espaço Geográfico em Análise, 35,* 68–94.

Porto, M. F. D. S. (2016). The tragedy of mining and development in Brazil: public health challenge. *Cadernos de Saúde Pública, 32*(2).

Pourhosseini, S. S., Ardalan, A., & Mehrolhassani, M. H. (2015). Key aspects of providing healthcare services in disaster response stage. *Iranian journal of public health, 44*(1), 111.

Salles, M. J. P., & Cavalini, L. T. (2012). Implementação da "estratégia de hospitais seguros frente a desastres" no Brasil: desafios e evidências empíricas. *Hygeia, 14*(8), 81–90.

Sena, A., Barcellos, C., Freitas, C., & Corvalan, C. (2014). Managing the health impacts of drought in Brazil. *International journal of environmental research and public health, 11*(10), 10737–10751.

Sena, A. R. M., Silva, E. L., & Corvalan, C. (2015). *Atuação do setor saúde frente a situações de seca* (p. 52). Brasília: OPAS, Ministério da Saúde.

Shoaf, K. (2014). Organizing the health sector for response to disasters. *Ciência & Saúde Coletiva, 19*(9), 3705–3715.

Silva, E. A. B., Suassuna, C. C. A., Furtado, M. F. R. G., & Bezerra, O. G. (2012). *Resiliência e vulnerabilidade de cidades brasileiras: lições aprendidas com os desastres da região serrana do Rio de Janeiro e da Zona da Mata de Pernambuco*, VI Encontro Nacional da ANPPAS, Belém, Pará, Brasil, 18 a 21 de setembro, pp. 22

Souza, R. F., & Nascimento, S. L. (2017). Doenças e Agravos no Contexto das Grandes Inundações Graduais no Estado do Amazonas-Brasil. *Hygeia: Revista Brasileira de Geografia Médica e da Saúde, 13*(26), 139.

Stanke, C., Kerac, M., Prudhomme, C., Medlock, J., Murray, V. (2013). Health effects of drought: A systematic review of the evidence. *PLoS currents, 5.*

Thouret, J. C. (2007). Avaliação, prevenção e gestão dos riscos naturais nas cidades da América Latina. In Y. Veyret (Ed.), *Os riscos: o homem como agressor e vítima do meio ambiente* (pp. 83–113). São Paulo: Contexto.

Tucci, C. E. (2007). Mudanças climáticas e impactos sobre os recursos hídricos no Brasil. *Ciência & Ambiente, 34,* 137–156.

Universidade Federal de Santa Catarina – UFSC/ Centro Universitário de Estudos e Pesquisas sobre Desastres - CEPED. (2013). *Atlas Brasileiro de desastres naturais 1991–2012* (2 Ed., pp. 126). Florianópolis: CEPED, UFSC.

Universidade Federal de Santa Catarina – UFSC/ Centro Universitário de Estudos e Pesquisas sobre Desastres - CEPED. (2016). *Relatório de danos materiais e prejuízos decorrentes de desastres naturais no Brasil: 1995–2014* (pp. 230). Florianópolis: CEPED UFSC.

Vázquez, E. L. (2001). Risk perception interactions in stress and coping facing extreme risks. *Environmental Management and Health, 12*(2), 122–133.

Wanderley, L. J., Mansur, M. S., Milanez, B., & Pinto, R. G. (2016). Desastre da Samarco/Vale/BHP no Vale do Rio Doce: aspectos econômicos, políticos e sócio-ambientais. *Ciência e Cultura, 68*(3), 30–35.

Xavier, D. R., Barcellos, C., Barros, H. D. S., Magalhães, M. D. A. F. M., Matos, V. P. D., & Pedroso, M. D. M. (2014a). Organização, disponibilização e possibilidades de análise de dados sobre desastres de origem climática e seus impactos sobre a saúde no Brasil. *Ciência & Saúde Coletiva, 19*(9), 3657–3668.

Xavier, R. D., Barcellos, C., & Freitas, C. M. (2014b). Eventos climáticos extremos e consequências sobre a saúde: o desastre de 2008 em Santa Catarina segundo diferentes fontes de informação. *Ambiente & Sociedade, 17*(4), 273–294.

Extreme Events in Mexico: Impacts on Public Health and Development

María E. Ibarrarán, Jerónimo Chavarría and Carmen Zúñiga

Abstract Scientific literature has established that climate change increases the frequency and intensity of extreme events. Extreme events, in turn, may cause disasters and impacts on health. This chapter presents an overview of the effects climate change is expected to have in Mexico through extreme events. In a nutshell, EE such as drought, forest fires, hurricanes and frosts have increased over recent years and further intensification is yet to be expected. Furthermore, these events have impacts, some health-wise, some economic, and others in terms of future economic performance. In the case of Mexico, many of these impacts have not been systematically measured. The goal of this paper is therefore to establish the links between extreme events in Mexico and the health impacts and other costs imposed, particularly those related to the loss of infrastructure and economic performance, and their impact on regional development.

Keywords Extreme events · Mexico · Health impacts · Vector-borne disease · Economic performance · Regional development

1 Introduction

Scientific literature has established that climate change increases the frequency and intensity of extreme events (IPCC 2014; Santiago et al. 2008; Aguirre et al. 2008; Trenberth 2001). Extreme events (EE) in turn may cause disasters and impacts on health. This chapter presents an overview of the effects climate change is expected to have in Mexico through extreme events. In a nutshell, EE such as drought, forest fires, hurricanes, and frosts have increased over recent years and further intensification is yet to be expected. Furthermore, these events have impacts, some health-wise, some economic, and others in terms of future economic performance. In the case of Mexico, many of these impacts have not been systematically measured. The goal of

M. E. Ibarrarán (✉) · J. Chavarría · C. Zúñiga
Instituto de Investigaciones en Medio Ambiente Xabier Gorostiaga SJ, Universidad Iberoamericana Puebla, Blvd. del Niño Poblano 2901, Reserva Territorial Atlixcáyotl, 72820 San Andrés Cholula, Puebla, Mexico
e-mail: mariaeugenia.ibarraran@iberopuebla.mx

R. Akhtar (ed.), *Extreme Weather Events and Human Health*,
https://doi.org/10.1007/978-3-030-23773-8_22

this paper is therefore to establish the links between extreme events in Mexico and the health impacts and other costs imposed, particularly those related to the loss of infrastructure and economic performance. This is a descriptive piece, so it illustrates how these issues are connected through the use of available data, mainly government reports and academic papers. The aim is to highlight the effects climate change is causing in Mexico and the need for additional policies to identify and contain impacts on health and regional development.

This chapter is divided into five sections. Section 2 discusses extreme events in Mexico, particularly drought, hurricanes, forest fires, and frost. It describes how they have evolved over time, where they occur, their frequency, intensity, and effects. Section 3 presents the main health impacts associated with extreme events and their evolution, as well as the policies to address such effects and their impacts on prevalence. Section 4 gathers some information on the costs that have been reported due to extreme events in Mexico and the impact that EE have on economic growth and regional development. The last section presents some conclusions and policy implications.

2 Intensity and Frequency of Extreme Weather Events in Mexico

Even though EE are expected to increase worldwide, a few long-run and broad studies have been found for Mexico (Álvarez and Martínez 2015). Between 1980 and 2005, 52 extreme events were reported, but over 46% happened in the last six years of that period (Santiago et al. 2008; Zúñiga 2007).

Based on the Emergency Events Database (EM-DAT) of the Centre for Research on the Epidemiology of Disaster (CRED), between 1900 and 2018, 50% of the disasters caused by extreme events were related to tropical storms, 15% due to riverine floods, 8% to flash floods, and 7% to cold waves. This can be seen in Fig. 1.

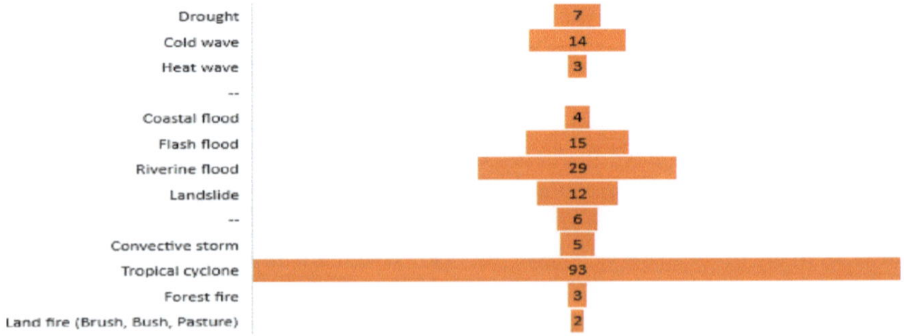

Fig. 1 Extreme events in Mexico (1900–2018). *Source* EM-DAT (2018)

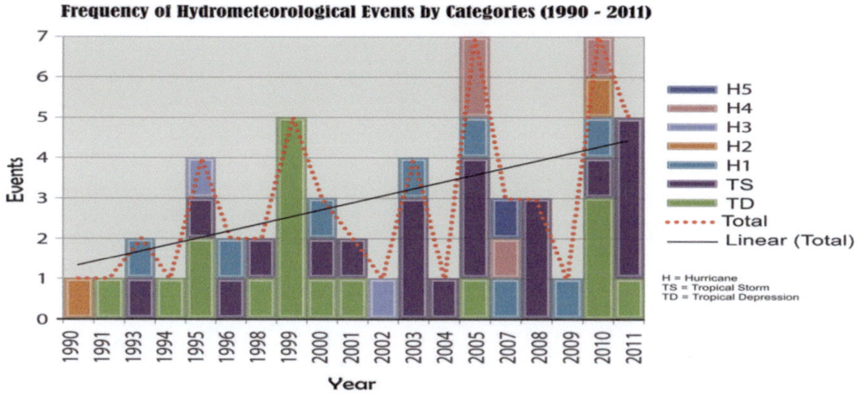

Fig. 2 Frequency of hydro-meteorological events by category. *Source* SMN (2013)

In the remaining part of this section, we look at the evolution of hurricanes, droughts, forest fires, and frost, and the type of effects they produce.

2.1 *Hurricanes*

The number of hurricanes has increased in Mexico since 1950. The damage they cause has also been on the rise (Ochoa et al. 2010; Blake et al. 2009; Jauregui 2003). They hit on both the Pacific and Gulf Coast. Figure 2 shows how the frequency and intensity, i.e., category, of hurricanes has increased in recent years (Baltazar 2017).

Finally, Fig. 3 shows the states that have suffered the most damage due to tropical storms during the 1970–2014 period (RioDoce 2015). Thereafter, tropical storms have become ever more frequent, causing severe flooding (Baltazar 2017; Ochoa et al. 2013).

2.2 *Drought*

As with other extreme events, IPCC models predict more intense and longer drought periods in northwestern Mexico. Forecasts show an increase in average temperature anywhere from 0.5 to 4.8 °C in the 2020–2100 period. Rainfall could decrease up to 15% in Winter and 5% in Summer, and the dry period could extend until the Fall (Sosa 2015). In any case, the twentieth century was plagued with drought, well into the 1990s. Figure 4 shows the states of the north that were the most affected during different periods (IMTA 2016).

Drought is the EE that causes more economic and social damage in Mexico, since it fosters migration and significant changes to ecosystems, such as desertification. As shown in Fig. 5, the frequency and intensity of drought has increased since 1970;

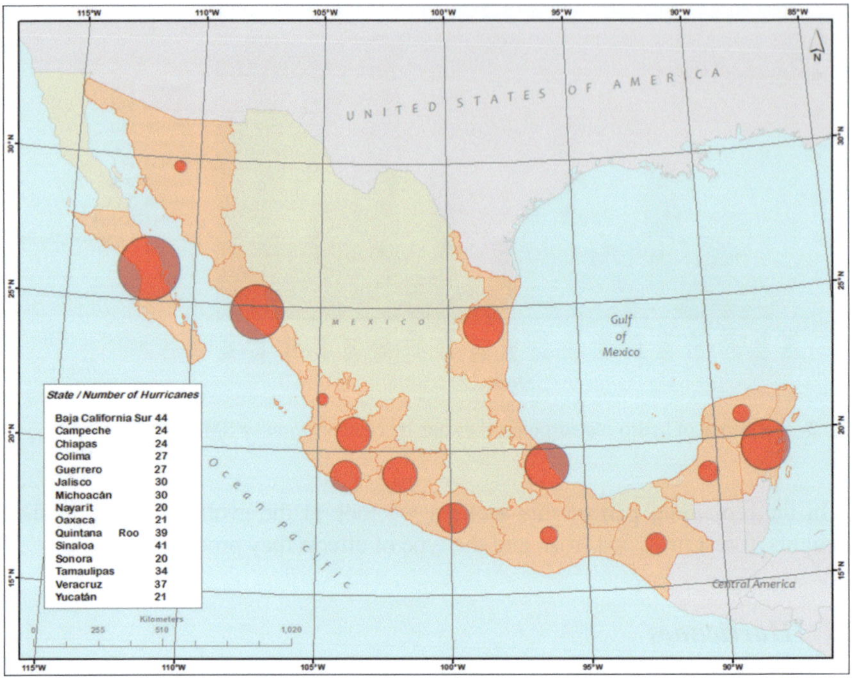

Fig. 3 Locations with higher impact of tropical storms, 1970–2014. *Source* RioDoce (2015)

2003, 2006, and 2011 were among the warmest in the last 70 years (IMTA 2016; Martínez 2016).

Recent drought spells are expected to change the distribution and abundance of native species, leading to the expansion of invasive species that can survive under water stress. This is the case in the Sonoran Desert where highly flammable grass has expanded, increasing the risk of forest fires (Álvarez and Martínez 2015). This will also affect agriculture and livestock, local food production, and increase pollution of water bodies, as well as lower water availability in an already stressed area, possibly leading to social conflict (Garcidueñas 2015).

2.3 Forest Fires

Climate change leads to conditions that favor forest fires and increase their destructive capacity. Forest fires are highly related to high temperature, heat waves, and an increase in desertification, that produce large areas full of dry vegetation that fuel fires (Flannigan et al. 2009). In Mexico forest fires have contributed to the loss of natural resources, human lives, and economic losses. Between 1970 and 2006 there have been about 7000 forest fires, affecting more than 222,000 ha. About 98% of

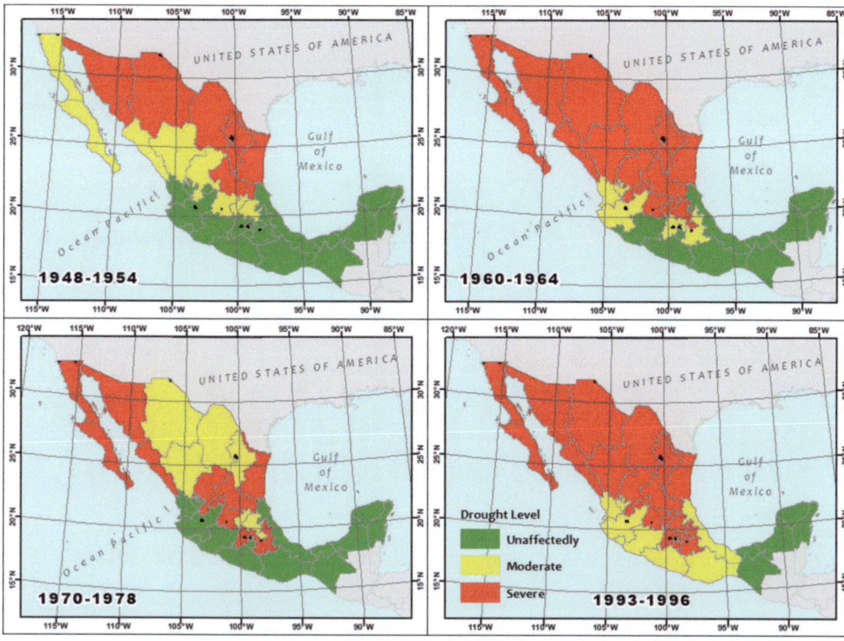

Fig. 4 Occurrence of drought in Mexico during different time periods. *Source* IMTA (2016)

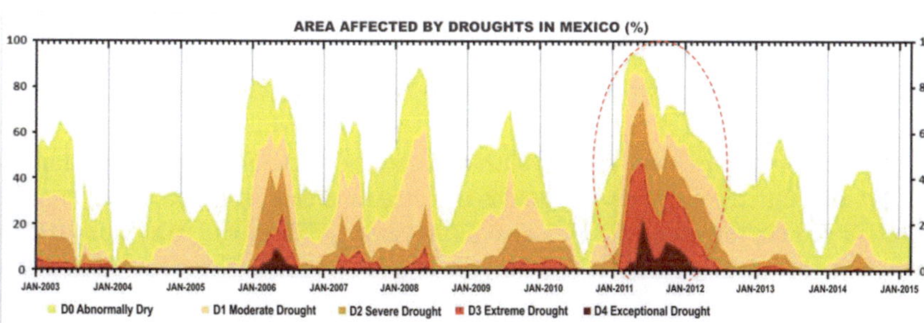

Fig. 5 Share of land affected by drought in Mexico, 2003–2015. *Source* IMTA (2016)

the fires are caused by human activities, such as slash-and-burn agriculture, that contribute with 41% of forest fires, 13% by smokers, and 11% by campfires; 13% are due to unknown causes (Ressl and Cruz 2012).

Even though forest fires are mostly caused by people, extreme climate conditions increase the frequency and intensity of fires (Villers and Hernández 2002). As Fig. 6 shows, an average of 42,000 additional hectares is lost each year to forest fires, increasing deforestation. The area burnt went from 158,000 to 280,000 ha/year from 1970 to 2013. The states with a larger number of forest fires are usually Oaxaca, Tlaxcala, Puebla, Mexico City, Morelos, Coahuila, Hidalgo, Jalisco, and Chiapas.

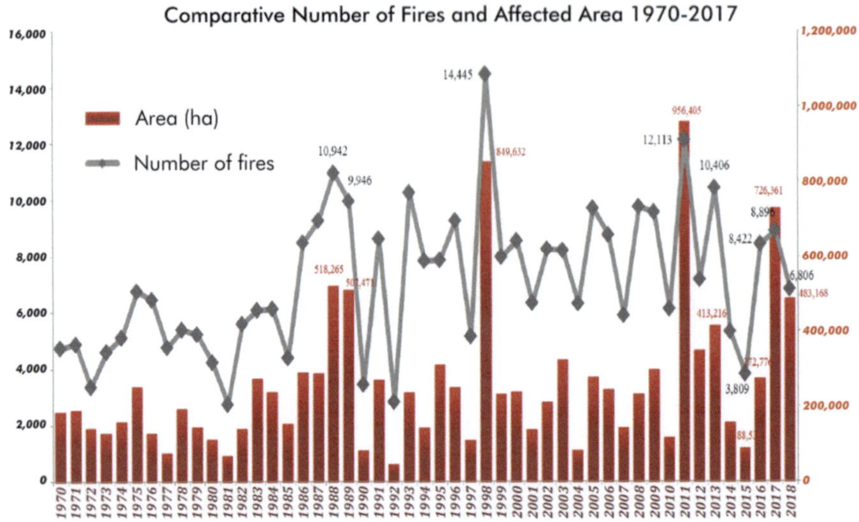

Fig. 6 Number of forest fires and affected area, 1970–2018. *Source* CONAFOR (2018)

2.4 *Frost*

Another extreme event associated with climate change is frost, that is related with cold fronts and can cause large economic losses and health impacts (Synder and De Melo-Abreu 2010). Among the most important of these events was one registered in 2011, affecting the northeastern states of Sonora, Sinaloa, and Chihuahua. In Sonora, 71 out of its 72 counties were affected by frost and 59 were declared under emergency. Agriculture was severely hurt and had to receive State compensation for their losses. Native ecosystems were also affected, and thus, natural carbon capture in second growth forests of the area also diminished (Álvarez and Martínez 2015).

Overall, this section has reviewed the type of extreme events that occur in Mexico, their location, and their trend. Section 3 will now look at the health impacts that such events cause and who gets affected.

3 Impact of Extreme Events on Health in Mexico

The fact that climate change alters environmental conditions has led to a host of health effects on the population. Studies show that morbidity and mortality are expected to increase because of higher temperatures that in turn increase ozone levels and other pollutants that affect air quality. Additionally, increasing concentrations of CO_2 enhance plant growth and the liberation of allergens related to asthma and other cardiovascular and respiratory disease (Fann et al. 2016; WHO 2014; Ize 2002). On the other hand, the geographical distribution of water-borne disease may change due to changes in climatic conditions as well (Martine et al. 2004; Beniston 2002). Floods related to hydro-meteorological events pollute fresh water increasing the risk of water-borne illness and allowing breeding grounds for vector-borne maladies (Balbus et al. 2016). The World Health Organization (WHO) estimates that for 2000, 2.4% of the total cases of diarrhea can be attributed to climate change and 6% of malaria in some middle-income countries (WHO 2002, 2003). In Mexico, specific human, social, and environmental conditions faced by different population groups such as children or the elder increases the severity of the effects they suffer (Berberian and Rosanova 2012; Haines and Patz 2004; Magaña and Gay 2002).

This section describes some of the effects of EE on health in Mexico in recent years. The first part is a brief overview of mortality and EE, and the second presents in detail the effect of EE on health in Mexico, focusing on vector-borne disease, heat strokes, and water- and air-borne disease.

3.1 Mortality

Statistical information on health outcomes associated with extreme events is quite limited in Mexico, and often figures are only available for the past 6 years. During the last seventy years, the number of EE-related disasters increased, predominantly since the eighties, as mentioned in Sect. 2. However, as shown in Fig. 7, the death toll due to storms and extreme temperatures was high from the 1950s through the 1970s, but then fell toward the year 2000, except for some years (EM-DAT 2018). Thus, specific policies that will be reviewed below seem to have allowed decoupling ocurrance of EE and mortality. This is an important finding because the fact that the number of deaths is significantly lower implies that Mexico has been improving its coping capacity to extreme events.

3.2 Morbidity

Climate change acts as an enabler of certain diseases that result from altering environmental conditions. Changing temperatures and precipitation patterns have expanded the breeding grounds for vector-borne diseases that are spread by mosquitoes, tics, and flies. These insects feed on blood that may be infected and transmit the disease to other organisms, propagating virus and bacteria. Water-borne diseases are also related to higher temperature and changes in water availability that allow for water pollution and thus diarrhea and similar ailments. Some other health impacts are due to air pollution, given that climate change produces, among other effects, atmospheric stability and thus poorer air quality (Ibarrarán et al. 2017). Finally, heatwaves themselves provoke heat strokes that have a significant impact on the elderly mainly.

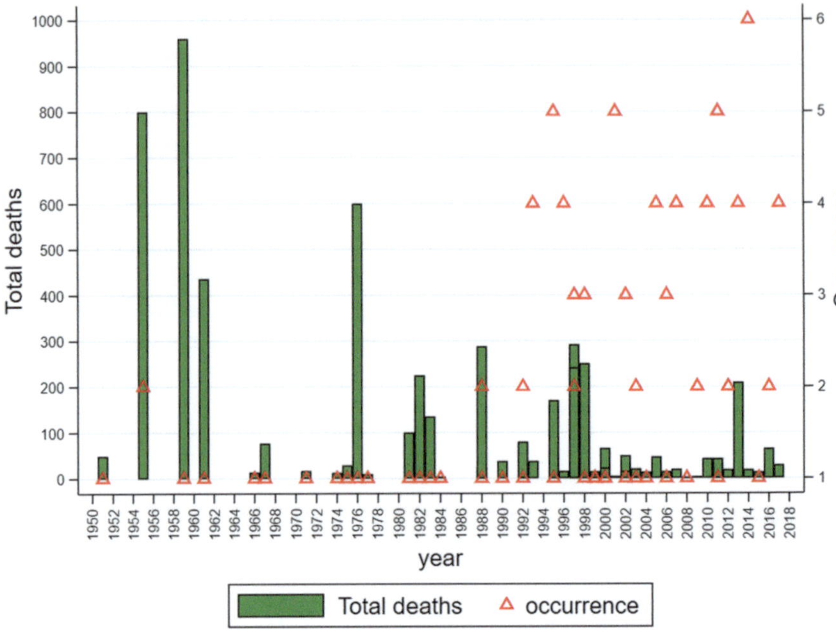

Fig. 7 Occurrence of EE (storms and extreme temperatures) and death toll, 1950–2017. *Source* Own elaboration based on EM-DAT (2018)

In this section we contrast the occurrence of health impacts with average temperature and precipitation patterns in Mexico. Figure 8 shows how these parameters have behaved in the past century and present an average for that period.

3.2.1 Vector-Borne Disease

The most common vector-borne ailments in Mexico are skin leishmaniasis, acute and chronic American Trypanosomiasis (Chagas disease), infection by zika virus, sickness produced by the Chikungunya virus, and dengue fever. Some statistics on their occurrence are presented below.

Chikungunya

This disease was found in 1952 in Tanzania, but has expanded to Africa, Asia, and the Americas, including Mexico (Secretaría de Salud 2014). It is transmitted by *Aedes aegypi*, the same mosquito that spreads dengue, but this ailment is usually not fatal even though it produces pain and rigidity in articulations and can last from months to years (PAHO 2014). Chikungunya was first diagnosed in Mexico in 2014, with 222 new cases. However, in 2015 there were 12,588 additional cases. In 2017 though, only 64 cases were reported. The months when this illness increases are also highly variable: In 2015 it was in July and August, that have relatively higher temperatures

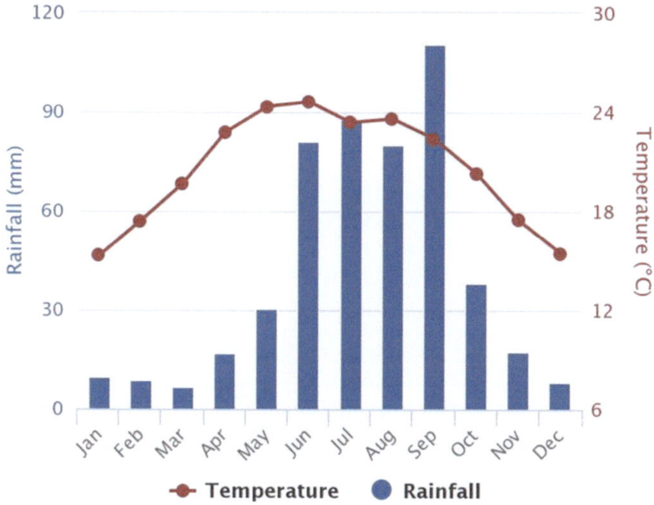

Fig. 8 Average monthly temperature and rainfall for Mexico from 1901–2015. *Source* Climate Change Knowledge Portal

and rainfall as shown in Fig. 7. However, in 2016 it was diagnosed mostly in January, and in 2017 in October and November, periods of relatively low temperatures and rainfall (DGIS 2017).

Dengue fever

This is among the most prevalent vector disease in Mexico. In 2015, about 61,700 new cases were reported. Most cases took place in August and October, the former being the warmest month in the year and the latter with enough temperature to ensure survival of the virus (DGIS 2017). On the other hand, 41% of these cases took place in 4 states that faced tropical storms and relatively higher temperatures of around 30 °C, namely Colima, Guerrero, Jalisco, and Nayarit, while 37% happened in 19 states that suffered frost and snow with temperatures around 2 and 6 °C (DGIS 2017). Distribution also varied, as can be seen in Fig. 9, that shows number of cases and concentration across Mexico for the 2012–2016 period.

3.2.2 Heat Strokes

Heat waves result from an increase in minimum temperatures, such that temperatures remain high even through the night. In Mexico heatwaves have become more intense and have affected particularly the northwest, but also central areas such as Mexico City. In 2017, there were 3820 new cases of heat strokes reported. The states with most cases were Sonora and Baja California, with 20 and 16%, respectively. The average temperature of these states was over 40 °C (DGIS 2017). No previous data on heat strokes were reported prior to 2017.

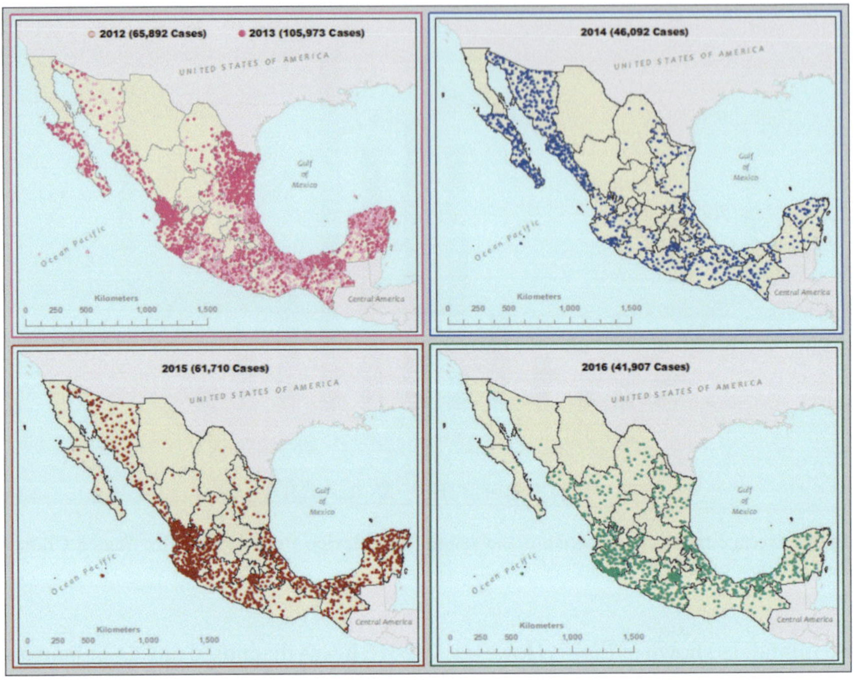

Fig. 9 Cases of dengue fever reported in Mexico. *Source* Own based on DGIS (2018)

3.2.3 Water- and Air-Borne Morbidity

Even though sickness due to polluted water and to low air quality is expected to increase with climate change, no hard evidence has been found of changes in water- or air-borne diseases flowing an extreme event. Mexico does not have a systematic reporting of these diseases in such a way as to establish a clear relation between extreme events and health outcomes in any of these two fields. However, due to policies directed at reducing these diseases, there has been a significant reduction of water-borne cases particularly.

3.3 Epidemiological Surveillance in Mexico

Overall, the health sector has had significant advances in controlling disease and promoting hygienic practices across the board (SEMARNAT 2017). On the other hand, the incidence of vector disease has fallen due to epidemiological surveillance from the Pan American Health Organization (PAHO) and the WHO that send epidemiological alerts to health sector in all countries when they detect sanitary risks to reduce the density of such vectors (SEMARNAT 2017).

Following international protocols, Mexico maintains a weekly epidemiological vigilance. Following the Integrated Management Strategy, EGI-Dengue, the Americas have strengthened the institutional response by a multi- and intersectoral approach through the dengue working group, GT-Dengue.

In Mexico in particular, the National Center for Prevention and Control of Disease (CENAPRECE), within the Ministry of Health, is in charge of leading and implementing crucial programs to reduce morbidity and mortality within the Mexican population. CENAPRECE has, among its prevention programs, one specially aimed at controlling vector-borne disease. Research related to this program has estimated that 60% of the Mexican territory presents favorable conditions for the appearance of these ailments, that means that about 50 million people are at risk, and this area is where most agrarian, industrial, fishing, oil-producing, and touristic centers are located (CENAPRECE 2018).

In summary, even though the evidence of the effects of EE on health is not systematically reported, this section, and the international literature for that matter, has shown that EE increase health outcomes, but that epidemiological surveillance has played a significant role containing these effects. On the other hand, some other impacts such as water-borne disease have fallen significantly due to sanitation. Evidence does not show the trends' change with EE.

The next section goes a step further and looks at the costs associated with EE. It reports what has been found related to health and other impacts that EE may have on development.

4 The Effect of EE on Development

As has been discussed in the previous sections, extreme events foster vector-borne disease and heat strokes. Eventually, both EE and health outcomes associated with such EE may eventually lead to fatalities. Both morbidity and mortality often affect the most vulnerable (Ibarrarán et al. 2010). Another set of impacts of EE is death, due to drowning and lack of access to food and water, as well as the destruction of assets. Disease and fatalities may have a direct impact on macroeconomic performance as well as on local and regional development.

4.1 Health Setbacks

The occurrence of an EE may have two sets of costs. The first one can be estimated through the out-of-pocket expenditures to face disease, or the resources public health has to invest to treat those affected. The other is more of a dynamic effect through time.

On the one hand, EE can have a direct impact on health, by causing short-term disease due to the presence of vector-borne disease, water pollution and scarcity, or traumatism because of destruction. Most of this information can be found only in a case-by-case basis, but it has seldom been collected and systematized so as to estimate an aggregate cost of extreme events on health. Some results may be found in Moreno (2006) that reports the occurrence of disease in Latin America related to EEs, but monetary costs are not provided for the different events. For Mexico, there is only anecdotal evidence that does not stand a rigorous analysis.

In the medium run, EE may have another set of effects on health and on economic growth. These long-term effects can play out as a setback in initial health conditions because of a disruptive effect on infrastructure or on public health that may be persistent in time. Mayer-Foulkes (2001) finds a Granger causality from health to economic growth for Mexico in the 1950–1995 period. In general studies find that an improvement in health conditions lead to higher economic growth. This was also found for Latin America (Mayer 2001; Mayer et al. 2001). For Brazil, health is related to the dynamics of income, education, school attendance, women's participation in the labor market, wages, and net fertility (Mayer-Foulkes 2000). Gyimah-Brempong and Wilson (2004) also find that the growth rate of per capital income has a positive and robust relation with health. Following this line of thought, impoverished health may lead to lower individual income and to negative macroeconomic effects such as lower growth. However, no research has been found on this topic for Mexico.

4.2 The Monetary Cost of Disasters

Another way of addressing the cost of EE is by looking at the economic losses that they impose. These include the economic losses caused by damage to transport and hydraulic infrastructure, agricultural areas, and housing; it also includes the expenditures needed to assist in emergencies (CENAPRED 2018). Results for Latin America show that economic losses have increased due to disasters. By and large, most of these costs in Latin America concentrate in South America. However, Mexico suffers about 25% of the overall cost of losses to the region (CEPAL 2011). Figure 10 shows the cost of disasters in the 2000–2015 period for Mexico.

The Ministry of the Interior has stated that in Mexico the hydro-meteorological factor generates 9 of every 10 disasters in the country. The cost of disasters has varied during the 2000–2016. The worst year was 2010, because of three hurricanes (CENAPRED 2016).

In 2015, 96.2% of damages and losses corresponded, in the first instance, to hydro-meteorological phenomena, followed by geological with 1.4%, and 1.4% of chemical origin. For its part, the sanitary-related damages (0.02%) and those due to social disruption (0.97%) as a whole reached 1% of the economic losses. Regarding hydro-meteorological phenomena, 44% of the impact, that is damage and losses, were caused by rain, while 28.4% by tropical cyclones, and 18.4% by floods (CENAPRED 2016).

Some results produced by CENAPRED are worth mentioning. For example, 56% of damage and losses between 2010 and 2014 have been on highway infrastructure,

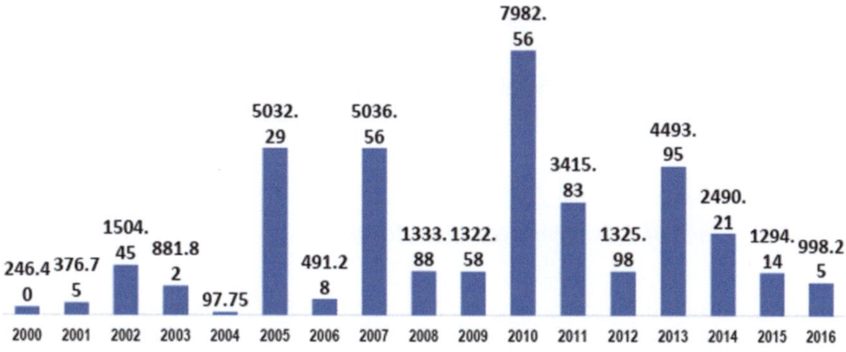

Fig. 10 Annual economic impact of disasters for the 2000–2015 period (Constant million dollars). *Source* CENAPRED (2016)

affecting trade and communication. On the other hand, 40% of the municipalities declared as affected by a disaster during that same period have a high degree of marginalization, and these EE increase their vulnerability as well as decrease their coping capacity in the future. However, the death toll in those four years went down significantly as a result of Early Warning Systems being installed in locations prone to tropical cyclones since 2000 (Constantino et al. 2011; Bitrán 2001).

To make the costs comparable throughout time, costs are also given as a share of GDP. Thus, for the case of Mexico, EE represent about 0.25 of a percentage point of GDP. The peak of 1985 corresponds to the earthquake that shook Mexico City, that is not an extreme weather event even though it did produce havoc. This is shown in Fig. 11.

Mexico is a large country, so most EE do not impose a systemic threat to the entire country. Nevertheless, they may have had a perceivable cost as a share of GDP. However, as Toya and Skidmore (2007) have found, losses from disasters are

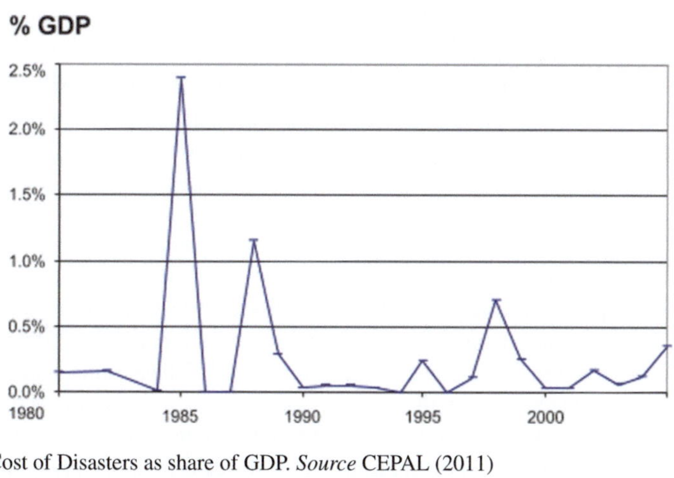

Fig. 11 Cost of Disasters as share of GDP. *Source* CEPAL (2011)

lower for countries with higher income and schooling, with an open economy, and with a more complete financial system and a smaller government. Mexico has some of these characteristics, and therefore, it faces a lower toll from EE. However, EE may still have a toll on future development, in particular regionally.

4.3 The Cost of Disasters on Development

In terms of economic growth, one may think that EE may generate a pattern of economic contraction, and as a result of the costs imposed by destruction, GDP would fall. However, due to reconstruction efforts, there might be an increase in spending, both government and private, and this brings back GDP growth. At this point growth will continue its trend, and in terms of macroeconomic indicators, the effect of the EE may basically disappear, especially for a country like Mexico where most disasters do not have systemic effects; that is, they do not impair the performance of the entire country, financially or infrastructure-wise.

None of this evidence, however, even if it makes intuitive sense has not been found. Cavallo et al. (2013) find that "only extremely large disasters have a negative effect on output, both in the short and long run… where radical political revolutions followed the natural disasters. Once these political changes are controlled for, even extremely large disasters do not display any significant effect on economic growth. It is also found that smaller, but still very large natural disasters, have no discernible effect on output."

Evidence is mixed and still under discussion. Loayza et al. (2012) find that disasters affect economic growth depending on the type of disaster and economic sectors it hits. Moderate disasters may even have a positive effect on some sectors, while severe disasters usually affect growth negatively. Finally, the less developed a country (or a region), the more sectors affected and the more meaningful the economic effects. Macroeconomic effects may be negative but small (Hochrainer 2009) and can be reversed by aid and the inflow of remittances.

All these arguments seem to fit with what has happened in Mexico. However, data was not enough to perform a statistically significant correlation analysis showing the impact of EE on future GDP growth, either at the national or at the regional levels.

Findings by Hochrainer (2009) can be used to analyze what happens with the poorest regions in Mexico. With particular EE, these different regions are expected to face diverse types of severe disruption, based on the resources they possess. For instance, Esquivel (2000) shows there is a link between geographical conditions, defined as climate and vegetation, and economic development for different Mexican states and regions. Geography is strongly correlated with income levels, life expectancy and education, and therefore economic growth and regional inequalities. States that are more prone to EE are in fact the ones with lower growth. These same states are often the ones that rely extensively on the primary sector for their economic activities and livelihood. Thus, when one of these states gets hurt, chances of a setback in future growth increases, becoming a poverty trap. This is the case for some areas and sectors of Chiapas, Oaxaca, Guerrero, Puebla, and Veracruz. Insurance to disasters is not universal in Mexico, let alone in the often poor affected areas, that additionally are highly dependent on natural resources and with a large informal sector that lives off primary activities. All of this makes them prone to suffer the most when EE hit.

Finally, EE may also impose a setback on the quality of life. This may be due to the fact that expenditure is diverted from high impact social expenditure. Even if a compensation is given to those affected, their lifestyle may be permanently hurt because of destruction. Moreover, often natural capital is destroyed and this is not reflected in an orthodox national accounting system, so GDP may actually rise showing economic growth even though natural capital is severely affected and a green version of GDP would fall. This is particularly true for the states that depend heavily on primary activities and their natural resources, and have a relatively marginalized population, such as the states in southern Mexico. Eventually extreme events may lead to migration (Runfola et al. 2016) and other impacts that are difficult to quantify.

5 Conclusions

Extreme events show an upward trend due to climate change, both worldwide and in Mexico. Both climate change in general and extreme events in particular lead to significant modifications in the natural environment that in turn may bring health impacts, in the form of morbidity and mortality. Even though there is enough evidence to state that EE are on the rise in Mexico, lack of systematic reporting of illness following EE prevents from obtaining strong results of the relation between disasters and health outcomes. On the other hand, costs of health impacts are not available either.

Another venue that was explored is the relationship between EE and development. However, data were not enough to perform a statistically significant correlation analysis showing the impact of EE on future GDP growth, either at the national or at the regional levels.

Some policy implications do arise. First, there is a need for reporting health outcomes related to EE. This is particularly needed to gain a better understanding of the broad effect of EE in the country. Second, costs have to be better estimated so that a broad valuation of the cost of EE in Mexico is readily available. This information may allow for better policy design both for prevention and for adaptation to climate change and its effects.

References

Aguirre, J., Brunet, M., & Jones, P. (2008). *Cambio climático regional y sus impactos. Cambios observados en los extremos climáticos de temperatura y precipitación en el estado de Veracruz,*

México a partir de datos diarios (pp. 447–456). Tarragona: Asociación Española de Climatología. Obtenido de Arcimís: https://repositorio.aemet.es/handle/20.500.11765/8657.

Álvarez, J., & Martínez, A. (2015). *Huracanes, sequías y heladas: eventos climáticos extremos en México*. Obtenido de http://web.ecologia.unam.mx/oikos3.0/, http://web.ecologia.unam.mx/oikos3.0/index.php/articulos/fracturacion-hidraulica/8-articulos/165-eventos-climaticos-extremos-y-sus-impactos-en-los-socio-ecosistemas-de-mexico12.

Balbus, J., Crimmins, A., Gamble, J. L., Easterling, D. R., Kunkel, K. E., Saha, S., et al. (2016). *Ch. 1: Introduction: Climate change and human health. The impacts of climate change on human health in the United States: A scientific assessment* (pp. 25–42). Washington, DC: U.S. Global Change Research Program. http://dx.doi.org/10.7930/J0VX0DFW.

Baltazar, E. (2017). México enfrenta extremos climáticos. *El Sol de Torreón*, p. 1. 28 May 2017.

Beniston, M. (2002). Climatic change: Possible impacts on human health. *Swiss Medical Weekly, 132*(1), 332–337.

Berberian, G., & Rosanova, M. (2012). Impacto del cambio climático en las enfermedades infecciosas. *Archivos Argentinos de Pediatría, 110*(1), 39–45.

Bitrán, D. B. (2001). *Características del impacto socioeconómico de los principales desastres ocurridos en México en el período 1980–99*. Sistema Nacional de Proteccón Civil, Centro Nacional de Prevención de Desastres.

Blake, E., Gibney, E., Brown, D., Mainelli, M., Franklin, J., & Kimberlain, T. (2009). *Tropical cyclones of the eastern north pacific basin, 1949-2006*. Asheville, NC: National Climatic Data Center.

Cavallo, E., Galiani, S., Noy, I., & Pantano, J. (2013). Catastrophic natural disasters and economic growth. *Review of Economics and Statistics, 95*(5), 1549–1561.

CENAPRED. (2016). http://www.cenapred.gob.mx/es/Publicaciones/archivos/340-NO.17-RESUMENEJECUTIVOIMPACTO2015.PDF.

CENAPRED. (2018). *Impacto socioeconómico de los principales desastres ocurridos en la República Mexicana en 2016*. México: CENAPRED.

Centro Nacional de Programas Preventivos y Control de Enfermedades (CENAPRECE). (2018). https://www.gob.mx/salud/cenaprece.

CEPAL. (2011). https://www.cepal.org/publicaciones/xml/9/33659/mexicocap3.pdf.

Climate Change Knowledge Portal. The World Bank Group. Mexico Dashboard. Climate baseline. http://sdwebx.worldbank.org/climateportal/countryprofile/home.cfm?page=country_profile&CCode=MEX&ThisTab=ClimateBaseline.

CONAFOR. (16 de agosto de 2018). *Reporte semanal de Incendios 2018*. Obtenido de www.gob.mx/conafor/, https://www.gob.mx/conafor/documentos/reporte-semanal-de-incendios.

Constantino, T., Roberto, M., Dávila, I., & Hilda, R. (2011). Una aproximación a la vulnerabilidad y la resiliencia ante eventos hidrometeorológicos extremos en México. *Política y cultura, 36*, 15–44.

DGIS. (2017). https://www.gob.mx/salud/acciones-y-programas/direccion-general-de-informacion-en-salud-dgis.

DGIS. (2018). https://www.gob.mx/salud/acciones-y-programas/panorama-epidemiologico-de-dengue-2018.

Esquivel, G. (2000). *Geografía y desarrollo económico en México*. Washington, D.C.: Inter-American Development Bank.

Fann, N., Brennan, T., Dolwick, P., Gamble, J. L., Ilacqua, V., Kolb, L., et al. (2016). *Ch. 3: Air quality impacts. The impacts of climate change on human health in the United States: A scientific assessment* (pp. 69–98). Washington, DC: U.S. Global Change Research Program. http://dx.doi.org/10.7930/J0GQ6VP6.

Flannigan, M., Krawchuk, M., De Groot, W., Wotton, M., & Gowman, L. (2009). Implications of changing climate for global wildland fire. *International Journal of Wildland Fire, 18*(5), 483–507.

Garcidueñas, P. (16 de abril de 2015). *7 consecuencias de una sequía*. Obtenido de Expok. Comunicación de Sustentabilidad y RSE: https://www.expoknews.com/7-consecuencias-de-una-sequia/.

Guha-Sapir, D. (2018). *EM-DAT: The emergency events database*. Universite catholique de Louvain (UCL)—CRED. www.emdat.be, Brussels, Belgium.

Gyimah-Brempong, K., & Wilson, M. (2004). Health human capital and economic growth in sub-saharan African and OECD countries. *The Quarterly Review of Economics and Finance, 44*, 296–320.

Haines, A., & Patz, J. (2004). Health effects of climate change. *JAMA, 291*(1), 99–103. Obtenido de https://jamanetwork.com/journals/jama/article-abstract/197911.

Hochrainer, S. (2009). *Assessing the macroeconomic impacts of natural disasters: Are there any?* Washington, D.C.: The World Bank.

Ibarrarán, M. E., Malone, E. L., & Brenkert, A. L. (2010). Climate change vulnerability and resilience: Current status and trends for Mexico. *Environment, Development and Sustainability, 12*, 365–388. https://doi.org/10.1007/s10668-009-9201-8.

Ibarrarán, M. E., Ortiñez, J. A., & Islas, I. (2017). Compounding factors: Air pollution and climate variability in Mexico City. In R. Akhtar & C. Palagiano (Eds.), *Climate change and air pollution: The impact on human health in developed and developing countries* (pp. 361–374). Berlin: Springer.

IMTA. (2016). *Taller para la Formulación de Bases de Políticas Provinciales de Gestión Integrada de Sequías. Política Nacional para la Sequía en México* (pp. 1–33). San Miguel de Tucumán, Argentina: Consejo Hídrico Federal.

IPCC. (2014). Impacts, adaptation and vulnerability. Summary for policymarkers. In B. C. Field, V. Barros, D. Dokken, K. Mach, & M. Mastrandrea (Eds.), *Global and sectoral aspects. Contribution of working group II to the fifth assessment report of the intergovernmental panel on climate change* (p. 1131). New York: Cambridge University Press.

Ize, L. I. (2002). El cambio climático y la salud humana. *Gaceta ecológica*, 43–52.

Jauregui, E. (2003). Climatology of landfalling hurricanes and tropical storms in Mexico. *Atmósfera, 16*(4), 193–204. Obtenido de http://www.ejournal.unam.mx/atm/Vol16-4/ATM16401.pdf.

Loayza, N. V., Olaberria, E., Rigolini, J., & Christiaensen, L. (2012). Natural disasters and growth: Going beyond the averages. *World Development, 40*(7), 1317–1336.

Magaña, V., & Gay, C. (2002). *Vulnerabilidad y adaptación regional ante el cambio climático y sus impactos ambientales, sociales y económicos* (pp. 7–9). México: Gaceta ecológica.

Martine, L., Pirard, P., Tillaut, H., Pascal, M., Vandentorren, S., Suzan, F., et al. (2004). The heat wave of August 2003: What happened? *La Revue du praticien, 54*(12), 1289–1297.

Martínez, P. (20 de septiembre de 2016). Los crecientes riesgos del clima extremo. *Milenio*, p. 1.

Mayer, F. D. (2001). The long-term impact of health on economic growth in Mexico, 1950-1995, *Journal of International Development, 13*(1), 123–126.

Mayer, D., Mora, H., Cermeño, R., Barona, A. B., & Duryea, S. (2001). Health, growth, and income distribution in Latin America and the Caribbean: A study of determinants and regional and local behavior. In *Investment in health, social and economic returns*. Pan American Health Organization Scientific and Technical Publication No. 582.

Mayer-Foulkes, D. (2000). A generalized fast algorithm for BDS-type statistics. *Studies in Nonlinear Dynamics and Econometrics, 4*(1), 45–50.

Mayer-Foulkes, D. (2001). The long-term impact of health on economic growth in Mexico, 1950–1995. *Journal of International Development, 13*(1), 123–126.

Moreno, A. R. (2006). Climate change and human health in Latin America: Drivers, effects, and policies. *Regional Environmental Change, 6*(3), 157–164.

Ochoa, C., Conde, C., Morales, M., Hayden, M., & Nava, E. (20 de julio de 2010). *Eventos extremos de precipitación en Veracruz: Impactos y efectos medioambientales*. Obtenido de aeclim.org: http://aeclim.org/wp-content/uploads/2016/02/0041_PU-SA-IX-2014-CA_OCHOA.pdf.

Ochoa, C., Welsh, C., Bonilla, E., & Morales, M. (2013). Fuentes de información sobre eventos hidrometeorológicos extremos en Veracruz de Ignacio de la Llave. Realidad, Datos y Espacio. *Revista Internacional de Estadística y Geografía, 4*(3), 66–73.

PAHO. (2014). *Chikungunya fact sheet*. https://www.paho.org/hq/index.php?option=com_content&view=article&id=8303:2013-hoja-informativa-chikungunya&Itemid=40023&lang=es.

Ressl, R., & Cruz, I. (2012). Detección y Monitoreo de Incendios Forestales mediante Imágenes de Satélite. *Biodiversitas, 100,* 12–13.

RioDoce. (15 de mayo de 2015). *Ciclones Tropicales*. Obtenido de http://riodoce.mx/, http://riodoce.mx/medio-ambiente/cerca-de-26-ciclones-se-preven-lleguen-a-mexico-14-podrian-ser-huracanes.

Runfola, D. M., Romero-Lankao, P., Jiang, L., Hunter, L. M., Nawrotzki, R., & Sanchez, L. (2016). The influence of internal migration on exposure to extreme weather events in Mexico. *Society & Natural Resources, 29*(6), 750–754.

Santiago, J., López, M., & López, S. (2008). Tendencias del Cambio Climático Global y los Eventos extremos Asociados. *Ra Ximhai: revista científica de sociedad, cultura y desarrollo sostenible, 4*(3), 625–633.

Secretaría de Salud. (2014). *Programa de Acción Específico Protección contra Riesgos Sanitarios 2013–2018*. México: Comisión Federal para la Protección contra Riesgos Sanitarios (COFEPRIS). http://www.cofepris.gob.mx/cofepris/Documents/QueEsCOFEPRIS/PAE%20180315.pdf.

SEMARNAT. (2017). Dirección General de Estadística e Información Ambiental, Julio, 2017.

Servicio Meteorológico Nacional (SMN). (2013). *Ciclones tropicales. Información histórica. Base de datos de ciclones tropicales que impactaron a México de 1970 a 2011*. CONAGUA. Consultado en: http://smn.cna.gob.mx/index.php?option=com_content&view=article&id=38&Itemid=46.

Sosa, F. (2015). Política del cambio climático en México: avances obstáculos y retos. Realidad, datos y espacio. *Revista internacional de estadíostica y geografía, 6*(2), 4–23. Obtenido de http://www.inegi.org.mx/rde/rde_15/doctos/rde_15_opt.pdf.

Synder, R., & De Melo-Abreu, P. (2010). *Protección contra las heladas: fundamentos, práctica y economía*. Cataluña, España: Organización de las Naciones Unidas para la Agricultura y la Alimentación (FAO).

Toya, H., & Skidmore, M. (2007). Economic development and the impacts of natural disasters. *Economics Letters, 94*(1), 20–25.

Trenberth, K. (2001). Climate variability and global warming. *Science, 293*(5527), 48–49.

Villers, L., & Hernández, J. (2002). *Incendios Forestales y la Variabilidad Climática en México*. CDMX: UNAM. Obtenido de http://observatoriogeograficoamericalatina.org.mx/egal11/Procesosambientales/Climatologia/08.pdf.

WHO. (2002). *The world health report 2002 - Reducing Risks, Promoting Healthy Life*. https://www.who.int/whr/2002/en/.

WHO. (2003). *Health topics. Climate chance*. http://www.who.int/globalchange/en/.

WHO. (2014). *Quantitative risk assessment of the effects of climate change on selected causes of death, 2030s and 2050s* (p. 2014). Geneva: World Health Organization.

Zúñiga, P. (2007). Protección civil y desastres naturales. In F. C. Tecnológico (Ed.), *Semanarios de Protección Civil y Desastres Inducidos por Fenómenos Naturales* (pp. 25–33). México: FCCyT.

The Impact of the Increase of the Level of the Sea in the Argentine Coastal Areas. Current Evidence and Future Scenarios

Daniel Oscar Lipp

Abstract The coastal area is now doubly threatened, due in the first place to the changes that affect the climate and the rise in sea level, which can of course reduce its size or alter it in a major way. As predicted, the rise in sea level will be the greatest impact. This report analyzes the Argentine coastal sector as a result of the increase in the mean sea level, describing the areas of greatest risk to the impact. The first of these impacts will be recorded in the Río de la Plata, causing an increase in the water level and favoring its erosive action. Large floods are envisaged along its entire coastline. The coastal area of Samborombón Bay is critical and is under the effects of widespread erosion. Likewise, there is a marked change in the Atlantic coastal belt that is experiencing an increase in flood levels, accelerated erosion, loss of wetlands, and intrusion of seawater into freshwater aquifers. According to studies carried out in the country, the retreat of the coastline in large sectors of Patagonia due to climate change has exceeded three meters in the span of 39 years. In the Río de la Plata, this trend also occurred. The analyses carried out in the Samborombón Bay indicate that the coastline has retreated some 50 m in the last 49 years (Codignotto et al. in Revista del Museo Argentino de Ciencias Naturales 13(2):135–138, 2011), hence our intention to address it.

Keywords Climate change in Argentina · Coastal erosion · Sea level increase · Río de la Plata · Samborombón Bay · Patagonian coast · Coast line

1 Introduction

After the entry into force of the Kyoto Protocol in 1997, and after an arduous negotiation, all countries made a joint commitment to reduce their annual greenhouse gas (GHG) emissions. But among other common obligations, the United Nations Framework Convention requires each state to periodically report on the inventory of

D. O. Lipp (✉)
Universidad de Buenos Aires, Buenos Aires, Argentina
e-mail: daniellipp@arnet.com.ar

Universidad Católica de Salta, Salta, Argentina

R. Akhtar (ed.), *Extreme Weather Events and Human Health*,
https://doi.org/10.1007/978-3-030-23773-8_23

its emissions and the status of its current situation with respect to climate change. All this is part of a National Communication that must be presented periodically before the United Nations Framework Convention on Climate Change (UNFCCC). It is clarified that this convention was not easy to sign since dissidents appeared that continue to this day. In the Second National Communication (SCN) of the Government of the Argentine Republic (2007), the Conference of the Parties identified important changes in the country of some trends of climatic variables, in comparison with what has been observed historically. This text highlights the climatic trends registered in most of the Argentine territory in the last three or four decades and, at the same time, points out that they are most likely related to global climate change. These trends have significantly affected natural systems and human activities and have required rapid adaptation. The most important are listed: (a) the increase in average annual rainfall in most of the countries, but especially in the northeast and in the western area surrounding the traditional wet region, (b) increase in the frequency of extreme rainfall in much of the east and the center of the country, (c) the increase of the temperature in the mountain range of Patagonia and Cuyo, together with a retreat of the glaciers, (d) the increase of the flow of the rivers and the frequency of the floods throughout the country, except in San Juan, Mendoza, Comahue, and northern Patagonia, and (e) the decrease in river flows of mountain origin in San Juan, Mendoza, and Comahue (Government of the Argentine Republic 2007, p. 93). In view of these impacts, which are very significant indeed, the author decided to contribute with a document whose interest focuses on the deterioration of Argentine coastal areas due to climate change. The study is part of an initiative whose main objective is to demonstrate the vulnerability acquired by the coastal area due to climate change, so that decision-makers at the national and local levels have a tool that allows them to take into consideration the relevant costs and benefits in their analyses. In this document, we analyze different effects of climate change in Argentine coastal areas, also known as impacts. The impacts that have been considered in this study are the floods that will suffer the coasts of the region due to: (a) rise in sea level (permanent flooding) and (b) extreme temporary events of flooding due to the combined action of tides, sea level, and overhang by waves (temporary floods). In this impact, factors such as the affected surface, the coastal population, and the ecosystems are studied. But the erosion of beaches that assumes serious consequences in our country will also be emphasized.

2 Changes in the Level of the Sea

Many predictions have been made about the sea level rise that will occur at the end of this century or in the next. There is a very high level of confidence, on the other hand, that the sea level rise rates can be several times higher or lower than the rise in the global mean sea level for periods spanning several decades, due to fluctuations in the sea level circulation of the ocean. For example, since 1993, regional rates for the Western Pacific are up to three times higher than the global average, while those for much of the Eastern Pacific are approaching zero or negative.

But what is meant by the average level of the sea, given the unavoidable evidence of its impact? The mean sea level is defined as the hourly average value (or smaller interval) of the recorded height of a tide gauge in a given period (one month, one year, or nineteen years), but with full tidal cycles. In this way, it is possible to eliminate the contribution of the tide in the calculation. In our country, the average level of the sea is materialized in practice by the zero of the scale of the Mar del Plata tide gauge. This materialization is official in the country and was established by the Military Geographical Institute (IGM) (Bergesio 2012, p. 6).

Because of the exponential growth of the population, economies, and urbanization in the past century, and in which it runs from it, the coastal areas have undergone great changes. At present, 10% of the world's population lives in coastal areas of elevation less than 10 m above mean sea level (McGranahan et al. 2007, pp. 17–37). On the other hand, there are 136 port cities with more than one million inhabitants around the planet where almost 400 million people live, and 10% of which are exposed to coastal flooding every 100 years. By way of example, we have Shanghai, China, with 10 million inhabitants. In Bangladesh, for example, a country with a large percentage of its territory at sea level, tens of millions of people are going to be displaced if sea levels rise to only 1 m (Nicholls and Cazenave 2010, pp. 1517–1520).

During the 1901–2010 period, the overall mean sea level rose 0.19 m (0.17–0.21 m). Since the middle of the century, the rate of sea level rise has been higher than the average of the previous two millennia. On the other hand, it is very likely that the average rate of average global elevation of sea level was 1.7 mm/year (1.5–1.9 mm) between 1901 and 2010 and 3.2 mm/year (2.8–3.6 mm) between 1993 and 2010 (IPCC 2014, pp. 44–46). The data collected with tide gauges and satellite altimeters coincide in that in this last period the rate was higher. It is also likely that high rates were recorded between 1920 and 1950 (Fig. 1). Today, the Fifth Assessment Report of the Intergovernmental Panel on Climate Change predicts an increase in the mean sea level until the year 2100 from 0.11 to 0.77 m, depending on the scenario considered. On the other hand, estimates of ascent of 3.8 mm/year were given by Nerem (1995, pp. 708–710) from studies with information from TOPEX/Poseidon satellites. However, it is noted that these predictions contain some uncertainties. Anyway, beyond the discrepancies on trends and values, the rise in sea level is categorical and was measured in various coastal locations around the world. Therefore, it is not a very encouraging scenario if the average level of the sea rises, and we do not deal with it during this century since a large part of the world population resides in this area with the aggravating circumstance of having an increasing tendency to choose it as a permanent habitat.

The mitigation of climate change is undoubtedly now of key importance if one wishes to avoid major catastrophes. Sea level rise is almost impossible to avoid, so it is urgent to evaluate adaptation measures. Even if global warming is mitigated, it will also be essential to adopt measures to adapt to sea level rise. This increase will cause more frequent coastal floods, changes in ecosystems, greater erosion in coasts, and salinization of surface and groundwater. Adaptation measures will be able to reduce

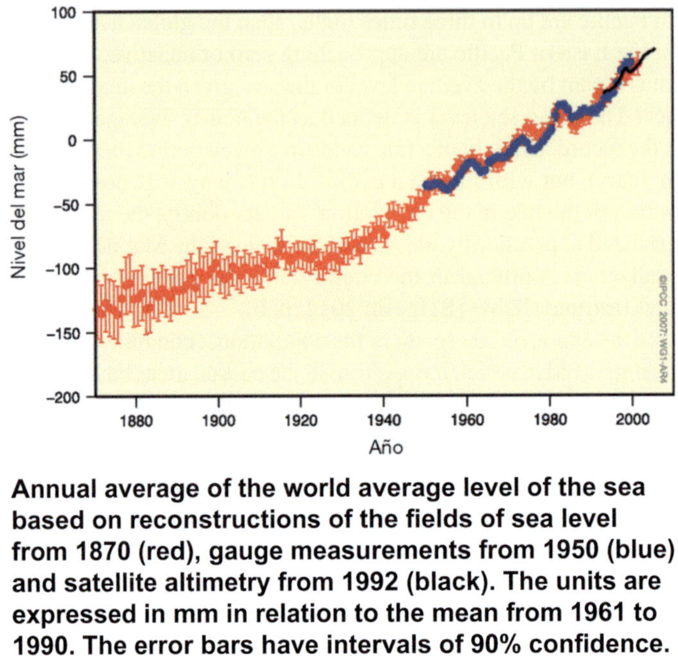

Fig. 1 Changes in the level of the sea. *Source* Report of Working Group I—Basis of Physical Science. http://www.ipcc.ch/publications_and_data/ar4/wg1/es/figure-ts-18.html

at least these damages by means of the planned zoning of the vulnerable coastal regions, also building cyclone protection centers, etc. In order to also ensure that nations have access to the information necessary to plan such measures, it will be necessary to continue advancing in the implementation of observation systems and improving climate system models and local resources to support decision-making.

In the Argentine coast, where we will now emphasize, the main data obtained from the tide gauges and whose statistical treatment allowed to recognize trends of sea level rise in the Port of Buenos Aires for the period 1905–2001 of 1.7 mm/year (Barros et al. 2005, pp. 26–27). This rise was very similar to the one registered more to the south, that is, outside the estuary of the Río de la Plata, and is within the average values observed in most of the coasts of the planet.

3 Class of Coastal Impacts

The impacts of climate change in coastal regions will manifest themselves in two different ways: erosion and flooding. To order our presentation, we will now deal with the floods and, in a subsequent examination, the erosion of coasts by sea level rise.

One of the biggest, if not the most critical, threats to coastal systems and low-lying areas of the world is the increase in flood events due to the effects of climate change, mainly due to the increase in mean sea level. But here, we must differentiate due to a double dimension of the problem. In general, when talking about coastal flooding, we are indicating an increase in the level of water that exceeds the usual, which ordinarily occurs, which causes serious damage to the socioeconomic and natural systems. This flood is temporary, of short duration, of almost always harmful effects, after which the water level returns to its usual situation. These floods are considered one of the phenomena of greatest impact in the world, due to the effect they cause in large densely populated territorial areas. According to the international database of disasters prepared by the CRED (OFDA-CRED Database), worldwide, 7,477,552 people were victims of floods during the period 1900–2001, while the total population affected reached figure of 2,379,092,236 inhabitants. Among the most important events are two floods in the Yangtze River, China, the first in 1931 that caused the loss of 3,700,000 lives and another more recently in August 1998, in which 238,973,000 inhabitants were affected (Garnica Pena et al. 2004, p. 24). These figures reflect in a tangible way the impact of this type of hazards, worldwide. Temporary floods are very complex environmental phenomena and include aspects of the most varied nature, such as climatic, hydrological, geological, geomorphological, and even social. All help to unleash these events, which although they are characterized by being brief, are of disastrous consequence for man.

However, the rise in the average level of the sea as a consequence of climate change implies the permanent flooding of land, especially in low-lying areas of the coast, leaving them flooded and thus losing their use. When talking about climate change, it is usual to analyze the rise in the mean sea level (sea level rise) as a predominant factor in coastal flooding. In this case, this type of flood, due to sea level rise, in a climate change scenario implies a permanent flood. Due to this double dimension of the problem, on the one hand, the flood of extreme events and with temporary duration (level of flood) and, on the other, the submergence or loss of soil due to sea level rise, it is necessary to contemplate both in the study.

4 Estuary of the Río de la Plata

In our country, the phenomenon of flooding lashes out in the most diverse environments; however, our purpose is to analyze it in coastal areas, which is what is discussed here. Let's start with the estuary of the Río de la Plata (Fig. 2).

The Río de la Plata is located on the east coast of South America, roughly between 34° and 36° south latitude and 54° 50′ and 58° 30′ west longitude, and constitutes the boundary between Argentina and Uruguay. Developing in the NW-SE direction, it reaches about 290 km in length. The main forcing of the dynamics of the Río de la Plata is the tidal wave that enters from the ocean, the discharge of the tributaries that enters the head of the river, and the wind field that acts on the whole surface of the river. The oceanic wave that arrives at the Río de la Plata has a regime of

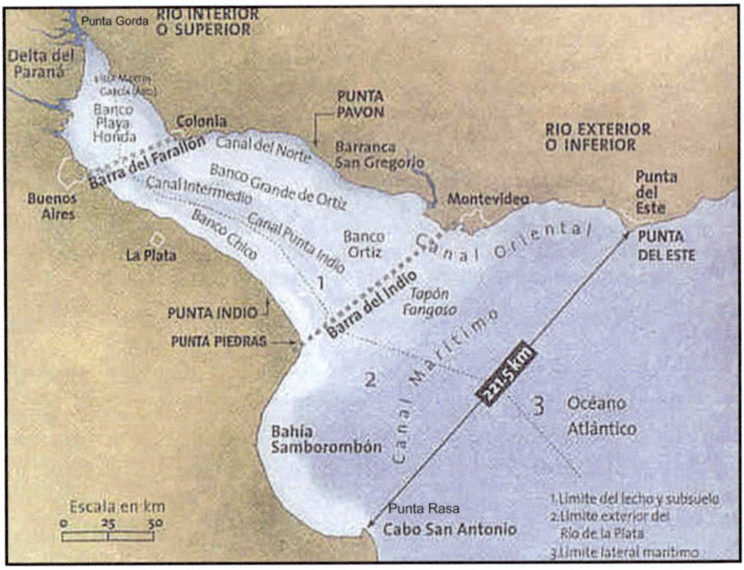

Fig. 2 Estuary of the Río de la Plata. *Source* http://irapescar.com/BODY/notas/todas_2010/03_10/cuendelpla/index.htm

diurnal inequalities, showing great differences between consecutive highs and lows. Its estuary, which is almost 50 km wide at the mouth of the Paraná and Uruguay, gradually widens to an imaginary line linking Montevideo (in Uruguay) to Punta Piedras (in Argentina). Its width here is 90 km, and it is the section known as Río de la Plata interior. From this point, the river becomes more saline to another imaginary line that connects Punta del Este (Uruguay) to Punta Rasa (Argentina), which is the limit of the Río de la Plata exterior, with a width of almost 200 km (Barros et al. 2005, pp. 23–38).

More than 97% of the water input to the Río de la Plata is contributed by the Paraná and Uruguay rivers, with the average flow of the Río de la Plata in the order of 23.000 m^3/s. The discharge of the Santa Lucia River, relatively close to Montevideo, is negligible with respect to the entire system, although it is an important local factor for the Uruguayan coasts. In the same way, the contributions of a series of small discharges that occur on the Argentine side are negligible.

The Río de la Plata has large metropolises on its banks, particularly Buenos Aires on its right bank and Montevideo on its left bank. Our special interest is now directed to the Metropolitan Region of Buenos Aires (RMBA, hereinafter), where more than thirteen million inhabitants live. Particularly for this sector of the Río de la Plata, the greatest impact that will come about due to climate change is the floods associated with extraordinary rainfall events. The floods in the RMBA are almost always tragic events of varying length and intensity.

In general, the area in which the agglomeration is located presents adequate conditions for the establishment of a large city. It has a soft relief, a good supply of ground and surface water, excellent agricultural quality soils, absence of great natural risks, and provision of materials suitable for construction. However, it must be said that the great growth experienced by the RMBA and its socioeconomic characteristics have resulted in the existence of major geo-environmental problems, some of which are very difficult to solve, such as floods. The disorderly and sustained growth of the city in the last century has taken place without the establishment of minimum guidelines of territorial order that took into account the characteristics of the physical environment.

The causes of flooding of the RMBA can be grouped into natural and anthropic. Within the first group, there are the frequent and intense rainfalls that plague the region, the existence of a poorly integrated drainage network due, among other factors, to the reduced slope of the land, the existence of waterlogged basses (bathed), the presence of a high phreatic layer, and the plug process exerted by the so-called sudestadas, in the mouths of the different streams that drain the RMBA. This does not allow water to run off the streams, which can overflow and flood the urban area. Within the second group of causes are the waterproofing produced by urbanization, the disappearance of green spaces and natural vegetation, the occupation of floodplain, lowlands, lagoons, and alluvial plains due to the growth of the city, the realization of infrastructure works (pipelines, dykes, etc.), obstruction of river courses, substantial modification of the coastline of the Río de la Plata, and the presence of communication routes that cross the courses and have not been built taking into account the frequent floods (Alcira et al. 2001, pp. 3–7).

5 Delta of the Paraná River

Given the importance of the Río de la Plata in global warming, researchers will now place special emphasis on two particularly threatened coastal areas of this great river course, which are the Paraná Delta, located at the headwaters of the river, and the wetland Samborombón Bay, on the coast of Buenos Aires.

The deltas constitute sites increasingly vulnerable to climate change as they are, in general, partially flooded areas. Of course, the Paraná Delta does not make an exception to the rule, and considering this condition, it is expected that future floods produced by the sudestadas of the Río de la Plata will be increasingly severe as the sea level rises, affecting its islands, defenses of the coastal zone, and the quality of the waters. With an area of 14,000 km^2 and a length of 320 km, the Paraná Delta is considered to be one of the largest in the world and the only one of its kind found entirely in a freshwater environment. It is born at the height of the city of Diamante, province of Entre Ríos, where the so-called predelta ends. It is divided into three large areas: the upper delta (from Diamante to Villa Constitución, province of Buenos Aires), the middle delta (from Villa Constitución to Puerto Ibicuy, province of Entre Ríos), and the lower delta (from Ibicuy port to the beginning of the great estuary of the Río de la Plata). It is located largely to the southwest of the province of Entre Ríos and northeast of the province of Buenos Aires. The constructive work of the river is constant, and it makes its way among the innumerable islands through a maze of streams and channels. The islands are depressed inside and with albardones on their banks, which determines the stagnation of the waters in the interior after the large floods. The main arms of the river are Paraná Pavón, Paraná Ibicuy, Paraná Guazú, Paraná de las Palmas, Paraná Miní and that limit island groups such as the islands of the Lechiguanas or the Ibicuy.

The advance of the delta is one of the most notorious facts. The current rate of advance, determined from the comparison of cartographies over a century, indicates average values of the order of many tens of meters per year. This vertiginous advance has led, in particular, to the appearance during the last century of all the islands that currently face the town of San Fernando (province of Buenos Aires). Associated to this, population and activities that are very vulnerable to the sudestadas and to the changes in the level of the waters have been established. The deltas are characterized for being low zones, with additional subsidence to the effect of the rise of the level of the sea and of course with a great ecological and economic importance. To quantify the rise in sea level in the Río de la Plata Delta, the flooded area and the affected population have been studied as a result of a rise in sea level to the level of 1 m and up to the level of 2 m in case of additional subsidences.

Table 1 shows the values obtained. For the purposes of comparison, we have contemplated in the analysis other deltas of Latin America and the Caribbean. As shown in the table, the deltas of the Río de la Plata and Río Magdalena would be the most affected in terms of population by a 1-m rise in sea level, while the former stands out for having the largest area of committed land. In contrast, Fig. 3 shows the level isolines for elevations 1 and 2 m of the delta of the Río de la Plata, which represent the scope of flood situations (United Nations 2015, pp. 39–40).

Table 1 Area and population in the most important deltas of Latin America and the Caribbean to the 1 and 2 m

Deltas	Between heights 0 and 1 m		Between heights 0 and 2 m	
	Flooded area (Has.)	Affected population (habs.)	Flooded area (Has.)	Affected population (habs.)
San Francisco	3403.62	60,516	12,005.84	108,275
Río de la Plata	79,824.22	103,180	130,481.38	148,472
Orinoco	3454.77	339	6373.20	365
Magdalena	16,901.40	103,962	36,292.86	128,334
Grijalva	28,928.93	27,490	65,401.78	37,897
Atrato	5588.19	6894	17,761.60	8131
Amazonas	7353.99	103,482	14,382.73	103,668

Source United Nations (2015) Effects of Climate Change on the Coast of Latin America and the Caribbean. Impacts. ECLAC. University of Cantabria. Institute of Environmental Hydraulics. Ed. Cepal. Project documents. http://www.cridlac.org/digitalizacion/pdf/spa/doc19588/doc19588-contenido.pdf

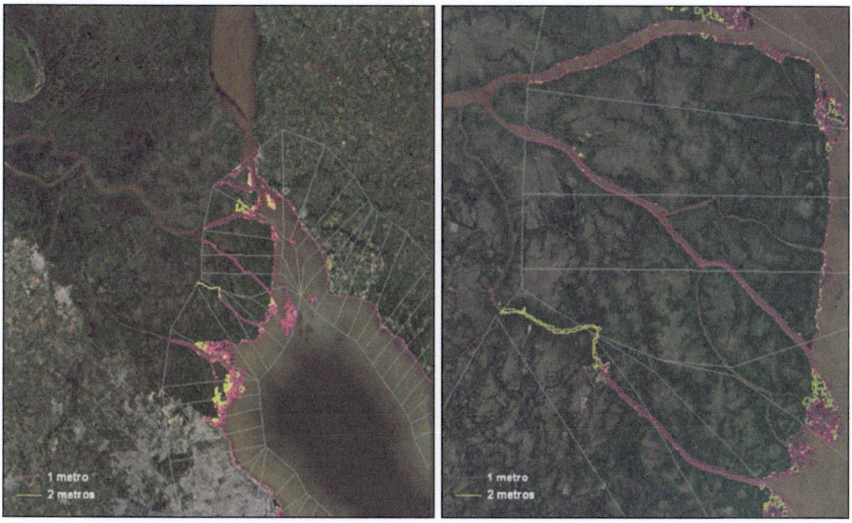

Fig. 3 Level isolines for 1 and 2 m in the delta of Río de la Plata. *Source* United Nations (2015) Effects of Climate Change on the Coast of Latin America and the Caribbean. Impacts (p. 40). ECLAC. University of Cantabria. Institute of Environmental Hydraulics. Ed. Cepal. Project documents. http://www.cridlac.org/digitalizacion/pdf/spa/doc19588/doc19588-contenido.pdf

6 Samborombón Bay

Another critical impact of the coastal zone of the Río de la Plata, which is the Samborombón Bay, will now be discussed. Coastal wetlands are one of the most productive ecosystems in the world, most of which are vulnerable to climate change. It is expected that they will experience impacts at the biophysical and socioeconomic level due to an increase in sea level. These impacts can modify the functions of the coastal wetlands and affect their dynamics, gravitating, of course, on the uses and resources exploited by the local communities. The Bay of Samborombón is one of the coastal wetlands of Argentina that would have the greatest impact due to climate change, especially in its southern zone, which would be the flooding area that would be covered by the rise of the sea. The biophysical effects that would be expected from the impact of climate change in Samborombón Bay are counted in hundreds, among which could be mentioned the alteration in the surface water quality of the lotic systems, the deterioration of the groundwater quality due to the marine intrusion, the loss of habitats, and the modification in the composition of plant and animal communities, among others. On the other hand, socioeconomic effects include the alteration of the anthropic cycle of water in the Bay (quantity and quality of surface and underground water due to different uses), the affectation of ports, such as that of Gral. Lavalle, the potential threat of health risk for the population, and impacts on the main productive activities in the Bay, such as cattle breeding, coastal commercial fishing, and hunting. For all these reasons, some mitigation measures have been proposed on the wetland to deal with climate change, among which can be mentioned the monitoring of the quality of surface and groundwater to plan the uses and the availabilities of the same, the study of the temporal and seasonal variability of the composition of the terrestrial and aquatic communities of the Bay, to demonstrate the tendency of the new trophic interrelations and the adaptation of the same ones to the new changes, the mapping of the redistribution of the micro-environments to estimate the new uses of the habitat by biota, the planned setbacks of the agricultural establishments located and the implementation of aquatic protected areas and corridors of terrestrial and aquatic fauna for the conservation of the Bay's biodiversity (Volpedo et al. 2005, pp. 342–343).

Figure 4 shows the permanent flooding area to which Samborombón Bay will be exposed by the year 2070. These sites are more vulnerable to the increase in the mean sea level because they are topographically low areas. Also indicated in the figure are areas that could be partially flooded.

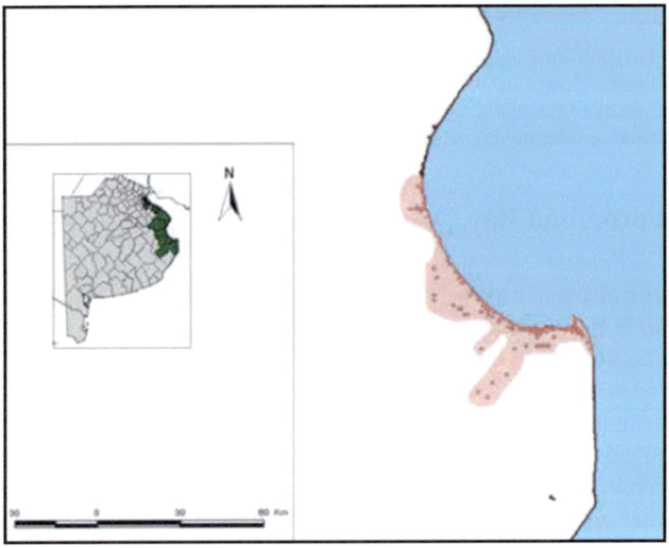

Observations:
In red: permanent flood area by the year 2070.
In pink: area that could be partially flooded.

Fig. 4 Samborombón Bay-Rió de la Plata: area of possible flooding. *Source* Prepared based on the report of the project "Impact assessments and adaptation to climate change" (2005). "Global Climate and the Coastal Areas of the Río de la Plata" Assessments of Impacts and Adaptation to Climate Change, Project N° LA 26. Barros (2005a)

7 Increase in Coastal Erosion

So far, we have paid attention to the coasts of the Río de la Plata from the town of Punta Gorda (33° 54′ 58″S, 58° 24′ 52″W), at the mouth of the Uruguay River, to Punta Rasa (36° 17′ 23″S, 56° 46′ 53″W), south of the Bay of Samborombón, in the province of Buenos Aires. Millions of people inhabit this place as two capitals; the country (Buenos Aires) and the province of Buenos Aires (La Plata) reside in this coastal area. In this document, evidence has already been presented that the most important environmental alterations produced by climate change will occur along this coastal strip. On the other hand, there is a lot of scientific work directly linked to this area because it is the most inhabited and will receive the greatest impact. The situation is very different with respect to what is presented from north to south, from Punta Rasa at the mouth of the Río de la Plata to 68° 36′ 38″W, on the Beagle Channel, southern end continental of the Argentine territory. This is where we will put the accent now. The Argentine coastal sector from the Punta Rasa town to the Beagle Channel presents several natural and anthropic phenomena of evolution. Here, we can distinguish three sectors (Fig. 5), each of which presents evidence of erosion and accumulation, which implies the loss of beaches, an attraction of tourist areas, which affects the economy of these places. On a total of 4.655 km coasts of said strip, 180 km is coasts of beaches with dunes, 4.175 km presents themselves with cliffs, and 300 km, the remainder, is mountain coasts. The coastal strip considered corresponds to a coast of passive margin, with an extensive continental shelf, which implies some tectonic stability (Codignotto et al. 1992, pp. 128–130). The beach coast with dunes, the most populated of spas and where the tourist most frequently accesses, is located in the province of Buenos Aires, between the town of Punta Rasa, at the mouth of the Río de la Plata, up to the Laguna de Mar Chiquita. A large part of the Argentine population visits this area in summertime, enjoying its waters. It is a coast, today very eroded anthropically, in constant construction by the contribution of sand that makes the sea to the continent, increasing the width of the beach and the height of the dunes. The causes that produce this phenomenon are the low height of the coast that in fact would correspond to the depressed region of the pampas and the marine current of northward direction called coastal drift. As we mentioned at the beginning, it is a coast that is essentially suitable for spa activities due to the extraordinary width of its beaches, its gentle slope, its fine sands, and the beauty of its dunes that accompany the sea in all its extension. Its waters are warm and free of malignant animal species like the living waters, which bother the tourist, and which appear further south. Unlike the coasts, which we will see later, with ravines and cliffs that drastically limit the width of the beaches and, consequently, their ability to house bathers, the dunes constitute an extension of the beach. For this set of enviable characteristics should be protected more; however, their situation is very vulnerable today. Its anthropic action, exacerbated recently by climate change, has generated unforeseen damages in coastal dynamics. The sector experiences an increase in episodic erosion, a change in water quality, and pollution in coastal aquifers. On the other hand, works are

Fig. 5 Argentina coastal regions. *Source* National Geographic Institute (IGN) http://www.ign.gob.sr/

generated that are harmful to the environment such as breakwaters, walls, moorings without proper planning of them.

The other sector of the Argentine coast exposed to climate change of 4.175 km of extension is almost entirely in Patagonia with a smaller strip in the province of Buenos Aires. It extends from the Laguna de Mar Chiquita to Cabo San Diego, on the eastern tip of Tierra del Fuego, within an area of cold and windy weather and unsuitable for the establishment of spas. The scenario is a coast of cliffs attacked by erosive processes that in some places adopt an almost perpendicular shape. At its base, there is a small rocky platform that slopes toward the sea and that lies just below the level of the waters. Therefore, the coast is high and cut to peak on the sea, which is separated in some cases narrow strips of coarse sand beach and gravels that are covered by high tide, which do not offer the minimum safety conditions, of course, for bathing and recreation. To the south of Carmen de Patagones, its waters are very cold and in some points, generally bays, ravines, or cliffs disappear in their entirety and the beaches are wider. This is a coast in continuous erosion. The fragments of rock that have been disintegrated are dragged out to sea, since the energy of the waves is excessive and does not allow the sand and gravel to form a beach. On the Patagonian coasts of heterogeneous rocky composition, the sea digs caves and canals, first taking the softer materials and then leaving the harder ones, which gives the coasts their own peculiar shapes. On the other hand, on the coasts of calcareous and clayey constitution, the sea produces cliffs of great verticality due to the even destruction of the constituent materials. At the foot of the ravines, rocks of different sizes accumulate that are also slowly crumbled to become sand, which is dragged mostly to the ocean floor where it is cemented and hardened over the centuries forming layers of new rocks sedimentary(.) Another portion of said sands is carried by the marine currents of the littoral (coastal drift) along hundreds of kilometers and deposited in the lowest places, or in front of the obstacles that oppose its march. In some sections of the Buenosairean coast, such as Miramar or Necochea, the sands have covered a large part of the ravines (Kokot 2004, pp. 715–717).

The phenomenon of erosion in our country has been described for decades by various authors. During the Pleistocene and part of the Holocene, there was a great contribution of sediments to the coastal zone due to the important load of the Patagonian rivers that reached that area. This contribution of sediments and gravels was distributed in the Patagonian area and formed a kind of protective shield making the erosive action of the waves and the sea less efficient. This large sedimentary contribution was mainly through the main emitters of Atlantic discharge, i.e., the large rivers. Result of this is the distribution of gravels that are observed along the entire Patagonian coast. However, currently in the main Patagonian rivers, hydraulic conditions for the transport of gravels are not given, since by reducing their flow, abundant in other periods, they configured a hydraulic system with very low transport capacity. Evidence of this is the thick mantle of alluvium present in the alluvial plains of the main Patagonian rivers. Currently, the coastal morphology is partly relict and originated from other climatic conditions that occurred in the recent past, while the current conditions present an evolutionary trend toward the erosion of the present geoforms.

The mountain coast, the last coastal sector that we would not mention, is developed along the Beagle Channel to the limit with Chile in a stretch of 300 km. It gets its name from the main feature of this coast, the continuous mountain and mountainous cords that frame it, which in some sectors fall to sink over the waters. It lacks beaches and due to the climatic conditions derived from the proximity of the Antarctic, it only allows as a recreational activity, navigation through the fueguine channels, famous for its panoramic beauty. The mountain coasts in Argentina constitute the least extensive stretch of the maritime territorial edge of the country. In this section, the conjunction of the sea with high mountains of snowy peaks is presented as an atypical feature of the landscape. The action of the waves, but especially the abrasive power of the glaciers, has formed a fringed margin where coastal accidents abound, both deep entrances and pronounced outcrops. Finally, the Argentine mountain coast concludes by transposing the international boundary with Chile. Finally, I would like to emphasize that an increase in the level of the waters of the entire Patagonian coast will undoubtedly influence the rate of coastal erosion. Therefore, the greatest attention in the area should be paid on the beaches, where there are, and the low coasts. On the other hand, on the coasts of hard rocks, generally characterized by coastal platforms or submerged cliffs, a rise in sea level will cause immersion in the former and a change toward land of the coastline. In soft rocks, immersion tends to accelerate erosion. The existing cliffs on open coasts and those that contain high proportions of clays seem more sensitive to erosion. However, in the face of rising sea levels, low-lying coasts may flood or recede due to erosion. The less-resistant coasts will allow a fast advance of the sea; on the other hand, the most-resistant ones will slow down this action. I must warn that several authors disagree that the rise in sea level is caused by an anthropic climate change but that this increase is essentially natural. The coasts of the maritime archipelagos and those of the Argentine Antarctic Sector complete our coastline and its appearance presents, in alternate form, cliffs and mountains.

8 Negative Impacts on Health

In an earlier text, the author has indicated the main infectious and respiratory *noxas* that are limited to the Argentine territory and aggravate, due to climate change, the potential area of transmission, and the mortality rate. If the current sanitary conditions are not modified, a rapid advance of these ailments is expected by the end of the century. The author in the present work will not repeat information already considered in another chapter (Akhtar and Palagiano 2018, pp. 405–418) rather he will enrich and condition the subject that has been discussed.

In the Second National Communication of the Government of the Argentine Republic to the parties to the United Nations Framework Convention on Climate Change (UNFCCC) of 2007 (Government of the Argentine Republic 2007, p. 114), as in the Third Communication National 2015 (Government of the Argentine Republic 2015, p. 137), the impacts on health that climate change would bring were considered important; however, no specific studies were available to serve as the basis for these analyses. In the two previous documents, but mainly in the Third Communication, it is pointed out that an increase in temperature and observed variations in the geographical distribution of rainfall highlighted some expected impacts on health mainly as a consequence of the expansion of the area of action of certain vectors transmitting diseases such as dengue, malaria, American trypanosomiasis or Chagas disease, and other infectious diseases such as schistosomiasis, fungal diseases such as *Paracoccidioides brasiliensis* or *aspergillus*, yellow fever, rickettsia, and other virosis. Other disorders and diseases in the Argentine Republic could also be due to variations in temperature, precipitation, and humidity levels, as well as extreme events such as droughts, floods, and storms.

Within the direct impacts of climate change, floods are the catastrophes that have caused the most damage in Argentina. These events have serious impacts on health since they tend to be favorable for the development of various diarrheal diseases and others transmitted by water. This risk has materialized in numerous coastal locations during the last decades, leading to outbreaks of infectious diseases, forced evacuations and the loss of livelihoods, and increased poverty. The most tragic cases occurred in the city of Santa Fe and La Plata, capital of the province of Buenos Aires (Argentina). Between February and April of 2003, intense rains occurred that exceeded 1000 mm in several localities of the northern Salado basin, which flows into the Paraná River in front of the city of Santa Fe. But these conditions were aggravated by the intense rainfall of the second half of April, which were decisive for the flooding of the city. Unfortunately, dozens of deaths and missing lives were regretted (Kullock 2007, p. 38).

On the other hand, more direct impacts such as heat waves can increase mortality rates, as happened recently in Europe. The human capacity to adapt to high temperatures could be exceeded in some regions of our country, making it impossible to perform outdoor work without adequate protection in the warmer times. Almeira and Rusticucci investigated the relationship between temperature and general mortality for the cities of Rosario and the city of Buenos Aires, finding an increase in mortality at both ends of their value, both for cold and heat (Almeira and Rusticucci 2014, p. 4). The heat waves have been increasing in Argentina. On December 13, 2013, an intense heat wave began in Argentina that lasted with few interruptions until almost half of January. It was the longest and most intense in the region, covering the center of Argentina, from Buenos Aires to Córdoba and Mendoza, with maximum temperatures above 40 °C and minimum temperatures above 24 °C. The distribution of electrical energy collapsed in many sectors of the metropolitan area of Buenos Aires, due to the consumption record due to the intense use of air conditioning equipment and the difficulties of the transformers to dissipate the heat. The risk of heat waves increased throughout the country between 1960 and 2010 due to the change in extreme thermal conditions, and this increase was greater in the north and the east of the country.

But climate change will aggravate other diseases in the country. It is estimated that the higher temperatures will favor the ailments, previously circumscribed to hot regions, such as dengue, malaria, and yellow fever. In Argentina, special attention was paid to two vector-borne diseases: dengue and malaria, both of which are typical ailments of the country's tropical and subtropical climates and where climatic scenarios point to an increase in mean and minimum mean temperatures. In addition, until now, it is the diseases that the Intergovernmental Panel on Climate Change identifies as most important. On the one hand, dengue is not considered an endemic disease in Argentina, but there have been several cases of reappearance of the disease: in 1998, in an epidemic that affected the province of Salta, and in the year 2000 when it returned to affect the provinces of Formosa and Misiones. In 2009, there were epidemics and autochthonous cases that spread to almost half of the country. The predicted temperature changes in the country allow us to suppose that both the presence of the transmitting vector of dengue (*Aedes aegypti*) and the same disease could spread south of the territory. Logically that would result in an increase in the population exposed to the disease.

Dengue fever is problematic in Argentina due to the presence of the vector (*A. aegypti*) and endemism in the bordering countries. In the north and northeast of the country, the risk of transmission of dengue is high throughout the year, while the risk of the center of the country is concentrated only in the summer months (Carbajo et al. 2001, p. 15). Figures 6 and 7 show the areas with different levels of risk of transmission of the disease and the population in danger. By the 2030s, the number of people at risk of contracting the disease could triple, while by the end of the century, the population at risk would be almost six times higher than at present, although it is estimated that only 0.22% of the population potential affected would contract the disease.

Malaria, on the other hand, is an endemic disease in some areas of northwestern Argentina, especially in some areas of Salta and Jujuy, where there is an almost invariable annual incidence, while in Misiones and in some areas of the northwest, like Tucumán and some areas of Salta and Jujuy, has epidemic characteristics. Although the incidence of malaria has decreased in the country, the density of the vector has increased, especially in the northeast of the country with the change of climatic variables. On the other hand, the distribution area of the *Anopheles darlingi* mosquito, one of the three vectors of the disease, is likely to expand, a circumstance that will require attention. Another vector disease that has shown sensitivity to climate change is Chikungunya fever, transmitted by *A. aegypti* mosquitoes, as well as dengue fever. This disease is native to Africa and currently has a vector transmission in South America. In Argentina, there have been imported cases of this pathology during 2008 and epidemiological surveillance plans have been activated given the presence of the vector in our country. Finally, the abundance of Leishmaniasis vector in the north of the country shows in some months of the summer an intimate association with temperature and in some cases with humidity (Government of the Argentine Republic 2015, p. 137).

Fig. 6 Geographical areas of Argentina with risk of dengue transmission. *Source* Carbajo et al. (2001)

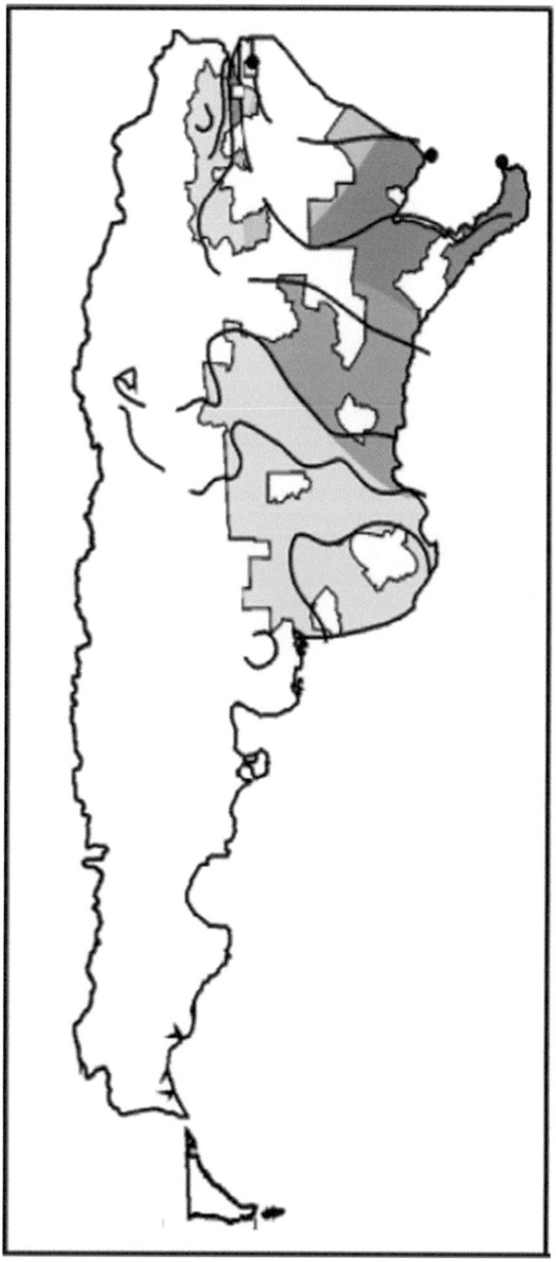

Observations: dark gray areas are the most affected by the disease.

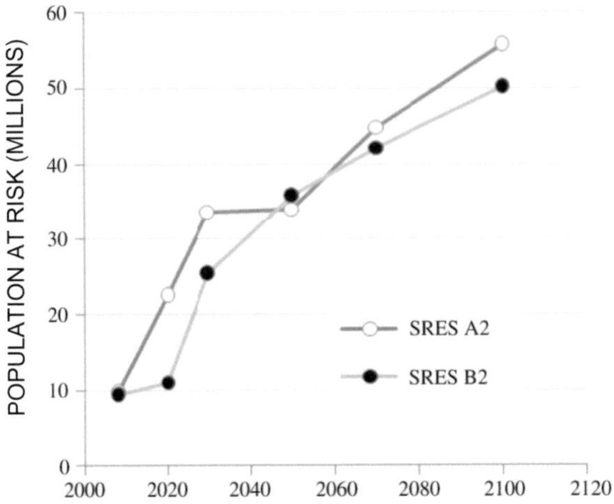

Observations : Population at risk of contracting Dengue in several periods considering the SRES A2 and SRES B2 emission scenarios. These scenarios are not climate forecasts, since each scenariois an alternative of how the future climate can behave.The emission scenarios (SRES) are grouped into four families (A1, A2, B1 and B2). Scenario A2 describes a very heterogeneous world based on the preservation of local identities. Economic development is regionally oriented, and per capita economic growth and technological change are very fragmented. In contrast, scenario B2 describes a planet where the emphasis is placed on local solutions aimed at social, economic and environmental sustainability. It is a world with a growing population but at slower rates than in the other scenarios, with intermediate levels of economic development, and a slow but varied technological change. The society is oriented towards environmental protection and social equity, and prioritizes the local and regional spheres.

Fig. 7 Population at risk of contribution dengue. *Source* https://www.ipcc.ch/pdf/special-reports/spm/sres-sp.pdf. The information comes from the number of inhabitants of the cities at risk. (*Source* Based on the information contained in ECLAC (2014))

9 Conclusions

The author has intended to analyze at first the changes that have occurred, and still occur, on the coasts of the Río de la Plata due to global warming, particularly due to the increase in water levels, due to the severity that this has reached in the coastal strip of said river. Many points of this coastline experienced an increase in flood levels, accelerated erosion, and loss of coastal wetlands. The extensive delta located in the springs of the Río de la Plata floods has affected entire populations and material goods, and as the storms become worse, it will increase. These frequently flooded areas will surely be aggravated by the effects of climate change.

According to the data provided by the tide gauge of Buenos Aires, the tendency of increase in water level is 1.7 mm/year (Buenos Aires Frente Al Cambio Climático 2014, p. 32). If this trend continues, the entire region and its adjacent areas will also be considerably affected by an increase in the frequency of flooding in low areas and increased erosion in some coastal areas. Finally, it should be noted that more than a third of the country's population is concentrated in this coastal sector.

Among the areas that have also been treated in the present work aggravated by global warming is the Ramsar Site Bahía Samborombón, and its surrounding area of influence. The coast of the bay, a escarpment of 0.60–0.90 m in height, would be receding at a rate of about 0.80 m per year, according to reports by several authors (Canziani et al. 2013, p. 158). This value was estimated by very recent field observations made at Punta Piedras and General Conesa, in the province of Buenos Aires. The rise in sea level will also impact on the discharge of the Samborombón River and the Salado River, as well as on the various drainage channels located on the site. Substantive effects on agricultural activities in the area are also expected, since the rise of water will reduce the area devoted to extensive cattle rearing, in addition to the redistribution of forage resources and the loss of natural pastures. The area of Samborombón Bay has a great extension of coasts (120 km) and has the peculiarity of being affected by ocean storms (sudestadas), which will surely bring with them great flood events.

Also, the Atlantic coast of our country was studied in this document to the limit with Chile, paying attention to three sectors: the coast of beaches with dunes, the coast of cliffs, and the coast of mountains. As the anthropic component on the coast of beaches already indicates, the greatest complexity is given here by the contribution of man due to the construction of high-cost works, without proper planning, generating erosion, and the destruction of dunes. This will contribute, of course, to greater vulnerability to climate change. It is important to keep in mind the harmful practices that man carries out in numerous spas in the country. This will only aggravate the future situation. The exploitation of sand, the destruction of dunes, the authorization of works near the shore, and inadequate infrastructure are some of the bad practices that enhance the erosive action caused by global warming.

On the other hand on the coastal strip where cliffs are evident is not exempt from those interventions, in recent years, there has been a growing occupation of space to urbanize it has generated an erosive phenomenon on the coast, accentuating the existing natural process in the area. But unlike the coast of the province of Buenos Aires, where the urbanization process was strongly motivated by beach and sun tourism, on this Patagonian coast urbanization is substantially linked to port, commercial and extractive activities. On the other hand, the intense action of the waves on the coastal outcrops, especially in the sectors of coast more exposed to breakers, helps to make this erosion process even more accentuated. Due to the orientation of this coastline, the area is mainly exposed to the action of waves coming from the east. The astronomical tide data are of fundamental importance to characterize the dynamics of the coastal environment, since the great height reached by them increases the exposure of the cliffs to the action of the waves.

As a final case, climate change may affect, if it has not already done so, the mountain coast located in the south of the country. This sector suffered with intensity the modeling produced by immense glaciers during the last glacial period. Geologically, the mountainous coasts of southern Argentina are related to the *Andes fueguinos* continuation of the Andes mountain range in the archipelago of Tierra del Fuego. However, the studies carried out on this coastal coast, related to the rise in sea level, are almost nil. The action of the waves will surely be forming a margin more bordered than the one that currently has outgoing and quite pronounced entrants.

References

Akhtar, R., & Palagiano, C. (Eds.). (2018). *Climate change and air pollution. The impact on human health in developed and developing countries* (430 p). Springer International Publishing. https://doi.org/10.1007/978-3-319-61346-8_1. https://www.springer.com/la/search?query=Rais+Akhtar&submit=Enviar+consulta, https://www.springer.com/la/book/9783319613451.

Alcira, K., Kullock, D., & Valdés, J. B. (Eds.). (2001). *Inundaciones en el Área Metropolitana de Buenos Aires*. Banco Internacional de Reconstrucción y Fomento. Banco Mundial. Disaster Risk Management Working Paper Series NO. 3. http://siteresources.worldbank.org/INTDISMGMT/Resources/buenosaires.pdf.

Almeira, G., & Rusticucci, M. (2014). Relations between mortality and temperatures in two big Argentineans cities. In *WCRP Conference for Latin America and Caribbean*, March 2014, Uruguay.

Barros, V. (2005a). *Global climate change and the coastal areas of the Río de la Plata*. Assessments of Impacts and Adaptations to Climate Change, Project N° LA 26.

Barros, V. (2005b). *El Cambio Climático Global* (2ª ed., p. 176). Libros del Zorzal: Buenos Aires.

Barros, V., Menéndez, A., Natenzon, J., Codignoto, J., Kokot, R., & Bischoff, S. (2005). *El Cambio Climático y la Costa Argentina del Río de la Plata* (44 p). Buenos Aires: Fundación Ciudad. http://www.fundacionciudad.org.ar/pdf/CClimatico_RdP.pdf.

Bergesio, L. I. (2012). *Precisiones Altimétricas. Comparación de Equipos y Métodos*. Buenos Aires: Universidad de Morón, Facultad de Ingeniería.

Buenos Aires Frente Al Cambio Climático. (2014). Agencia De Protección Ambiental (52 p). Ministerio de Ambiente y Espacio Público. Noviembre 2014, http://www.buenosaires.gob.ar/sites/gcaba/files/bs_as_cc_2014_corregido.pdf.

Canziani, O. F., Canziani, P. O., Cirelli, A. F., Codignotto, J. O., Gimenez, J. C., Giraut, M. A., & Volpedo, A. V. (2013). Aportes para abordar la adaptación al cambio climático en la Bahía Samborombón. *Boletín Técnico, 257*. Fundación Vida Silvestre Argentina.

Carbajo, A. E., Cardo, M. V., & Vezzani, D. (2012). Is temperature the main cause of dengue rise in non-endemic countries? The case of Argentina. *International Journal of Health Geographics, 11*, 26.

Carbajo, A. E., Schweigmann, N., Curto, S. I., de Garín, A., & Bejarán, R. (2001). Dengue transmission risk maps of Argentina. *Tropical Medicine and International Health, 6,* 170–183.

Codignotto, J. O. (2015). Capítulo 3: *"Acreción costera"*. En H. E. Zaixso & A. L. Boraso (Eds.), La Zona Costera Patagónica Argentina. Volumen II. Comunidades Biológicas y Geología (pp. 153–166). Editorial Universitaria de la Patagonia. Comodoro Rivadavia. Versión digital.

Codignotto, J. O., Dragani, W. C., Martin, P. B, Campos, M. I., Alonso, G., Simionato, C. G., & Medina, R. A. (2011). Erosión en la bahía de Samborombón y cambios en la dirección de los vientos, provincia de Buenos Aires, Argentina. *Revista del Museo Argentino de Ciencias Naturales, 13*(2), 135–138. http://www.scielo.org.ar/scielo.php?script=sci_arttext&pid=S1853-04002011000200004&lng=es&tlng=es.

Codignotto, J. O., Kokot, R. R., & Marcomini, S. C. (1992). Neotectonism and sea - level changes in the coastal zone of Argentina. *Journal of Coastal Research, 8*(1), 125–133.

Comisión Económica para América Latina y el Caribe (CEPAL). (2014). *La Economía del Cambio Climático en La Argentina*. Primera Aproximación. – Colección Documentos de proyectos. Naciones Unidas. https://www.cepal.org/es/publicaciones/35901-la-economia-cambio-climatico-la-argentina-primera-aproximacion.

D'Onofrio, E., & Fiore, M. (2002). *Report on the Tide Dynamics of The Río de la Plata*. Proyecto: Assessment of Impacts and Adaptation To Climate Change (AIACC) LA 26, Buenos Aires.
Fiore, M., D'Onofrio, E., Grismeyer, W., & Mediavilla, D. (2008). *El ascenso del nivel del mar en la costa de la provincia de Buenos Aires*. Ciencia Hoy. V. 18 Nro. 106 agosto-septiembre 2008.
Garnica Pena, R. J., & Alcantara A, I. (2004). Riesgos por inundación asociados a eventos de precipitación extraordinaria en el curso bajo del río Tecolutla, Veracruz. *Investigaciones Geográficas*, México, n° 55, 23–45. http://www.scielo.org.mx/scielo.php?pid=S0188-46112004000300003&script=sci_abstract.
Gobierno de la República Argentina. (2007). *Segunda Comunicación Nacional de la República Argentina a la Convención Marco de las Naciones Unidas sobre Cambio Climático*. Buenos Aires. Octubre. http://unfccc.int/resource/docs/natc/argnc2s.pdf.
Gobierno de la República Argentina. (2015). *Tercera Comunicación Nacional de la República Argentina a la Convención Marco de las Naciones Unidas sobre Cambio Climático.* Buenos Aires. Noviembre. http://3cn.cima.fcen.uba.ar/docs/3Com-Resumen-Ejecutivo-de-la-Tercera-Comunicacion-Nacional.pdf.
IPCC (Grupo Intergubernamental de Expertos sobre el Cambio Climático). (2014). *Cambio climático 2014*. Informe de síntesis. Resumen para responsables de políticas. http://www.ipcc.ch/pdf/assessment-report/ar5/syr/SYR_AR5_FINAL_full_es.pdf, http://www.ipcc.ch/home_languages_main_spanish.shtml.
Kokot, R. R. (2004). Erosión en la costa patagónica por cambio climático. *Revista. Asociación Geológica Argentina, 59*(4) Buenos Aires oct./dic. 2004. http://www.scielo.org.ar/scielo.php?script=sci_arttext&pid=S0004-48222004000400018.
Kokot, R. R. (2009). *Cómo los cambios en el nivel del mar afectarán las costas argentinas?* Primeras Jornadas Interdisciplinarias de la Universidad de Buenos Aires sobre Cambio Climático, Buenos Aires (Argentina).
Kullock, D. (2007). *Sistema Urbano en Programa Nacional de Adaptación y Planes Regionales de Adaptación.* Actividad habilitante para la Segunda Comunicación Nacional. Buenos Aires, Fundación e Instituto Di Tella.
McGranahan, G., Balk, D., & Anderson, B. (2007). The rising tide: Assessing the risks of climate change and human settlements in low elevation costal zones. *Environmental and Urbanization, 19*(1), 17–37.
Naciones Unidas. (2015). *Efectos del Cambio Climático en la Costa de América Latina y el Caribe. Impactos*. CEPAL. Universidad de Cantabria. Instituto de Hidráulica Ambiental. Ed. Cepal. Documentos de proyectos. http://www.cridlac.org/digitalizacion/pdf/spa/doc19588/doc19588-contenido.pdf, www.cepal.org/ddsah.
Nerem, R. S. (1995). Global mean sea level variations from TOPEX/POSEIDON altimeter data. *Science, 1268*:708–710.
Nicholls, J., & Cazenave, A. (2010). Sea-level rise and its impacts on coastal zones. *Science, Special Section, 328*. https://doi.org/10.1126/science.1185782.
Pittau, M., Sarubbi, A., & Menéndez, Á. (2003). *Análisis del avance del Frente del Delta del Río Paraná*. III Congreso Argentino de Ingeniería Portuaria. Asociación Argentina de Ingenieros Portuarios (AADIP).
Volpedo, A. V., Yunes, T., & Cirelli, A. F. (2005). *El humedal mixohalino de Bahía Samborombón: conservación y perspectivas*. En: Humedales Fluviales de América del Sur. Hacia un manejo sustentable. Fundación Proteger y UICN, Santa Fe, Argentina.

Extreme Events and Health in Mumbai, India

Hem H. Dholakia, Ishita Jalan and Amit Garg

Abstract India's urban future is under threat from a changing climate. Evidence over the last twenty-eight years shows an increasing trend in the frequency and intensity of extreme events. With over 300 extreme events, the majority of which occurred since 2005, estimated damages are around US$78.8 billion. There is a significant toll of these extreme events with over 1.33 billion people being affected. Studies on future climate projections indicate that as the frequency and intensity of extreme events increase, cities may experience nonlinear health impacts. In addition to increasing the burden of disease, it will put enormous strain on financial resources as well as health systems. These effects are exemplified in the context of the mega-city, Mumbai. Urban planning choices have left the city vulnerable to flooding and heat waves. There exists a substantial burden of both communicable and non-communicable diseases in Mumbai. A single, unprecedented rainfall event, in 2005, resulted in over 200,000 patients being treated in a single tertiary care hospital of the city. These findings indicate the need for cities to proactively plan for responding to extreme events. There remain significant opportunities to design cities that are resilient to extreme events and to protect the health of people.

Keywords Extreme events · Disasters · India · Mumbai · Mega-city · Health

1 Introduction

There is a clear understanding that India's cities are in a state of flux. Opportunities for employment, a better standard of living as well as access to health services are some of the drivers of increasing urbanisation. It is expected that by 2050, India's urban population will double and be reaching nearly 814 million persons (UN 2014). This is also the time when the country will transition to a predominantly urban

---

H. H. Dholakia (✉) · I. Jalan
Council on Energy, Environment and Water, New Delhi, India
e-mail: hem.dholakia@ceew.in

A. Garg
Indian Institute of Management, Ahmedabad, India

R. Akhtar (ed.), *Extreme Weather Events and Human Health*,
https://doi.org/10.1007/978-3-030-23773-8_24

population. This means that present choices surrounding urban development will disproportionately continue to influence human health in India, now as well as in future.

Increased urbanisation is known to influence health positively as well as negatively. Urban areas concentrate people, thereby exposing a large population to disaster risk. These risks are compounded in developing countries, where population growth outpaces that of the formal local economy (Butsch et al. 2016), resulting in inadequate infrastructure and service provision to citizens. As a result, settlements spawn on wastelands of the city, near drains and marshy areas, that are devoid of any services or basic facilities. These settlements are mostly occupied by the urban poor. This factor enhances vulnerability of people in the event of a disaster (Revi 2008). It also facilitates the rapid spread of several epidemics. The geography of most cities, i.e. proximity to coasts and riverbanks, makes them particularly vulnerable to climate change (Rosenzweig et al. 2010). Urban planning choices and materials used, e.g. paved and impermeable roads, promote surface run-off, adding pressure to drainage systems (Mishra et al. 2015). This may pre-dispose urban areas to flooding. These surfaces also retain and radiate heat adding to the urban heat-island effect.

At the same time, several studies suggest that cities may confer an ‘urban health advantage’ that results in urban health indicators often being better than its rural counterpart (Vlahov et al. 2005). Several hypotheses attempt to explain this advantage. One explanation is that the proximity to wealth may have spillover effects to benefit economically weaker sections of society. The presence of affluent and prominent society members may exert a certain ‘pressure’ on government agencies to provide basic civic services within reach of more disadvantaged urban residents (Lynch et al. 2004). For instance, Wen et al. (2003) find that measures of neighbourhood health have positive associations with affluence after adjusting for neighbourhood-level poverty, income inequality as well as education. A second explanation is that cities offer access to markets for food, housing, education and health care. This gives urban residents an edge compared to non-urban residents, to access better nutrition, knowledge and health services, despite an inequitable distribution of these goods (Panel on Urban Population Dynamics 2003). Urban areas also provide the potential for social cohesion, social movements and political mobilisation on account of density of people. In addition to creating a public voice to demand more resources, social support is known to play an instrumental role in contributing to an environment that is conducive to health. The health status of populations in urban areas, then, is an outcome of the complex influence of these factors.

The urban transition will happen against the backdrop of a changing climate. Human activities have already resulted in warming of 1 °C since pre-industrial times, and there is strong evidence that it will reach 1.5 °C between 2030 and 2052 at current rates (IPCC 2018). Global mitigation efforts will determine the extent of warming, and in a worst-case scenario, average temperatures may exceed 4 °C by the end of the century (IPCC 2018). There is considerable evidence to demonstrate that a changing climate is driving changes in the frequency, intensity, spatial extent, duration, timing of weather and climate extremes (IPCC 2012). As a consequence, the world over, we are witnessing unprecedented climate extremes. Increased warming will result in

nonlinear increases in adverse health outcomes—this is true for India's cities as well. It is within this broader context that we provide a perspective on climate extremes and linkages to health in India. We further draw upon the experience of one mega-city, Mumbai, to provide a deeper insight into the connections between outliers and health.

2 A History of Extremes

To understand trends of extreme events such as floods, storm, heat waves as well as drought events, we analysed data from EM-DAT[1] database for the period 1990–2018 (Fig. 1). Over this period, we find that a total of 335 extreme events were recorded, 60% of which were related to flooding and 26% to storms. The balance events were attributed to heat waves and droughts. It is estimated that across these events, a total of approximately 1.33 billion people were affected. The direct economic losses incurred were about US$78.8 billion (EM-DAT). However, the actual damages were probably an order of magnitude higher as indirect losses (e.g. loss of employment, disability, disease, migration, etc.) were not accounted for. Some studies estimate that extreme events alone cost India an average of US$5–6 billion a year in damages (Garg et al. 2015).

We find a clear upward trend in the total number of extreme events over the last 28 years, suggesting that the frequency of extreme events is increasing over time (Fig. 1). For instance, forty-nine flood events occurred in just three years between 2005 and 2007. This is a significant change when considering the fact that only 44

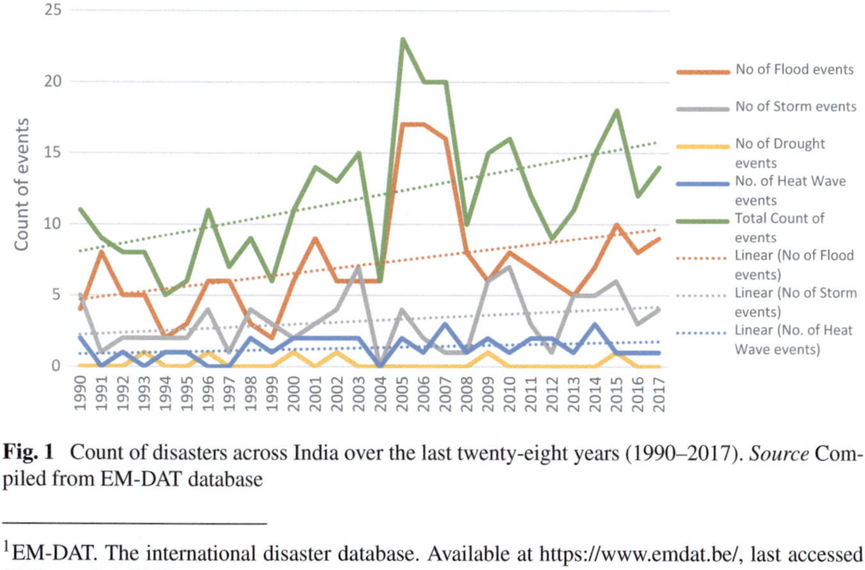

Fig. 1 Count of disasters across India over the last twenty-eight years (1990–2017). *Source* Compiled from EM-DAT database

[1]EM-DAT. The international disaster database. Available at https://www.emdat.be/, last accessed 22 November 2018.

such events were recorded for the decade from 1990 till 1999. A similar trend is observed for heatwaves as 2015–2017 have been recorded as the three consecutive hottest years in the last century. While eight states experienced temperatures above 45 °C in 2015, this number rose to twelve states in the year 2016. Some of the highest temperatures were observed in Churu in Rajasthan and Telangana (both recording 48 °C) as well as Allahabad and Vidarbha (recording 47 °C) in 2015. Just the last three years have taken a large health toll resulting in death of 2812 people (EM-DAT). As death registration systems, in India, are weak, a large number of cases do not get recorded. The reported heat-wave deaths are therefore an underestimate. In addition to an increase in frequency, the lived experience indicates that the intensity of these disasters is also increasing over time.

3 Future Projections of Extreme Events

3.1 Precipitation

Future projections suggest that extremes of rainfall as well as heat are likely to increase across India. For instance, Ali et al. assessed future climate projections across 57 urban locations based on data from 1901 till 2010 (Ali et al. 2014). They found that the majority of the cities studied (33 of 57) were likely to experience increases in one-day maximum mean rainfall from 2036 to 2060, while the remaining would experience declines of up to 27%. These findings are indicative of flood risk in certain cities across west-central India and drought risk in north-western and north-eastern India. These trends are corroborated by other studies (Garg et al. 2015) that suggest that projected increases in extreme precipitation may cause damage to urban infrastructure as well as the impact on crop production.

Heavy rainfall and flooding can trigger widespread health impacts through water-borne and vector-borne diseases. This is especially a concern in tropical and sub-tropical regions (McMichael et al. 2005) such as India. A study in Chennai analysed the relation between extreme precipitation and gastrointestinal-related illness. They found a significant relation between extreme precipitation and gastrointestinal-related, particularly in children below the age of five (Bush et al. 2014). In addition to epidemics, flooding can displace people and may also lead to loss of life.

3.2 Heatwaves

Temperatures across India are projected to increase by 3.3–4.8 °C by 2080s relative to pre-industrial times under changing climate (Chaturvedi et al. 2012). A strong increase in intensity, duration and frequency of heatwaves is projected under different climate change scenarios in India. This finding is common across several different studies. For instance, Murari et al. (2015) used future temperature projections across three climate scenarios (Representative Concentration Pathways 2.6, 4.5 and 8.5). They found a strong prevalence of heat waves towards the latter half of the century. For Southern India, where the phenomenon is seldom observed, projections for heat waves were severe along with west and east coast. Impacts were projected to be lower under an ambitious global mitigation regime.

Recently, Mazdiyasni et al. (2017) have demonstrated that changes in heat-wave characteristics across two time periods (1960–1984 and 1985–2009) are associated with an increased probability of deaths in India. The northern region of India is at most risk for adverse health impacts due to heat (Patz et al. 2014). Heat is also known to cause cardiac arrest and respiratory illness (UN-HABITAT 2014), thereby exacerbating these diseases. Further, increased morbidity and loss of productivity are other major implications of heat. This is particularly important from the angle of outdoor workers who will also face economic loss as a result (Kjellstrom et al. 2009).

Increases in temperature will change the distribution of vector-borne diseases (Dhiman et al. 2010). For several colder or high-altitude regions, warming may increase the transmission window for malaria and dengue, whereas it may restrict this window in other parts. For example, Himachal Pradesh and north-eastern states can experience increase in the malaria transmission window. For southern states of India where malaria and dengue are common due to their warm and humid climate, a decrease in number of cases might take place due to higher temperature. For cities like Delhi, a 2 °C temperature rise might increase the transmission for dengue by one week, whereas 4 °C temperature rise might shrink this period.

There is a clear linkage between changing extremes (precipitation and heat) and urban health outcomes. This has been shown across the world and more recently in the context of India. In the next section, we exemplify, in the context of Mumbai, some of the more general points discussed previously. After a short introduction, we will discuss the impacts of heat waves as well as flooding on urban health in the context of Mumbai.

4 Mumbai: A Case Study

Mumbai, with an area of 4355 km^2, had an estimated population of 12.5 million people (Census 2011), making it the fifth largest mega-city in the world. The city is the financial capital of India, a base for many industries and multi-national companies, contributing 5% to the GDP of India (Hallegatte et al. 2010). As a coastal city, Mumbai has evolved from seven islands and reclaimed land to form one mass of land with varying topology (Satterthwaite et al. 2007). At places, the elevation is just 1 m above mean sea level and 1.5 m below high tide level such as the Bandra-Kurla complex, Wadala and Worli (Patankar et al. 2010). Based on this information, it is not hard to imagine that Mumbai is vulnerable to flooding. In addition to this, land-use change has also contributed in making the city prone to extreme events of climate change. Over the last 60 years, Mumbai's built-up area has expanded from 12 to 52%, and the mangrove forest cover has shrunk from 235 to 160 km^2 in parts of Mumbai and Navi Mumbai (Bhagat et al. 2006). This rapid encroachment of nature is leading to an environmentally sensitive city by eliminating ecosystems which act as barriers to extreme events and calamities. Along with infrastructure and ecology, risk to a city is also determined by social, political and economic dimensions (Butsch et al. 2016). Nearly 55.67% of the population resides in slums or informal housing in Mumbai, mostly located on disaster-prone areas such as creek sides, low-lying and marshy zones or on hill slopes (Bhagat et al. 2006) (Fig. 2).

Income inequality is another driver of vulnerability in the Mumbai. A household survey of 549 slum dwellers found that 76.6% of households earned less than 12,000 INR per month. Of these, about 50% of households earned less than 9000 INR (Raghu et al. 2018). Disasters can impact economic stability of the poor by damaging their habitation, health and livelihoods. For example, Patankar and Patwardhan (2016) show that households with income less than INR 5000 per month suffered an average loss of INR 37,000 during the Mumbai floods in 2005. Damages were roughly five times their monthly income, whereas for middle-income households' damages were roughly three times their average monthly income.

4.1 Health Status of People

Given the size of the city and the number of people, Mumbai has a fairly large burden of disease in terms of infectious as well as non-communicable diseases. Data from official sources indicate that while there has been a decline in malaria cases (21,939 in 2012 and 11,607 in 2017), dengue has increased (4867 in 2012 to 17,771 in 2017) by nearly 265% while diarrhoea cases have remained more or less the same (99,827 in 2012 to 100,643 in 2017) (Praja 2017). Further, there is an enormous disease burden due to non-communicable diseases as well. These contribute to the top ten causes of death in Mumbai (Table 1).

Flooding remains a significant risk factor for diarrhoeal disease, leptospirosis and hepatitis due to contamination of water. Acute respiratory infections are known to be associated with overcrowding, while changes in habitat and behaviour influence the spread of vector-borne diseases such as malaria and dengue.

4.2 Extreme Rainfall in Mumbai

The flood event of 26 July 2005 in Mumbai was in many ways unprecedented. The city experienced 944 mm of rainfall in less than 24 hours. In contrast, the historical average annual rainfall of Mumbai City has ranged between 2050 and 2300 mm (Gupta 2007). Economic damages were estimated to be around US$2 billion, and the official death toll recorded was 447 persons. Immediately following the floods, a

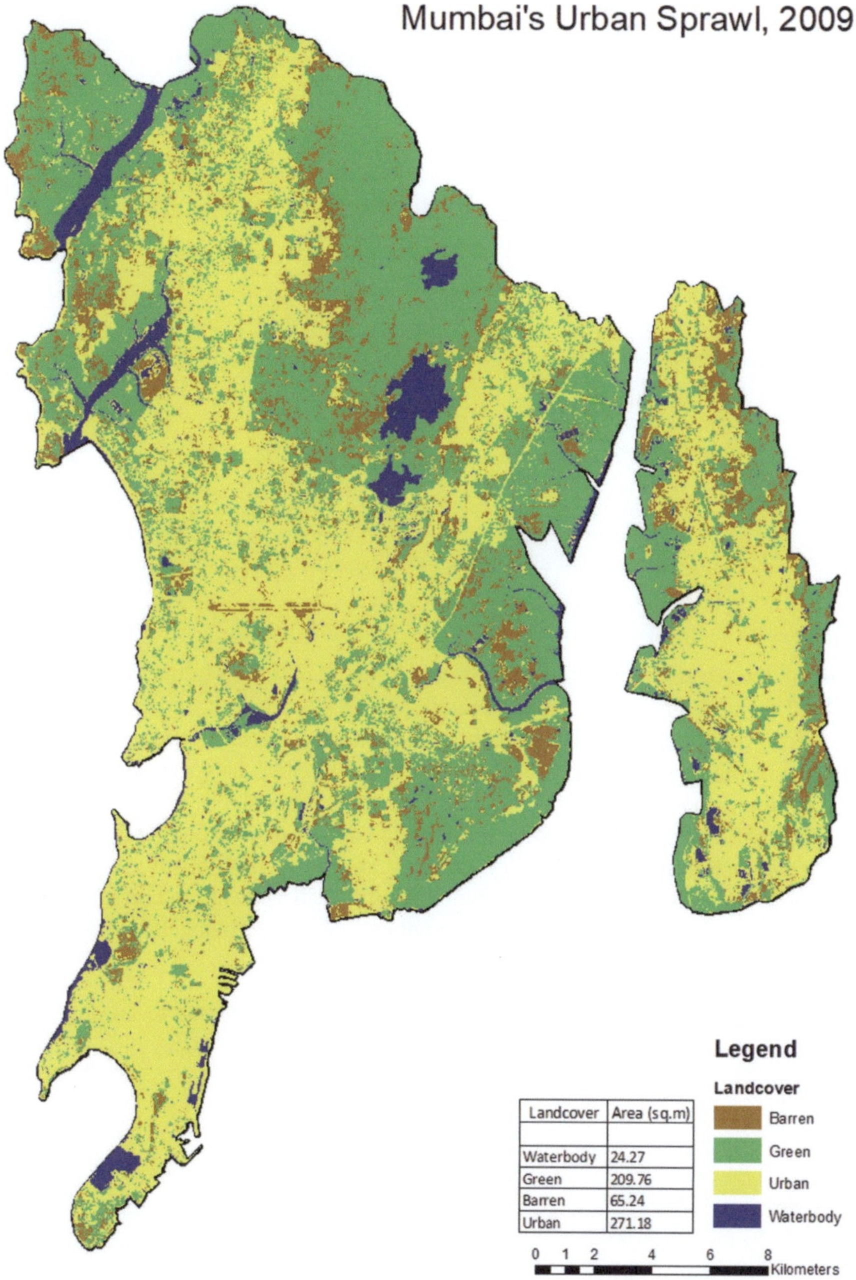

Fig. 2 Land-use map for Mumbai city (2009). *Source* CEEW compilation using Google Earth and QGIS

Table 1 Top ten causes of death in Mumbai

Cause of death	2011–12	2012–13	2013–14	2014–15	2015–16[a]	2016–17[a]
Acute myocardial infarction	10,475	9897	10,187	10,263	8955	8961
Other forms of heart diseases	7690	7488	7507	8781	6696	5659
Septicaemia	6024	5611	5650	6014	5117	3787
Tuberculosis	8375	7170	7319	6501	5181	6472
All other ischaemic heart diseases	4590	4375	4366	4554	4298	4249
All other hypertensive diseases	3541	3585	4118	4604	3998	3407
All other diseases of the respiratory system	3934	4078	4131	4336	3674	3674
Pneumonia	4072	3330	2937	3215	3129	3034
Renal failure	3734	3431	3377	3308	3065	3160
Diseases of the liver					2859	2751

Source Praja (2017)

spike in the adverse health outcomes was witnessed. A single tertiary care hospital (Kind Edward Memorial Hospital) treated over 12,000 patients, and the community outreach programme addressed health needs of nearly 200,000 persons (Table 2). It is conceivable that the health toll was significantly higher; however, in the absence of a central data repository, this figure is difficult to determine.

The impact of the 2005 episode was compounded by major flooding that occurred in the Mithi River. This river is 14 km long and flows from north towards west-central Mumbai to finally drain in the Arabian Sea. It is also a major storm water drainage channel of Mumbai. Encroachment by informal settlements, industrial and

Table 2 Cases treated at King Edward Memorial Hospital after the July 2005 floods

Disease	Outpatient attendance	Indoor patients
Malaria	5514	746
Leptospirosis	626	282
Dengue fever	157	58
Undifferentiated fever	6325	1157
Community patients treated	200,516	–

Source Compiled from Supe and Satoskar (2008)

municipal waste dumping on the banks and inside the river as well as the construction of roads on flood plains were responsible for affecting the carrying capacity of Mithi (Ranger et al. 2011). In addition, suburban Mumbai lacked underground drainage infrastructure and reverse flow of water into the city during the high tide further inhibited drainage, thereby exacerbating the floods (Stecko and Barber 2007). The low-lying areas around the Mithi River remained flooded even after the majority of the city returned back to normal functioning.

The worst impacts were the people from informal habitations who comprised 70% of the population residing on banks of Mithi River (Parthasarathy 2016). Unsanitary conditions along with contamination of drinking water resulted in 1318 cases of gastroenteritis, 194 cases of hepatitis, 406 malaria cases and 197 cases of leptospirosis in a single week from 29 July till 5 August 2005 (Patankar and Patwardhan 2016). Water-borne diseases and gastrointestinal infections led to 248 deaths. An increase in the vector-borne diseases, especially malaria, was observed as following the disaster (Butsch et al. 2016).

While the scale and impact of the 2005 floods remain unmatched, smaller and more localised floods in the city, particularly in the northern suburbs and in central and western parts of the city, are now a regular feature during the annual monsoon season (June–September) (Supe and Satoskar 2008). On 29 August 2017, more than 200 mm rain fell within 12 hours in Mumbai flooding the city and blocking all the modes of transport including railways and flights. Other extreme rainfall events include June 2015 with rain gauge in Santa Cruz measuring 283.4 mm rainfall in 25 h and July 2009 when 293.1 mm of rainfall was recorded by the Santa Cruz station and 152.4 mm rainfall in Colaba in 30 hours.[2] It is clear that the city remains vulnerable to flooding events.

4.3 *Heatwaves and Coldwaves*

Temperature extremes such as heat waves and cold spells have been studied to a lesser extent in India. Dholakia et al. (unpublished) studied the heat as well as coldwaves for Mumbai between 2005 and 2012. As there is no universally accepted definition of heat or coldwaves, they looked at two or more consecutive days where temperatures exceeded 97th, 98th or 99th percentile of the historical temperature distribution (1951–2005) for heatwaves and 1st, 2nd and 3rd percentiles for coldwaves, respectively. The study found that between 2005 and 2012, Mumbai had ten heatwaves and three coldwaves with an average duration of 2.3 days. Mortality was found to increase with the intensity of heat waves (Fig. 3). Our findings suggest that planning interventions at a local level for temperature extremes are important.

[2]https://www.livemint.com/Politics/4FCQcYA57XwzwBqXNMNzaP/Heavy-rains-throw-life-out-of-gear-in-Mumbai.html.

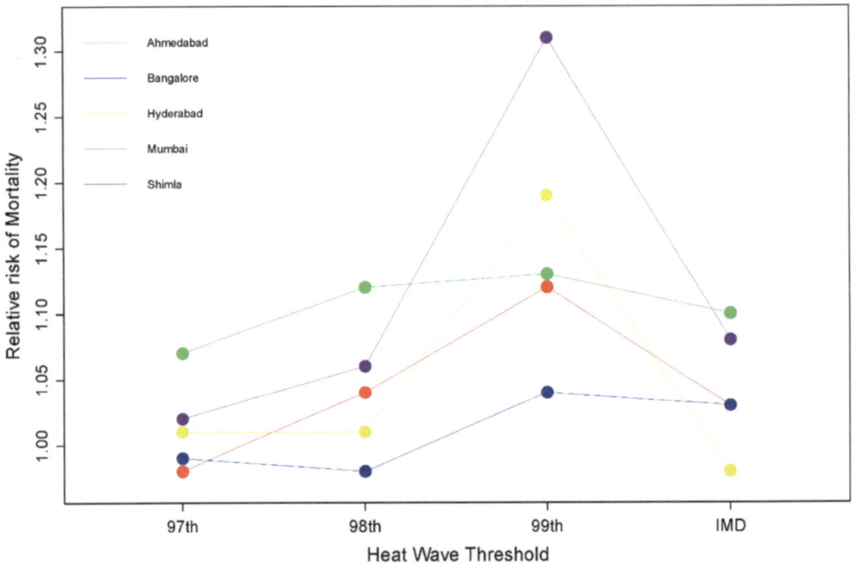

Fig. 3 Changes in relative risk of mortality with heat-wave intensity. *Source* Dholakia, Bhadra and Garg (unpublished). The x-axis shows the heat-wave thresholds based on the 97th, 98th, 99th percentile of temperature distribution. IMD refers to the definition of heatwaves by the Indian Meteorological Department

4.4 Future Projections

For Mumbai, findings from studies that look at future projections of extreme events are similar to those for India. For example, Ranger et al. quantified the hazard due to rainfall in Mumbai using the PRECIS model. The PRECIS is a high-precision model for regional climate, modelling based on Hadley Centre climate models (HadCM3). In the A2 SRES scenario, it was projected that there will be an increase in the mean temperature by 3.6 °C and in mean rainfall by 6.5% by 2080. Further, they downscaled data from the PRECIS model at the city level using a weather generator (WXGEN). Results based on a 200-year simulated record of rainfall indicated increases in extreme rainfall for all the return periods, especially for shorter return periods.

Hallegatte et al. (2013) in their research estimated the future flood losses in major coastal areas. While Kolkata ranked 14 in terms of annual average loss incurred due to flood, Mumbai ranked 5th in the list. Annual average losses, due to flooding for Mumbai, were estimated at US$6 billion or 0.49% of the gross domestic product (Hallegatte et al. 2013). Future increases in heat-related mortality are projected by studies across all climate scenarios.

5 Conclusions

There is a clear recognition that extreme events are linked to adverse health outcomes. Though studies linking extreme events and health are sparse in India, the few that exist establish this link at the national level as well as in the context of Mumbai. Additional research in this area is important given that India already faces a large burden of disease. Disasters are likely to exacerbate current burdens and possibly result in new ones. Most studies thus far look at specific events (e.g. 2005 Mumbai floods) or use surveys or modelling approaches to understand current and future impacts. Garnering evidence through cohort studies would help establish causality between extreme events and health outcomes. Very often, the focus is on pathologies (e.g. malaria, dengue, leptospirosis) other than mental health, which also contribute to the disease burden. It is estimated that 20–40% of the affected population suffer from mild psychological distress and 30–50% from moderate to severe distress (WHO 2019).

One possible reason that disease burdens are not adequately captured within the Indian context is that data are fragmented. Despite a common framework, data are often spread across registries of different hospitals. Whereas there is some attempt to assimilate data from the government hospitals, data from the private sector hospitals are seldom available. This means that prevalence and incidence numbers may be underestimated.

Finally, in India there remains a lack of warning systems for extreme events. Whereas there is a well-developed system in place for cyclone warnings, systems for other extreme events are lacking. It is only of late that cities like Ahmedabad have developed a heat-health warning system. Though preliminary evidence shows that it may have had a positive impact on heat and health, there remains the need for monitoring and evaluation. The generation of evidence will ultimately feed into policy and decision-making around preventing the health impacts of disasters.

Acknowledgements The authors would like to thank Ms. Tanushree Garg for her assistance with developing the land-use map for Mumbai.

References

Ali, H., Mishra, V., & Pai, D. S. (2014). Observed and projected urban extreme rainfall events in India. *Journal of Geophysical Research: Atmospheres*, 12,621–12, 641. https://doi.org/10.1002/2014jd022264.

Bhagat, R. B., Guha, M., & Chattopadhyay, A. (2006). Mumbai after 26/7 deluge: Issues and concerns in urban planning. *Population and Environment, 27,* 337–349.

Bush, K. F., et al. (2014). Associations between extreme precipitation and gastrointestinal-related hospital admissions in Chennai, India. *Environmental health perspectives, 122,* 249–254.

Butsch, C., Kraas, F., Namperumal, S., & Peters, G. (2016). Risk governance in the megacity Mumbai, India—A Complex Adaptive System perspective. *Habitat International*. https://doi.org/10.1016/j.habitatint.2015.12.017.

Census of India. (2011). Office of the Registrar General of India. New Delhi.

Chaturvedi, R. K., Joshi, J., Jayaraman, M., Bala, G., & Ravindranath, N. H. (2012). Multi-model climate change projections for India under representative concentration pathways. *Current Science, 103,* 12.

Dhiman, R. C., Pahwa, S., Dhillon, G. P. S., & Dash, A. P. (2010). Climate change and threat of vector-borne diseases in India: Are we prepared? *Parasitology Research, 106,* 763–773.

Garg, A., Mishra, V., & Dholakia, H. H. (2015). *Climate change and India: Adaptation gap*. Ahmedabad: Indian Institute of Management.

Gupta, K. (2007). Urban flood resilience planning and management and lessons for the future: A case study of Mumbai, India. *Urban Water Journal, 4,* 183–194.

Hallegatte, S., et al. (2010). Flood risks, climate change impacts and adaptation benefits in Mumbai. *OECD Environment Working Papers*. https://doi.org/10.1787/5km4hv6wb434-en.

Hallegatte, S., Green, C., Nicholls, R. J., & Corfee-Morlot, J. (2013). Future flood losses in major coastal cities. *Nature Climate Change, 3,* 802–806.

IPCC. (2012). *Managing the risks of extreme events and disasters to advance climate change adaptation*. Intergovernmental Panel on Climate Change.

IPCC. (2018). *Global warming of 1.5 °C.* Intergovernmental Panel on Climate Change.

Kjellstrom, T., Holmer, I., & Lemke, B. (2009). Workplace heat stress, health and productivity—An increasing challenge for low and middle-income countries during climate change. *Global Health Action, 2,* 1–6.

Lynch, J., et al. (2004). Is Income Inequality a Determinant of Population Health? Part 1. A Systematic Review. *The Milbank Quarterly, 82,* 5–99.

Mazdiyasni, O., AghaKouchak, A., Davis, S. J., Madadgar, S., Mehran, A., Ragno, E., et al. (2017). Increasing probability of mortality during Indian heat waves. *Science Advances, 3*(6). Available from: http://advances.sciencemag.org/content/3/6/e1700066.abstract.

McMichael, A. J., Woodruff, R. E., & Hales, S. (2005). Climate change and human health: Present and future risks. *Lancet, 367,* 859–869.

Mishra, V., Ganguly, A. R., Nijssen, B., & Lettenmaier, D. P. (2015). Changes in observed climate extremes in global urban areas. *Environmental Research Letters, 10*.

Murari, K., Ghosh, S., Patwardhan, A., Daly, E. & Salvi, K. (2015). Intensification of future severe heat waves in India and their effect on heat stress and mortality. *Regional Environmental Change*. https://doi.org/10.1007/s10113-014-0660-6.

Panel on Urban Population Dynamics. (2003). *Cities transformed: Demographic change and its implications in the developing world*. National Academies Press.

Parthasarathy, D. (2016). Decentralization, pluralization, balkanization? Challenges for disaster mitigation and governance in Mumbai. *Habitat International, 52,* 26–34.

Patankar, A., & Patwardhan, A. (2016). Estimating the uninsured losses due to extreme weather events and implications for informal sector vulnerability: A case study of Mumbai, India. *Natural Hazards, 80,* 285–310.

Patankar, A., Patwardhan, A., Andharia, J., & Lakhani, V. (2010). *International Workshop on Climate Change and Vulnerability Assessment and Urban Development Planning for Asian Coastal Cities*.

Patz, J. A., Frumkin, H., Holloway, T., Vimont, D. J., & Haines, A. (2014). Climate change: Challenges and opportunities for global health. *JAMA—Journal of American Medical Association, 312,* 1565–1580.

Praja. (2017). *The state of health of Mumbai*. Available from: https://praja.org/praja_docs/praja_downloads/Report%20on%20The%20STATE%20of%20HEALTH%20of%20MUMBAI.pdf.

Raghu, A., Ernst, J., Sahariah, S. A., Gandhi, M., & Murphy, G. (2018). Healthcare choices in Mumbai slums: A cross-sectional study. *Wellcome Open Research*, 1–17. https://doi.org/10.12688/wellcomeopenres.13127.1.

Ranger, N., et al. (2011). An assessment of the potential impact of climate change on flood risk in Mumbai. *Climatic Change, 104,* 139–167.

Revi, A. (2008). Climate change risk: An adaptation and mitigation agenda for Indian cities. *Environment and Urbanization, 20,* 207–229.
Rosenzweig, C., Solecki, W., Hammer, S. A., & Mehrotra, S. (2010). Cities lead the way in climate-change action. *Nature, 467,* 909–911.
Satterthwaite, D., Huq, S., Pelling, M., Reid, H., & Lankao, P. R. (2007). Adapting to climate change in urban areas: the possibilities and constraints in low and middle income nations. Human Settlements, 58.
Stecko, S., & Barber, N. (2007). Exposing vulnerabilities: Monsoon floods in Mumbai, India. Case study prepared for Revisiting Urban Planning: Global Report on Human Settlements, 1–14.
Supe, A., & Satoskar, R. (2008). Health services responses to disasters in Mumbai sharing experiences. *Indian Journal of Medical sciences, 62,* 242.
UN. (2014). *World urbanization prospects*. Geneva: United Nations.
UN-HABITAT, & UNESCAP. (2014). Quick guide for policy makers: Pro-poor urban climate resilience in Asia and the Pacific.
Vlahov, D., Galea, S., & Freudenberg, N. (2005). The urban health 'advantage'. *Journal of Urban Health: Bulletin of the New York Academy of Medicine, 82,* 1–4.
Wen, M., Browning, C., & Cagney, K. (2003). Poverty, affluence, and income inequality. Neighborhood economic structure and its implications for health. *Social Science and Medicine*, 843–860.
WHO. (2019). *Disasters and mental health*. Available from: http://www.searo.who.int/bangladesh/disastersandmentalhealth/en/.

Future Outlook

Conclusion and Suggestions

Rais Akhtar

Abstract Having analyzed various case studies from regions encompassing all continents, there cannot be two opinions that climate change will lead to an increase in the number of extreme weather events, including increased frequency and intensity of hurricanes/typhoons/cyclones, heat waves, sea-level rise, flooding, droughts and wild fires. Extreme heat conditions in Australia and deadly cold snap to US states due to Polar vortex in January 2019 due to climate change are important instances of extreme weather events.

Having analyzed various case studies from regions encompassing all continents, there cannot be two opinions that climate change will lead to an increase in the number of extreme weather events, including increased frequency and intensity of hurricanes/typhoons/cyclones, heat waves, sea-level rise, flooding, droughts and wild fires. Extreme heat conditions in Australia and deadly cold snap to US states due to Polar vortex in January 2019 due to climate change are important instances of extreme weather events.

1 Paris Climate Agreement and Temperature Rise

The major objective of Paris climate Agreement is to strengthen the global response to the threat of climate change by keeping a global temperature rise this century well below 2 °C above pre-industrial levels and to pursue efforts to limit the temperature increase to 1.5 °C. IPCC asserts that emissions of carbon dioxide (CO_2) caused by human activity must reach "net zero" by 2050 to keep the average rise in global temperatures at 1.5 °C above pre-industrial levels to reduce catastrophic climate-change risk on populations (IPCC 2018).

R. Akhtar (✉)
Health Management Research, IIHMR Delhi, International Institute of Health Management Research, New Delhi, India
e-mail: raisakhtar@gmail.com

R. Akhtar (ed.), *Extreme Weather Events and Human Health*,
https://doi.org/10.1007/978-3-030-23773-8_25

However, according to Mark Lynas, "humans are now transferring 10 billion tonnes of carbon from the earth's crust—in the form of combusted coal, oil and gas—into the atmosphere each year" (Lynas 2019). Thus, the targets agreed in Paris on cutting emissions based on NDCs would not be enough even if there were larger and more ambitious cuts after 2030 (Storrow 2018).

The just-published report from British Met Office predicts "that from 2019 to 2023, we will see temperatures ranging from 1.03 to 1.57 °C above the 1850–1900 level, with enhanced warming over much of the globe, especially over areas like the Arctic. It emphasizes that until we reduce greenhouse gas emissions, we can expect to see upward trends in global averaged temperatures."(McGrath 2019).

Warmer ocean temperatures and higher sea levels are expected to intensify their impacts. Eight of the 10 costliest hurricanes on record in the USA have occurred since 2004. Hurricanes Katrina (2005) and Sandy (2012) cost $125 billion and $65 billion, respectively. Hurricanes Andrew (1992) and Ike (2008) cost $27 billion each. The cumulative cost of the 16 separate billion-dollar weather events in the USA in 2017 was $306.2 billion, breaking the previous cost record of $214.8 billion in 2005. Frequency and intensity vary from basin to basin (https://www.c2es.org/content/hurricanes-and-climate-change/ 2017).

An important driver of the increased cost of hurricanes is increasing development in coastal areas. US coastal populations grew by nearly 35 million people between 1970 and 2010. Coastal counties account for nearly 40% of the total US population. As more development occurs in coastal vulnerable areas—regardless of climate change—the more likely the damage will grow.

Presently, about 42% of the world's population lives within 100 km of the coast. As populations and economic activity in the coastal areas grow the region faces more vulnerability due to sea-level rise. According to Science Daily, rising seas could result in 2 billon climate refugees by 2100. (Science Daily 2017). Studies also reveal that sea-level rise flooding has caused a total home value loss of $7.4 billion since 2005 in the states of Florida, South Carolina, North Carolina, Virginia and Georgia (Morad 2018).

Reducing greenhouse gas emissions is, therefore, a long-term solution to reduce the risk of the strongest storms in the future. Communities can also bolster their resilience to the impacts of hurricanes by:

- Preserving coastal wetlands, dunes, and reefs to absorb storm surges (currently, the emphasis has been on the construction of green infrastructures that can help mitigate flooding by absorbing rain and storm water).
- Replenishing beaches and improve infrastructure that affords coastal protection, such as seawalls (vertical walls as well as sloping structures on the banks).
- Elevating vulnerable buildings to reduce flood damage.
- Designing structures to be resilient to high winds and flying debris.
- Enacting policies that discourage development and high population density in vulnerable areas.

- Preparing prior to a storm's arrival by boarding windows, clearing property of potential flying debris, and having an evacuation plan is being currently in practice in the Florida region.

If global temperatures rise by up to 3 °C, above their preindustrial levels, the risk of extreme events could grow by as much as fivefold in certain parts of the world. Overall, up to 60% of locations across North America, Europe, East Asia and parts of southern South America would likely see at least a threefold increase in various extreme events, according to a study published in the journal *Science Advances* (Harvey 2018).

Events like record-setting heat, extreme rainfall and drought will happen more frequently around the world even if global climate targets are met, new research suggests, and missing those targets could make the risk even worse.

2 Case Studies

Case studies in the book suggest important measures with a focus on adaptation in order to minimize the impacts of extreme weather events. Nevertheless, even developed economies, including USA, EU, Japan and Australia, become vulnerable due to extreme weather events. Like cold blast in northern USA, and hottest January 2019 in Australia, flooding in Europe and heat wave in Japan are contemporary examples of extreme events.

Measures include that at local levels, multi-sectoral programmes to reduce extreme weather impacts may need to be integrated with wider poverty reduction strategies, combining agriculture development with economic support to increase the resilience of the most disadvantaged communities. At national levels in developing countries, strong multi-ministry action by governments could usefully combine economic, agricultural, and other programmes to drive sustainable national programmes to improve countries' resilience to climate change. At a global level, donors and multilateral agencies need to prioritize the most vulnerable, even when political/policy environments may not be the easiest to work in.

Floods and excessive rainfall are associated with increased incidences of various diseases, especially water-borne, vector-borne, and food-borne diseases. Mitigation of these negative effects remains a critical topic for developing society. In addition to focusing on the health effects of extreme temperature events, the public and private sectors in developing countries may be required to increase their attention on the health consequences associated with water-related extreme events with a focus on water availability, access, and water pollution.

Heat health warning systems are important for reducing the harmful effects of extreme weather and climate events. Government and stakeholders require appropriate mechanism to effectively improve health and address the social determinants of health equity in developing countries. France's heat alert system that commenced after terrible European heat wave of 2003 has since won plaudits from the UN, which has recommended that other countries refer to the European nations as a model for preventing such disasters. Besides, NRDC Ahmedabad Heat Action Plan 2017 is relevant for developing countries.

Understanding and incorporation of more explanatory variables of Urban Heat Island (UHI) from both meteorological and urban factors is required to identify their pivotal association with UHI intensification in the context of developing countries. With such acquired knowledge of the UHI phenomenon and related underlying mechanisms, urban planners, designers, and decision-makers can perform more evidence-based decision-making to create, reform, or rejuvenate climate-friendly sustainable cities for enhanced livability in future. Furthermore, such scientific knowledge will provide valuable input and tools for policy-makers to make the right policy choices for sustainable healthcare solutions.

Besides extreme storms and weather pose a great economic threat: Analysis of the probability of extreme storms in different parts of a country would be very valuable in identifying the costs they will impose and setting priorities for investments in adaptation. To the extent that the fields of climatology and oceanography can shed light on this issue, investments in further research will be essential. Policy-makers must order a more detailed analysis of the impacts of extreme storms on large urban areas, particularly metropolitan cities. In particular, work that can consider the macroeconomic or multiplier implications of the loss of key urban infrastructure will be essential to set priorities for investments in adaptation or strategies to minimize the harm caused by such storms. The private sector must anticipate these problems as well; every company whose business relies on transportation through the harbour or airport should create plans for continuity of its operations in the face of such extreme storms.

For building resilience, in particular, it is recommended to reorient the institutional structure of the national government, strengthen technical and financial support to local governments, and recognize the importance of NGOs and community-based initiatives.

The magnitude of the cyclone in some regions and its associated impacts raised the question of the influence of climate change and sea-level rise in the frequency and intensity of these extreme events. Although new studies focus on the attribution of cyclones to climate change, it is still difficult to precisely quantify the influence anthropogenic climate change has on cyclones. As mentioned earlier, the high number of people seeking shelter in the evacuation centres increased the risk of secondary outbreak of diarrheal, acute respiratory diseases, and vector-borne diseases Although some cases were observed in the most affected areas or in urban environment, the monitoring and treatment of these cases prevented the generation of large outbreaks of these diseases. Malnutrition was closely monitored as well as vector-borne disease and campaigns of reduction of mosquito populations were conducted in the months after the cyclone.

Records of natural disasters, and their social and economic losses, show the need for understanding the dimension of different types of disasters which require inter-disciplinary and inter-sectorial action for their proper management and prevention. Particularly, the health sector must operate directly on the socio-environmental determinants of disasters to concretely assist in mitigating damage—a responsibility for which environmental health surveillance is fundamental. It is necessary to advance knowledge regarding the populations and places in the country that is most vulnerable to disasters particularly in coastal zones. Furthermore, policies need to work both horizontally and vertically with regard to protection and civil defense in the coming decades, with a focus on reducing the high social and economic costs associated with disasters in developing countries.

As we understand that global warming imposes important changes in climate and weather, with increasing temperatures and levels of drought, Australians experience a higher risk of suffering from extreme wildfires and intense heat waves as happened in January 2019. The many impacts will probably be distributed unequally throughout society. Infants, the elderly, labourers, people with pre-existing conditions, lower socioeconomic status, or belonging to indigenous groups may be at higher risk of suffering from smoke and heat-related illnesses while those living at the urban–rural interface are at higher risk of being directly affected by wildfires. Geographically, population exposure and density differ around Australia, and studies suggest that the Southeast of Australia has higher risk of mortality and disease from wildfire and smoke-related impacts while South Australia presents the highest heat-related mortality rates, although most fatalities have concentrated in Victoria, NSW, and South Australia. There are some important aspects that have had relatively limited consideration, such as the possible synergistic adverse health impacts of simultaneous exposure to fire smoke and heat waves, and the more extreme personal exposures of professionals who work outside during heat waves and bushfire air pollution events. Necessary steps should be taken at federal, state, and local levels to ensure the safety of communities and minimize the loss of resources, lives, and wellbeing, especially focussing on the sectors of society that are more vulnerable due to their physical locations, social and economic disadvantage, or higher risk occupational groups. Required adaptation actions are urgently needed to be implemented. As we know adaptation essentially means anticipating the impact of climate change induces extreme weather events and making cities and rural areas resilient in the face of 1.5 °C rise in global warming over pre-industrial levels by the 2030s.

In the context of India and other developing countries, there is need to adopt environmental governance in the process of economic development by focusing on climate-proof infrastructure to minimize the impact of extreme weather events such as heat waves, flooding and water stress.

Co-benefit analysis is extremely important for policy-makers. In this regard, co-benefit by reducing short-lived climate pollutants is also necessary to be studied in a regional perspective both in developed and in developing countries. For this purpose, development of impact models that incorporate temperature and air pollutants is the first step. Although some researchers controlled for temperature in evaluating air pollution effect, deeper understanding of the inter-relation between the air pollutants and temperature is necessary for co-benefit analyses (Hanaoka and Masui 2018). Current trends in the increase in temperature and the occurrence of heat waves in USA, Europe, Australia as well as in developing countries like China and India, present enormous opportunities for research on co-benefit analysis.

Coastal regions are directly affected by recurrent flooding, loss of biodiversity, and storm damage exacerbated by sea-level rise. To adequately prepare for the impacts, individual citizens, community leaders, and policymakers must address the causes of climate change and sea-level rise while identifying the multi-sector risks associated with the consequences. At present, research has established the health and socio-economic impacts of sea-level rise on coastal communities in developed countries, yet the capacity to address the challenges in more developing locations are still limited. This hinders the assessment of coastal impacts and future predictions for developing counties. Therefore, further research may continue to synergize across disciplines and geopolitical boundaries to address these deficiencies.

In some regions, the health consequences of thunderstorm asthma may be prevented with adequate measures by meteorological forecast, identifying most vulnerable areas, and by the correct use of patients of adequate anti-allergic and antiasthma therapy and avoiding to be outdoor at the start of a storm during pollen season.

References

Hanaoka, T., & Masui, T. (2018). Co -benefit reductions of short-lived climate pollutants and air pollutants by 2050 while achieving the 2 degree target in Asia. *Journal of Sustainable Development of Energy, Water and Environment Systems, 6*(3), 505–520.

Harvey, C. (2018). Extreme weather will occur more frequently worldwide and if countries do not meet the Paris climate agreement goals, the risks will be even greater. *Scientific American*, E.E. News, February 15.

https://www.c2es.org/content/hurricanes-and-climate-change/. (2017). Hurricanes and climate change, Centre for Climate and energy Solutions.

IPCC. (2018). *Special Report: Global Warming of 1.5 °C: Summary for Policy Makers*. Geneva: IPCC.

Lynas, M. (2019). *Climate change: The more we know, the worse it seems*, CNN News, January, 26.

McGrath, M. (2019). *Climate change: World heading for warmest decade*, BBC, 6th February, 2019.

Morad, R. (2018). *As sea levels rise, home values sink, FORBES*. https://www.forbes.com/sites/reneemorad/2018/07/31/study-as-sea-levels-rise-home-values-sink/#5dbf58455029.

Science Daily (2017). Cornell University. Rising seas could result in 2 billion refugees by 2100. *Science Daily*, 26 June. www.sciencedaily.com/releases/2017/06/170626105746.htm.

Storrow, B. (2018). Global CO_2 emissions rise after Paris climate agreement signed, *Scientific American*, E.E.News, March 24.

Author Index

R. Akhtar (ed.), *Extreme Weather Events and Human Health*,
https://doi.org/10.1007/978-3-030-23773-8

Subject Index

R. Akhtar (ed.), *Extreme Weather Events and Human Health*,
https://doi.org/10.1007/978-3-030-23773-8

MIX
Papier aus verantwortungsvollen Quellen
Paper from responsible sources
FSC® C105338

If you have any concerns about our products,
you can contact us on
ProductSafety@springernature.com

In case Publisher is established outside the EU,
the EU authorized representative is:
Springer Nature Customer Service Center GmbH
Europaplatz 3, 69115 Heidelberg, Germany

Printed by Libri Plureos GmbH
in Hamburg, Germany